Wastewater Treatment and Sludge Management Systems - The Gutter-to-Good Approaches

Edited by Başak Kılıç Taşeli, Eduardo Jacob-Lopes, Leila Queiroz Zepka and Mariany Costa Deprá

Published in London, United Kingdom

Wastewater Treatment and Sludge Management Systems - The Gutter-to-Good Approaches
http://dx.doi.org/10.5772/intechopen.1001656
Edited by Başak Kılıç Taşeli, Eduardo Jacob-Lopes, Leila Queiroz Zepka and Mariany Costa Deprá

Contributors
Adedayo O. Adebusuyi, Adeyemo Festus Adeniyi, Ahmadi Borhan, Alexis Rubén Bracamontes-Ruelas, Alfonso Mata Bermúdez, Amadou Diao, Andile Mkhohlakali, Ariana Aisa, Atashkar Sheida, Avner Ronen, Başak Kılıç Taşeli, Büşra Çayir, Carlos Alberto Méndez Cuesta, Chetan Prakash Sharma, Cuauhtemoc Pérez González, Dessy Ariyanti, Dina Lesdantina, Dongwon Ki, Eduardo Jacob-Lopes, Erick Cuauhtemoc Pérez Montiel, Fakoro Souleymane Dia, Farida Diyah Hapsari, Filicia Wicaksana, Firas Fayssal, Harvinder Singh Saini, Hong-Kwan Choi, Hossini Hooshyar, Ibrahima Niane, James Tshilongo, José Rafael Irigoyen-Campuzano, Kamar T. Oladepo, Katayoon Nematpour, Leila Queiroz Zepka, Liliana Reynoso-Cuevas, Luis Arturo Torres-Castañon, Manoj Kumar, Maria D. Ciuca, Mariany Costa Deprá, Massahi Dariush, Massahi Tooraj, Mezie Chinazor Angus, Miguel Angel Zavala Sánchez, Miguel Martell Mendoza, Mohammadhassan Gholami-Shabani, Mokgehle R. Letsoalo, Mpfunzeni C. Raphulu, Napo Ntsasa, Nyiko M. Chauke, Oche Sunday Oche, Olawale Theophilus Ogunwumi, Oluwademilade Obisakin, Oluwanishola A. Azeez, Opeyemi K. Olayanju, Parnoon Kimya, Radu C. Racovita, Se-Won Park, Senghane Mbodji, Sharafi Maryam, Shin Young Kang, Tumelo Mogashane, Wei Gao, Ze Zhu

First published in London, United Kingdom, 2025 by IntechOpen
IntechOpen is the global imprint of INTECHOPEN LIMITED, registered in England and Wales, registration number: 11086078, 167-169 Great Portland Street, London, W1W 5PF, United Kingdom

For EU product safety concerns: IN TECH d.o.o., Prolaz Marije Krucifikse Kozulić 3, 51000 Rijeka, Croatia, info@intechopen.com or visit our website at intechopen.com.

British Library Cataloguing-in-Publication Data
A catalogue record for this book is available from the British Library

Wastewater Treatment and Sludge Management Systems - The Gutter-to-Good Approaches
Edited by Başak Kılıç Taşeli, Eduardo Jacob-Lopes, Leila Queiroz Zepka and Mariany Costa Deprá
p. cm.
Print ISBN 978-0-85466-743-7
Online ISBN 978-0-85466-742-0
eBook (PDF) ISBN 978-0-85466-744-4

Meet the editors

Editor Prof. Başak Kılıç Taşeli is a professor at the Environmental Engineering Department at Giresun University in Türkiye. She is an expert with 27 years of experience in protected area management, environmental management, waste prevention, renewable energies, and zero waste. She is an environmental engineer and has completed her M. Sc. in soil pollution due to thermal power plants and her Ph.D. in environmental sciences-biodegradation of toxic organic compounds at Middle East Technical University in Türkiye. She teaches graduate-level sustainability courses concerning environment, energy and sustainability; biofuels: biomass conversion systems; biogas production and usage; sustainable operation of treatment plants and undergraduate-level courses covering topics such as water supply; wastewater disposal; environmental and engineering ethics; environmental ecology, sludge treatment; unit operations; industrial wastewater treatment. She has more than 25 years of experience working in national/international environmental projects on water quality monitoring and conservation of specially protected areas, design and operation of wastewater treatment plants, environmental impact assessment, management plans, carrying capacity and Turkish Grand National Assembly Research and Investigation Commission Reports on various environmental pollution issues. Dr. Taşeli has excellent knowledge of zero waste, ecological sanitation and greenhouse gas emissions calculations. Additionally, she has strong knowledge of climate change–greenhouse gas emissions relationships, greenhouse gas emission calculations, wastewater treatment plant's sustainability and their effect on climate change, and strong knowledge of protected area management and International environmental standards (ISO 14001, EMAS). Finally, she has 10 years of experience working on EU-funded Erasmus + projects, coordinating several projects as a promoter and partner.

Prof. Dr. Eduardo Jacob-Lopes is currently an Associate Professor in the Department of Food Technology and Science at the Federal University of Santa Maria in Brazil. He has more than 15 years of teaching and research experience. He has coordinated or coordinates more than 50 research projects and/or technological developments financed by public funding agencies and private initiatives. He has published over 800 scientific publications/communications, including books, book chapters, original research papers, research communications in national and international conferences, and patents. He is a member of the editorial boards of 10 journals and acts as a reviewer for several national and international journals. His research interests focus on a multidisciplinary approach to applied phycology and its intersection with bioprocess engineering and sustainable engineering.

Prof. Dr. Leila Queiroz Zepka is currently an Associate Professor in the Department of Food Technology and Science at the Federal University of Santa Maria in Brazil. She has more than 15 years of teaching and research experience. She has published over 700 scientific publications/communications, including books, book chapters, original research papers, research communications at national and international conferences, and patents. She is a member of the editorial boards of 5 journals and acts as a reviewer for several national and international journals. Her research interests include biotechnology, with an emphasis on microalgae-based products.

Dr. Mariany Costa Deprá is currently a Researcher in the Department of Food Technology and Science at the Federal University of Santa Maria, Brazil. She has published over 150 scientific publications/communications, including book chapters, original research papers, research communications at national and international conferences, and patents. She acts as a technical and scientific consultant for companies and international scientific journals. Her research interests include sustainability metrics and indicators, with an emphasis on bioprocess engineering.

Contents

Preface **XI**

Section 1
Introduction **1**

Chapter 1 **3**
Introductory Chapter: Tackling Challenges of Sludge Management Systems – How to Drive a Critical Paradigm Shift in the Sustainability Era?
by Mariany Costa Deprá, Leila Queiroz Zepka and Eduardo Jacob-Lopes

Section 2
Wastewater Treatment – Past and Future Perspectives **7**

Chapter 2 **9**
Embodied Carbon and Greenhouse Gas Emissions from a Full-Scale Advance Wastewater Treatment Plant
by Büşra Çayir and Başak Kılıç Taşeli

Chapter 3 **29**
Continuous Tubular Photoreactor with TiO_2 Nanoparticles Catalyst for Batik Industrial Wastewater Treatment
by Dessy Ariyanti, Dina Lesdantina, Ariana Aisa, Farida Diyah Hapsari, Filicia Wicaksana and Wei Gao

Chapter 4 **49**
Application of Advanced Oxidation Processes for Domestic and Industrial Wastewater Treatment
by Alexis Rubén Bracamontes-Ruelas, José Rafael Irigoyen-Campuzano, Luis Arturo Torres-Castañon and Liliana Reynoso-Cuevas

Chapter 5 **69**
Novel Approaches for Sustainable Management of Chromium Contaminated Wastewater
by Manoj Kumar and Harvinder Singh Saini

Chapter 6 95
Trends in Systematic Techniques for Pollutants Monitoring in the Environmental Water Systems
by Napo Ntsasa, Andile Mkhohlakali, Tumelo Mogashane, James Tshilongo and Mokgehle R. Letsoalo

Chapter 7 113
Membrane Filtration for Wastewater Treatment – Fouling Mitigation
by Chetan Prakash Sharma, Ze Zhu and Avner Ronen

Chapter 8 143
Metal Organic Frameworks Used as Antibiotic Removal Agents in Water
by Miguel Martell Mendoza, Carlos Alberto Méndez Cuesta, Miguel Angel Zavala Sánchez, Erick Cuauhtemoc Pérez Montiel, Alfonso Mata Bermúdez and Cuauhtemoc Pérez González

Chapter 9 171
Tin Oxide Materials for Industrial Wastewater Treatment: Promising Adsorbents and Catalyst
by Nyiko M. Chauke and Mpfunzeni C. Raphulu

Chapter 10 187
Wastewater Treatment Approaches for the Removal of Antidepressant Residues
by Radu C. Racovita and Maria D. Ciuca

Chapter 11 205
Review of Anaerobic Baffled Reactor for Treatment of Industrial Wastewater in Nigeria
by Opeyemi K. Olayanju, Kamar T. Oladepo, Oluwanishola A. Azeez, Adedayo O. Adebusuyi and Oluwademilade Obisakin

Chapter 12 221
Biological Wastewater Treatment
by Olawale Theophilus Ogunwumi, Adeyemo Festus Adeniyi, Mezie Chinazor Angus and Oche Sunday Oche

Chapter 13 261
Reuse of Wastewater as Non-Conventional Water: A Way to Reduce Water Scarcity Crisis
by Mohammadhassan Gholami-Shabani and Katayoon Nematpour

Chapter 14 295
Modeling of a Septic Tank Waste Treatment System for the Production of Biogas
by Ibrahima Niane, Amadou Diao, Fakoro Souleymane Dia and Senghane Mbodji

Section 3
Sludge Management Systems – The Gutter-to-Good Approaches **315**

Chapter 15 **317**
Good Practices in Sludge Management for Non-Annex Mediterranean Countries
by Firas Fayssal

Chapter 16 **331**
Advancing Sustainable 3D Printing: Harnessing the Potential of Wastewater Sludge Incineration Ash for Composite Material Development and Practical Applications
by Dongwon Ki, Shin Young Kang, Se-Won Park and Hong-Kwan Choi

Chapter 17 **349**
Future of Sludge Management
by Massahi Tooraj, Hossini Hooshyar, Parnoon Kimya, Atashkar Sheida, Sharafi Maryam, Massahi Dariush and Ahmadi Borhan

Preface

Wastewater treatment plants are crucial in reducing the environmental impacts of discharging untreated wastewater into natural water bodies. However, in pursuit of long-term ecological sustainability, the objectives of wastewater treatment must extend beyond pollution control to include energy and resource savings, as well as waste reduction. One significant aspect of wastewater treatment is the management of sewage sludge. This unavoidable residue is primarily a byproduct of phase change, and although it constitutes only about 1–2% of the treated waste, it contains the majority of the original pollution.

Sewage sludge, the solid residue generated during wastewater treatment, is an unavoidable byproduct, even with proper management and treatment. However, the treatment and disposal of sludge present significant challenges, including high treatment costs, the presence of new pollutants, potential health risks, and concerns over public acceptance. Traditional disposal methods—such as storage and incineration—have raised environmental and societal concerns, prompting the exploration of alternative sludge management approaches.

Effective sludge management requires a comprehensive, systematic solution that balances environmental effectiveness, social acceptability, and economic feasibility. With increasing sludge production worldwide due to population growth, urban development, and industrial expansion, there is a growing need to treat and manage sludge to minimise its negative environmental impacts.

Investments in sustainable solutions and reliable, safe wastewater treatment facilities are vital for any industrial or municipal sector aiming to minimise environmental footprints and use natural resources wisely. Sludge management systems have become one of the biggest challenges for wastewater treatment plants. This is because improper disposal of sludge not only contributes to land and water pollution but also poses significant biological risks, thereby endangering public health.

In the past, sludge management strategies in wastewater treatment plants focused primarily on disposal. However, future perspectives point towards the intelligent and sustainable reuse of sludge. Managing sludge sustainably has become a critical issue for modern societies due to its increasing production, challenges in the proper placement of treatment and disposal facilities, and the need to meet quality requirements. There is still a need for a paradigm shift around sludge—from viewing it solely as waste to recognising it as a valuable biological resource.

The purpose of bringing together the chapters in this book is to foster transformative connections between leading academic innovations and decision-makers from industries, creating meaningful, sustainable strategies and insights to address and overcome the current challenges in sludge management systems.

The editor would like to sincerely thank the co-editors, Dr. Eduardo Jacob-Lopes, Dr. Mariany Costa Deprá, Prof. Leila Queiroz Zepka, and the authors for their invaluable contributions to this book.

Başak Kılıç Taşeli
Professor,
Department of Environmental Engineering,
Giresun University,
Giresun, Türkiye

Eduardo Jacob-Lopes, Leila Queiroz Zepka and Mariany Costa Deprá
Federal University of Santa Maria,
Santa Maria-RS, Brazil

Section 1

Introduction

Chapter 1

Introductory Chapter: Tackling Challenges of Sludge Management Systems – How to Drive a Critical Paradigm Shift in the Sustainability Era?

Mariany Costa Deprá, Leila Queiroz Zepka and Eduardo Jacob-Lopes

1. Introduction

Water is life's bottleneck on the planet. It is vital for the development of all living organisms. We, humans, generate wastewater every day. However, extravagant industrialization, population growth, and urbanization - in recent years - acted as a driving force behind the unprecedented increase in the scarcity of water resources [1]. Given these aspects, wastewater treatment is, undoubtedly, a naive and utopian need to try to curb the cascading effects of anthropogenic impacts on nature and biodiversity.

However, even if current wastewater treatment technologies are highly efficient in removing pollutants, a significant amount of organic matter, nutrients, and other macro and micropollutants result from this process; there is the formation of a solid residue known as sludge [2]. Sludge - this unavoidable residue, although representing only about 1 to 2% of the treated effluent, contains most of the original pollution, making it a pressing thorn in the side [3]. Currently, world sludge generation is on a relentless expansion curve; however, environmental quality standards for its management are becoming stricter, and economic pressures require low-cost solutions to alleviate the increased load of rapid production [4].

Accordingly, sustainable solutions and investments in reliable and safe wastewater treatment plants are crucial for any industrial and municipal segment that intends to minimize its environmental footprint and use natural resources wisely [5]. Sludge management systems have become one of the biggest challenges for wastewater treatment plants. This is because, in practice, beyond polluting the terrestrial and aquatic atmosphere, improper disposal of sludge puts public health at risk, due to its high biological risk [6]. Thus, treating and properly disposing of this waste becomes decisive. However, it is estimated that post-treatment processing costs for proper final disposal may represent more than 50% of the operating costs of a conventional plant [7].

In a historical context, the sludge management solutions adopted in wastewater treatment plants, in the past, were directed mainly toward its disposal; while the perspectives of the future point to its reuse intelligently and sustainably. In this way, extracting the best of this residue makes it possible to explore it in different areas, whether they are applied from agricultural activities (fertilizers) to their use as alternative energy sources, through the most varied processes such as, for example, anaerobic digestion and combustion [8]. Even so, when correlating and analyzing the concepts of past and future, it is verified how complex and controversial it is to consolidate the gutter-to-good approaches.

At the same time, innovation adopted in the sludge management sector is very slow for the ambitious sustainability goals that our society has ratified. The management of sludge in a sustainable way is one of the critical issues of modern society due to the growing increase in its production, difficulties in the proper location of treatment and disposal works, and compliance with quality requirements, which thus require a rethinking of its management [9].

Given these aspects, the sustainability era has reached a turning point. Unless we intervene sharply, current environmental reports will remain exactly what they are, data. Thus, in order to transform the tangible information, we have today into actionable insights, it is necessary to abandon old certainties and embark on new solutions [10].

Notwithstanding, the future of sludge management systems will also depend on what and if policymakers are ready to implement these measures in order to protect the quality of this by-product and the regulatory framework to support a wide variety of reuses [11]. Currently, there is still a need for a paradigm shift around sludge in order to start by not considering it as waste but starting to see it as a bioresource in itself. Hence, researchers have struggled with these issues and provided clear scientific evidence that the time has come – or even passed – to review sludge management systems [12]. Therefore, directing attention to what is valuable and reconsidering aspects of how to protect it, rather than looking for alternatives how to get rid of it, is imperative.

2. Final remarks

We believe that sublime connections can solve real-world problems. As we combine pioneering researchers from the academy with key actors in decision-making in industrial and municipal sectors in the search for greater positive impacts on the ecosystem, they are consummated in a crucial step in paving the paradigm shifts of the new sustainable era applied to waste sludge management. Therefore, our aim in bringing together the next chapters found in this book is to enable transformative connections between leading academic innovations with an international community of decision-makers in industries, creating meaningful sustainable strategies and insights to face and overcome the current challenges of sludge management systems.

Acknowledgements

The authors would like to thank the National Council for Scientific and Technological Development (CNPq) – Brazil, and the Coordination of Improvement of Higher Education Personnel (CAPES), Finance Code 001.

Conflict of interest

The authors declare that they have no known competing financial interests or personal relationships that could have appeared to influence the work reported in this chapter.

Author details

Mariany Costa Deprá, Leila Queiroz Zepka and Eduardo Jacob-Lopes*
Bioprocess Intensification Group, Federal University of Santa Maria (UFSM), Santa Maria, RS, Brazil

*Address all correspondence to: ejacoblopes@gmail.com

References

[1] Sohail NF, Iftikhar R, Saleem S. Microalgal treatment of high-nutrient wastewater using twin layer cultivation system. Journal of Environmental Chemical Engineering. 2023;**11**(2):109248. DOI: 10.1016/j.jece.2022.109248

[2] Arias A, Alvarino T, Allegue T, Suárez S, Garrido JM, Omil F. An innovative wastewater treatment technology based on UASB and IFAS for cost-efficient macro and micropollutant removal. Journal of Hazardous Materials. 2018;**359**:113-120. DOI: 10.1016/j.jhazmat.2018.07.042

[3] Spinosa L, Doshi P. Re-thinking sludge management within the sustainable development goal 6.2. Journal of Environmental Management. 2021;**287**:112338. DOI: 10.1016/j.jenvman.2021.112338

[4] Liu X, Huang H, Iqbal A, Chen J, Zan F, Jiang C, et al. Sustainability analysis of implementing sludge reduction in overall sludge management process: Where do we stand? Waste Management. 2022;**152**:80-93. DOI: 10.1016/j.wasman.2022.08.013

[5] Ramadan BS, Miftahadi MF, Suryawan IWK, Liang Z, Zhang J, Septiariva IY, et al. Circular economy from a water and wastewater management perspective. In: Sustainable and Circular Management of Resources and Waste Towards a Green Deal. Amsterdam, Netherlands: Elsevier; 2023. pp. 79-91. DOI: 10.1016/B978-0-323-95278-1.00003-6

[6] Poinen P, Bokhoree C. Sludge management practices: Drivers, opportunities and implications for small island developing states. Journal of Water Process Engineering. 2022;**48**:102860. DOI: 10.1016/j.jwpe.2022.102860

[7] Gomes LA, Gonçalves RF, Martins MF, Sogari CN. Assessing the suitability of solar dryers applied to wastewater plants: A review. Journal of Environmental Management. 2023;**326**:116640. DOI: 10.1016/j.jenvman.2022.116640

[8] Bagheri M, Bauer T, Burgman LE, Wetterlund E. Fifty years of sewage sludge management research: Mapping researchers' motivations and concerns. Journal of Environmental Management. 2023;**325**:116412. DOI: 10.1016/j.jenvman.2022.116412

[9] Flores-Alsina X, Ramin E, Ikumi D, Harding T, Batstone D, Brouckaert C, et al. Assessment of sludge management strategies in wastewater treatment systems using a plant-wide approach. Water Research. 2021;**190**:116714. DOI: 10.1016/j.watres.2020.116714

[10] Zhang X, Liu Y. Resource recovery from municipal wastewater: A critical paradigm shift in the post era of activated sludge. Bioresource Technology. 2022;**363**:127932. DOI: 10.1016/j.biortech.2022.127932

[11] Greene C. Briefing Note on Sludge Management. Waste Water Treatment – Sludge Management. Brussels, Belgium: EurEau; 2022. Available from: https://www.eureau.org/resources/briefing-notes

[12] Orhon D, Sözen S. Reshaping the activated sludge process: Has the time come or passed? Journal of Chemical Technology & Biotechnology. 2020;**95**(6):1632-1639. DOI: 10.1002/jctb.6290

Section 2

Wastewater Treatment – Past and Future Perspectives

Chapter 2

Embodied Carbon and Greenhouse Gas Emissions from a Full-Scale Advance Wastewater Treatment Plant

Büşra Çayir and Başak Kılıç Taşeli

Abstract

This chapter provides a valuable data regarding greenhouse gas emissions and embodied carbon emissions associated with wastewater treatment plants, and their implications for climate change. We have assessed greenhouse gas emissions, resulting in 91.05 tCO_2eq for CH_4 and 0.343 tCO_2eq for N_2O. Indirect greenhouse gas (GHG) emissions predominantly stem from electricity usage, comprising 15.20 tCO_2eq. The cumulative GHG emissions for the plant were estimated at 106.59tCO2eq. The examination of the structure of WWTP revealed that ready-mix concrete and steel contribute to the highest amount of embodied carbon, approximately 397.48 and 286.79 $tonCO_2e$, respectively. The major contributor to embodied carbon emissions is external walls and facade, which accounts for 39.83% of the total carbon, and emits 750 $tonCO_2e$. The major contributors of resource type to embodied carbon emissions are paints, coatings, and lacquers and ready-mix concrete for external walls and floors that account for 32.73% and 24.08% of the total carbon, respectively, and emit 620 and 450 $tonCO_2e$.The construction material stage (A1-A3) constitutes the highest proportion of embodied carbon (1371.01 $tonCO_2e$), followed by the construction/installation stage A5 (453.96 $tonCO_2e$) and transportation to the building site A4 stage (58.18 $tonCO_2e$). Cumulatively, this accounts for a total of 1883 tCO_2e.

Keywords: advance wastewater treatment, greenhouse gas emissions, embodied carbon emissions, methane emissions, nitrous oxide emissions, carbon dioxide emissions, sustainability

1. Introduction

The Global Carbon Project has announced its 2023 carbon budget, the annual scientific assessment of the global carbon cycle. Data shows that without the necessary acceleration in global efforts to decarbonise the economy, CO_2 emissions from fossil fuels are expected to reach a historic peak in 2023. According to data provided by the Global Carbon Project, if current emission rates persist, the remaining carbon budget will continue to dwindle unless robust climate actions are taken. There is a 50%

probability that the 1.5°C warming threshold will be surpassed permanently in approximately 7 years. This projection signifies a delay compared to the timeline outlined in the latest Intergovernmental Panel on Climate Change (IPCC) reports. Projections indicate a 1.1% increase in fossil carbon emissions in 2023 compared to 2022, amounting to 36.8 billion tons of CO_2. Emissions are anticipated to rise across all fuel types, including coal, oil, and gas. Despite a global slowdown in the growth of CO_2 emissions and reductions observed in some nations over the past decade, progress remains insufficient to avert the perils of climate change [1].

Turkey ranks 15th among the world's largest emitters, with 400 million tons of carbon emissions recorded in 2022. Coal emerges as the primary contributor to Turkey's emissions, accounting for 167.2 million tons of CO2, followed by the oil, gas, and cement sectors. While China, the United States, and India lead in global emissions, regional emission trends exhibit significant variations. Projections indicate an uptick in emissions for India (8.2%) and China (4.0%) in 2023, whereas the EU (−7.4%), the USA (−3.0%), and the remainder of the world (−0.4%) are expected to witness declines [1].

As a pivotal metric for evaluating sustainability, the carbon footprint assumes a critical role in comprehensively gauging the environmental impact of wastewater treatment plants (WWTPs) on climate change. The analysis of carbon footprint has garnered widespread adoption globally, underscoring the imperative of mitigating greenhouse gas (GHG) emissions from WWTPs in the collective pursuit of achieving global carbon neutrality. GHG emissions pose significant adverse ramifications on the environment, amplifying global climate change dynamics through the operation of the open carbon cycle [2, 3]. These emissions encompass a spectrum of gases known as the 'basket of six' GHGs, comprising carbon dioxide (CO_2), methane (CH_4), nitrous oxide (N_2O), hydrofluorocarbons (HFCs), perfluorocarbons (PFCs), and sulfur hexafluoride (SF6), quantified in tonnes of carbon dioxide equivalent (CO_2e). Both CO_2 and CO_2e are commonly denoted as carbon emissions. Among these, CO_2 emerges as the predominant GHG, constituting approximately 76% of total global emissions [4].

The evaluation of the greenhouse effect induced by prominent greenhouse gases (GHGs) such as carbon dioxide (CO_2), methane (CH_4), and nitrous oxide (N_2O) within wastewater treatment operations is predicated upon their Global Warming Potentials (GWPs). Over a century-long period, the warming impact of one ton of methane and nitrous oxide is equated to 25 and 298 tons of CO_2, respectively. Nitrous oxide (N_2O) emerges as a significant contributor to the greenhouse effect, possessing a lifespan of 114 years and exerting a 298-fold stronger influence on global warming in comparison with carbon dioxide. Furthermore, it is anticipated to serve as a potent agent in ozone layer depletion [5]. Human activities, including agriculture, wastewater management, and various industrial processes, contribute to the elevation of atmospheric N_2O levels. In the realm of wastewater treatment, renowned for its substantial energy consumption, approximately 5.6% of total N_2O emissions originate, positioning it as the fourth largest emitter subsequent to agriculture, energy production, and other industrial pursuits, according to data from the China National GHG Inventory. Additionally, ample evidence indicates that substantial quantities of N_2O are generated during biological nitrogen removal processes, with emissions escalating due to the proliferation of wastewater treatment plant (WWTP) construction and the imposition of more stringent discharge standards in recent years. The genesis of N_2O in wastewater treatment systems stems from diverse microbial reactions occurring during both nitrification and denitrification processes [6].

On one side, implementing sewage treatment holds the potential to mitigate direct environmental impacts. However, it introduces new challenges, primarily through heightened energy usage for operating treatment facilities and maintaining wastewater infrastructure. In many developing countries, there exists a dichotomy between prioritizing energy management and water management, with little emphasis placed on understanding the intricate interplay between water, energy, and greenhouse gas emissions [7]. Thus, there arises a pressing need to scrutinize energy consumption and its carbon footprint within wastewater infrastructure to equip policymakers and regulators with informed decision-making tools.

The aggregate energy consumption within wastewater treatment facilities encompasses both operational energy and the energy embodied in infrastructure and chemical utilization [8]. Embodied energy encompasses the energy expended in all upstream processes, encompassing raw material extraction, processing, manufacturing, and transportation, requisite for producing materials employed in wastewater treatment and construction [9]. While various studies scrutinize operational factors impacting energy intensity across diverse treatment technologies, they frequently overlook infrastructure components [10, 11]. It has been noted by Gallego-Schmid and Tarpani [12] that among 43 life cycle assessment studies of wastewater treatment plants, only 14 accounted for the construction phase. This disparity might arise from the labor-intensive nature of analysing wastewater infrastructure and the limited accessibility of inventories. Ideally, comprehensive project reports should provide validated civil estimates to precisely quantify the quantity of building materials in any treatment infrastructure. This construction inventory holds significance as environmental impacts attributed to construction reportedly exceed 15% of the total impacts [13, 14]. Furthermore, the impact of construction in wastewater treatment heavily depends on the treatment technology and system boundaries [15].

Due to the energy-intensive characteristics of wastewater systems, the total energy consumption is expected to escalate in tandem with population expansion and the consequent rise in wastewater production. Despite the predominance of cradle-to-grave analyses in the existing literature on wastewater treatment systems, there has been limited focus on comprehensive gate-to-gate system analyses, which can provide significant insights into energy efficiency during the operational phase of a plant. It is crucial to establish a comprehensive database of building materials employed in the construction of wastewater treatment plants, especially considering its neglect in the majority of studies concerning plants in the developing and underdeveloped nations.

In the study conducted by Chen et al. [16], it was found that cement, steel, brick, lime, and linoleum are the five primary materials exhibiting significant embodied energy and carbon. Collectively, these materials account for approximately 93.1% of the total embodied energy and roughly 95.7% of the total embodied carbon. Notably, steel and cement are highlighted for their widespread application in construction and substantial embodied energy, largely attributed to the mining operations inherent in their manufacturing processes [10].

The life cycle assessment (LCA) approach, a pivotal methodology mandated by the International Organization for Standardization (ISO) 14040/14044, is widely employed for the calculation and evaluation of the embodied carbon in buildings [17, 18]. Substantial endeavours have been undertaken in recent years to establish formal standards and recognized industry reports aimed at facilitating embodied carbon assessments [19]. Environmental Product Declarations (EPDs), introduced in 2014 in compliance with EN 15804, play a fundamental role in elucidating the environmental implications of various building materials and regulating the process of embodied carbon assessments [20, 21]. EPDs serve as an increasingly valuable source of environmental data within the construction products market, aiding in the assessment of environmental performance and enabling product comparisons for procurement decisions during the later stages of building design [22].

Furthermore, various software programs are available for embodied carbon assessment, with OpenLCA and SimaPro being among the most notable. However, Li [23] suggests that some LCA tools may not provide comprehensive results for full life cycle carbon emissions, and many tools rely on region-specific data and building norms, necessitating careful consideration of their applicability beyond their country of origin.

This study examines the greenhouse gas emissions linked with the operation of a treatment plant based on advanced activated sludge process technology. It delves into the emissions stemming from both electricity usage and overall plant activities. Primary data concerning wastewater inflow, total energy consumption, and materials employed across various unit processes within the treatment plant were systematically collected and subjected to rigorous analysis. Direct greenhouse gas emissions originating from the plant were estimated using the methodology delineated by the Intergovernmental Panel on Climate Change (IPCC). Additionally, the embodied energy contained within the materials utilized for the construction and ongoing operations of the treatment plant was calculated, leveraging country-specific data to ensure accuracy and relevance.

2. Materials and methods

2.1 Study area

The district located in the southern part of the Marmara Region covers an approximate area of 430 km^2. Characterized by highly rugged terrain, the district has an elevation of 253 m. Six small streams flow from its lands. The most mountainous sections of the district lie to the east and north.

The population targeted for the wastewater treatment facility designed to treat the district's wastewater is 20,000 for the year 2050 and 15,000 for the year 2035. A sewage network comprising 54 km of pipeline and 1875 manholes has been constructed. The total construction area of the facility is 10,000 m^2. The analyzed waste water treatment of 3678 m^3/day in 2020 is based on the activated sludge process technology, which employs advanced biological treatment, and includes coarse and fine screens followed by anaerobic phosphorus tanks. Nitrification-denitrification processes are carried out along with organic carbon removal, and redox meters and oxygen meters are utilized to monitor anoxic-oxic conditions in the tanks. The sludge obtained from the final settling tanks is dewatered using mechanical dewatering equipment. Cationic polyelectrolytes are employed as sludge conditioners for optimal conditioning of the dewatering units. UV disinfection is employed before discharged to receiving body.

2.2 GHG emissions calculating principles for a full-scale advance wastewater treatment plant

2.2.1 Methane (CH_4) emission calculating principles

Calculating CH_4 emissions involves several steps, including estimating the organically degradable material in domestic wastewater, determining the methane emission factor for domestic wastewater, and then estimating CH_4 emissions.

The amount of direct methane emissions is influenced by the quantity of degradable carbon present in the wastewater and sludge, along with an emission factor. Eq. (1) (derived from Eq. 6.2 of [24]) illustrates this relationship, where the emission factor (EF) depends on the maximum CH_4 producing potential (Bo) and the methane correction factor (MCF). It is recommended to use a Bo value of 0.6 kg CH_4/kg BOD removal, with an uncertainty range of ±30%, and an MCF value of 0.05.

The estimation of TOW in 2020 (total organics in wastewater in the inventory year, expressed in kg BOD/yr) is based on the human population and the BOD generation per person. TOW can be calculated using Eq. (2) (derived from Eq. 6.3 of Ref. [24], which quantifies the total organically degradable material in domestic wastewater.

CH_4 emission factor for each domestic wastewater treatment/discharge pathway or system:

$$EF_j = B_0 \bullet MCF_j \tag{1}$$

Where EF_j = emission factor, kg CH_4/kg BOD, j = each treatment/discharge pathway or system, B_0 = maximum CH_4 producing capacity, kg CH_4/kg BOD, MCF_j = methane correction factor (fraction).

Total organically degradable material in domestic wastewater:

$$TOW = P \bullet BOD \bullet 0.001 \bullet I \bullet 365 \tag{2}$$

Where TOW = total organics in wastewater in inventory year, kg BOD/yr, P = country population in inventory year, (person), BOD = country-specific per capita BOD in inventory year, g/person/day, 0.001 = conversion from grams BOD to kg BOD, I = correction factor for additional industrial BOD discharged into sewers (for collected the default is 1.25, for uncollected the default is 1.00).

The general equation for estimating CH_4 emissions from domestic wastewater is derived from Eq. (3) (Eq. 6.1 of [24]: total CH_4 emissions from domestic wastewater).

Total CH_4 emissions from domestic wastewater;

$$CH_4 \text{ emissions} = \left[\sum_{ij}(U_i \bullet T_{ij} \bullet EF_j)\right] \bullet (TOW - S) - R \tag{3}$$

Where CH_4 emissions = CH_4 emissions in inventory year, kg CH_4/yr, TOW = total organics wastewater in inventory year, kg BOD/yr, EF_j = emission factor, kg CH_4/kg BOD, S = organic component removed as sludge in inventory year, kg BOD/yr, U_i = fraction of population in income group *i* in inventory year, $T_{i,j}$ = degree of utilization of treatment/discharge pathway or system, *j*, for each income group fraction *i* in inventory year, İ = income group: rural, urban high income and urban low income, j = each treatment/discharge pathway or system, R = amount of CH_4 recovered in inventory year, kg CH_4/yr.

2.2.2 Nitrous oxide (N_2O) emission calculating principles

Estimating nitrogen in effluent and determining the emission factor and emissions of indirect N_2O emissions from wastewater are integral steps for calculating N_2O emissions. These emissions are associated with the microbial conversion of nitrogen compounds in wastewater and can occur either from treatment plants or from wastewater after effluent disposal into water bodies, lakes or the sea. The emission factor for domestic wastewater nitrogen effluent is typically taken as 0.005, which refers to the default value recommended by IPCC [24]. The factor 44/28 is applied for converting kg N_2O–N into kg N_2O. A simplified equation is provided in Eq. (5). The emission factors of N_2O were assessed by considering nitrogen loads in the influent of the SBR WWTP.

Total nitrogen in the effluent is calculated by Eq. (4).

$$N_{EFFLUENT} = (P \bullet Protein \bullet F_{NPR} \bullet F_{NON-CON} \bullet F_{IND-COM}) - N_{SLUDGE} \quad (4)$$

Where $N_{EFFLUENT}$ = total annual amount of nitrogen in the wastewater effluent, kgN/year, P = human population, Protein = annual per capita protein consumption, kg/person/yr, F_{NPR} = fraction of nitrogen in protein, default = 0.16, kg N/kg protein, $F_{NON\text{-}CON}$ = factor for non-consumed protein added to the wastewater, $F_{IND\text{-}COM}$ = factor for industrial and commercial co-discharged protein into the sewer system, N_{SLUDGE} = nitrogen removed with sludge (default = zero), kg N/yr.

N_2O emissions from wastewater effluent:

$$N_2O\ Emissions = N_{EFFLUENT} \bullet EF_{EFFLUENT} \bullet 44/28 \quad (5)$$

Where N_2O Emissions = N_2O emissions in inventory year, kg N_2O/yr, $N_{EFFLUENT}$ = nitrogen in the effluent discharged to aquatic environments, kgN/yr, $EF_{EFFLUENT}$ = emission factor for N_2O emissions from discharged to wastewater, kgN_2O-N/kgN, 44/28 = the factor 44/28 is the conversion kg N_2O-N into kg N_2O.

2.2.3 Carbon dioxide (CO_2) emission calculating principles

Two primary factors contributing to CO_2 production from wastewater treatment plants are the treatment process type and electricity consumption. During anaerobic treatment, BOD5 in wastewater is either converted to CO_2 and CH_4, or a portion enters the biomass and is subsequently converted to CO_2 and CH_4 through endogenous respiration. Additional sources of carbon dioxide emissions arise from sludge digesters and the combustion of digestion gas. In aerobic processes, CO_2 is generated through the decomposition of organic substances. Since CO_2 emissions from wastewater treatment plants are biogenic in nature, they are excluded from national total emissions and are not considered in the IPCC Guidelines. Their biogenic origin implies that they are part of the natural carbon cycle and originate from the food chain, passing from plants to animals and humans, or from natural atmospheric CO_2 sources [25].

2.2.4 Emission calculations associated with electricity usage

GHG emissions generated due to electricity usage are calculated by multiplying total MWh used annually by WWTP and emission factor 0.91 tCO_2e/MW as suggested by Ref. [24].

3. Embodied carbon emission calculating principles for a full-scale wastewater treatment plant

As a case study, advance wastewater treatment plant located in the southern part of the Marmara Region of Türkiye was selected. The selected wastewater treatment

Type of material	Name of material	Unit	Consumption
Concrete	Concrete with C8/10 compressive strength	m^3	175
Concrete	Concrete with C30/37 compressive strength	m^3	1888
Concrete	Plain reinforced concrete mould with plywood	m^2	15,213
Steel	Fine-ribbed flooring	ton	58
Steel	Thick ribbed flooring	ton	160
Whitewash	Tar whitewash	m^2	1207
Whitewash	Lime whitewash	m^2	298
Paint	Solvent-based epoxy paint	m^2	389
Paint	Silk matte paint	m^2	771
PVC	PVC door, window	kg	1708
PVC	Pipe	m	74
Brick	Brick	m^2	627
Tile	Tile	m^2	506
Ceramic	Ceramic flooring and wall covering	m^2	503
Insulation	Waterproofing with polymer bitumen covering	m^2	3361
Joint	Construction joint	m^2	572
Iron	Various iron manufacturing	ton	3
Wood	Wood	m^2	593
Glass	Glass	m^2	80
Handle	Handle	m^2	947
Steel pipe	Stainless steel pipe	kg	71,147
Steel flange	Stainless steel flange	kg	638

Table 1.
Material quantity in the WWTP.

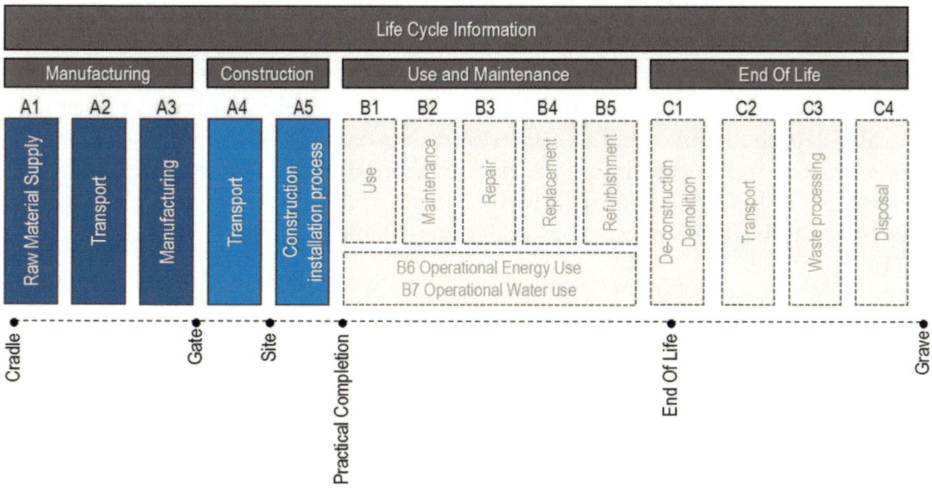

Figure 1.
Life cycle stages [26].

plant is built on an area of 10,000 m^2. The types and quantities of materials used in the construction of the WWTP are taken from the plant's final report and are detailed in **Table 1**. Determining embodied carbon entails quantifying the volume of carbon emissions associated with the production, transportation, manufacturing, and disposal of a product or material throughout its entire lifecycle. To elucidate the environmental impact of an asset, its lifecycle is segmented into stages and modules according to the specifications delineated in BS EN 159785 for buildings (depicted in **Figure 1**), as referenced by Gibbons [26].

A life cycle assessment (LCA) may encompass one or multiple of these stages. The minimum stages required in the LCA depend on its objectives. A cradle-to-grave assessment spans from manufacturing to disposal (A–C). Cradle-to-gate assesses the manufacturing process (A1–A3). When applied to building construction, a cradle-to-gate-type assessment includes stages A4 and A5, also known as upfront carbon, capital carbon, or cradle-to-practical completion. Cradle-to-cradle entails a cradle-to-grave LCA wherein the disposal method is recycling. Gate-to-gate involves an LCA that covers only one part of the lifecycle, such as the construction stage A product's whole life carbon (WLC) incorporates all stages (A–D) [26].

For this research, the system boundary is defined from the production stage (A1–A3), transportation stage (A4), and construction installation stage (A5). The A1–A3 phases pertain to the fabrication of raw materials (A1), production and assembly (A2), and transportation to the manufacturer's site (A3).

4. Results

4.1 Estimated full-scale advance WWTP GHG emissions

4.1.1 Estimated methane emissions

Eqs. (1)–(3), along with the default maximum CH_4 producing capacity (B0) for domestic wastewater of 0.6 kg CH_4/kg BOD, is employed for the calculation of methane emissions from advanced WWTP, as outlined in Ref. [24]. Additionally, (Ti, j) values provided in Table 6.5 of IPCC's "Suggested values for urbanization and degree of utilization of treatment discharge pathway or method (Ti,j) for each income group for selected countries" are utilized in the calculation process [25]. Total organics in WWTP influent is calculated as 157,668 kgBOD/yr. **Table 2** provides the estimated organically degradable material in domestic wastewater. Methane emission factor for domestic wastewater and estimated emissions from domestic wastewater are presented in **Tables 3** and **4**, respectively.

We have assessed greenhouse gas emissions, resulting in 91.05 tCO_2eq for CH_4.

4.1.2 Estimated nitrous oxide emissions

Eqs. (4) and (5), alongside the values specified in **Table 5**, are utilized for N_2O emission calculations. **Table 6** gives calculated N in effluent by using Table 6.11 of IPCC, Ref. [24]: N_2O methodology default data was used for calculations for estimation of nitrogen in effluent.

Table 7 provides the estimation of emission factor and emissions of indirect N_2O emissions from wastewater.

Population (a)	Degradable organic matter (kg/BOD/cap/yr) (b)	Correction factor (c)	Total organics in WWTP influent (kgBOD/yr) (d = a*b*c)
10,094	15.62	1.00	157,668

(c) Correction factor for industrial BOD discharged into sewers (uncollected default value is 1.00).

Table 2.
Organically degradable matter in domestic wastewater.

Max. methane producing capacity, B_0 ($kgCH_4/kgBOD$) (a)	Methane correction factor (MCF) (b)	Emission factor (EF_j) ($kgCH_4/kgBOD$) (c = a*b)
0.6 (aerobic treatment)	0.05	0.03

Table 3.
Methane emission factor for domestic wastewater.

Income group	Pop. income group fraction (U) (a)	Fraction of degree of utilization (T_{ij}) (b)	Emission factor (EF_j) ($kgCH_4/kgBOD$) (c)	TOW (kgBOD/yr) (d)	Sludge removed (S) (kgBOD/yr) (e)
Rural	0.32	0.58	0.03	157,668	0
Urban high income	0.68	0.96			
Urban low income	0.00	NA			
Total	1	0.77			

Methane recovered and flared (R) ($kg/CH_4/yr$) (f)	Sludge removed (S) (kgBOD /yr) (e)	Net methane emissions ($kg/CH_4/yr$) g = [(a*b*c)*(d-e)]-f	GWP for CH_4 (h)	Total CO_2e ($kgCO_2e/yr$) i = g*h	Total CO_2e (tCO_2e/yr) (j)
0	0	3642	25	91053.27	91.05

Table 4.
Methane emissions from domestic wastewater.

In our study, we have assessed greenhouse gas emissions, resulting in 91.05 tCO_2eq for CH_4 and 0.343 tCO_2eq for N_2O. It is widely recognized that CH_4 emissions stem from methanogenic activity occurring within anaerobic tanks, whereas nitrifying and denitrifying bacteria generate N_2O emissions during the denitrification–nitrification process within tanks.

4.1.3 Estimated emissions from electricity consumption

Subsequently, GHG emissions resulting from the consumption of electricity are computed using an emission factor of 0.91 tCO_2e Mwh^{-1}, as per are presented in **Table 8** [24].

Definition		Default value	Range
$EF_{EFFLUENT}$	Emission factor (kg N_2O-N/kg-N)	0.005	0.0005–0.25
P	Number of people in country	Country specific	±10%
Protein	Annual per capita protein consumption	Country specific	±10%
F_{NPR}	Fraction of nitrogen in protein	0.16	0.15–0.17
$F_{NON\text{-}CON}$	Non-consumed protein adjustment factor	1.1 for countries with no garbage disposals 1.4 for countries with garbage disposals	1.0–1.5
$F_{IND\text{-}COM}$	Co-discharge factor for industrial nitrogen into sewers. Higher for countries with significant fish processing plants.	1.25	1.0–1.5

Table 5.
N_2O methodology default data [24].

Pop. (a)	Annual per capita protein consumption (protein) (kg/person/yr) (b)	Fraction of nitrogen in protein (F_{NPR}) (kgN/kg protein) (c)	Fraction of non-consumption protein ($F_{non\text{-}cons\cdot}$) (d)	Fraction of industrial and commercial co-discharged protein ($F_{ind\text{-}com\cdot}$) (e)	N removed with sludge (N_{sludge}) (f)	Total N in effluent (kgN/yr) g = (a*b*c*d*e)-f
10,094	0.052	0.16	1.4	1.25	0	146.97

Table 6.
Estimation of nitrogen in effluent.

Indirect GHG emissions predominantly stem from electricity usage, comprising 15.20 tCO_2eq. The wastewater treatment plant exhibited relatively lower energy consumption, primarily due to two reasons: Firstly, gravity-based wastewater collection obviates the need for pumping energy, and secondly, the absence of wet-well pumping in treatment plants owing to embedded kinetic energy in wastewater.

The cumulative GHG emissions (combining direct and indirect emissions) for the plant were estimated at 106.59tCO_2eq as given in **Table 9**.

4.2 Embodied carbon emissions from a full-scale wastewater treatment plant

In this study, the life cycle assessment system boundary is delineated to encompass the production stage (A1–A3), transportation stage (A4), and construction installation stage (A5). Cradle-to-gate (A1–A3) analysis encompasses the impacts of a material or product that is prepared for shipment to the construction site, incorporating factors such as raw material extraction, transportation, and manufacturing emissions. Information embodied carbon, materials use efficiency, and global warming potential in terms of tCO_2e for lifecycle stages is presented in **Figure 2** and summarized in **Table 10**.

N in eff. kgN/ yr. (a)	Emission factor (kg N_2O-N/ kgN) (b)	Conversion factor of kg N_2O-N into kg N_2O-N 44/28 (c)	Emission from WWTPs (default as zero)(d)	Total N_2O emissions/ yr.(e)	GWP for N_2O (f)	Total CO_2e (kg CO_2e/yr) (g)	Total CO_2e (t CO_2e/ yr)(h)
146.97	0.005	1.57	0	1.15	298	342.7	0.343

Table 7.
Estimation of emission factor and emissions of indirect N_2O emissions from wastewater.

Total Mwh used yearly (a)	Emission factor tCO_2e/Mwh (b)	Emission tCO_2e (c = a*b)
16.70	0.91	15.20

Table 8.
Electricity consumption of the wastewater treatment plant.

Year	CH_4 (tCO_2e)	N_2O (tCO_2e)	Electricity usage (tCO_2e Mwh^{-1})	Total (tCO_2e)
2020	91.05	0.343	15.20	106.59

Table 9.
Total operational carbon emissions of a full-scale wastewater treatment plant.

The examination of the building revealed that ready-mix concrete and steel contribute to the highest amount of embodied carbon, approximately 397.48 and 286.79 tonCO_2e, respectively. As depicted in **Figure 2**, the construction material stage (A1–A3) constitutes to the highest proportion of embodied carbon (1371.01 tonCO_2e), followed by the construction/installation stage A5 (453.96 tonCO_2e) and transportation to the building site A4 stage (58.18 tonCO_2e).

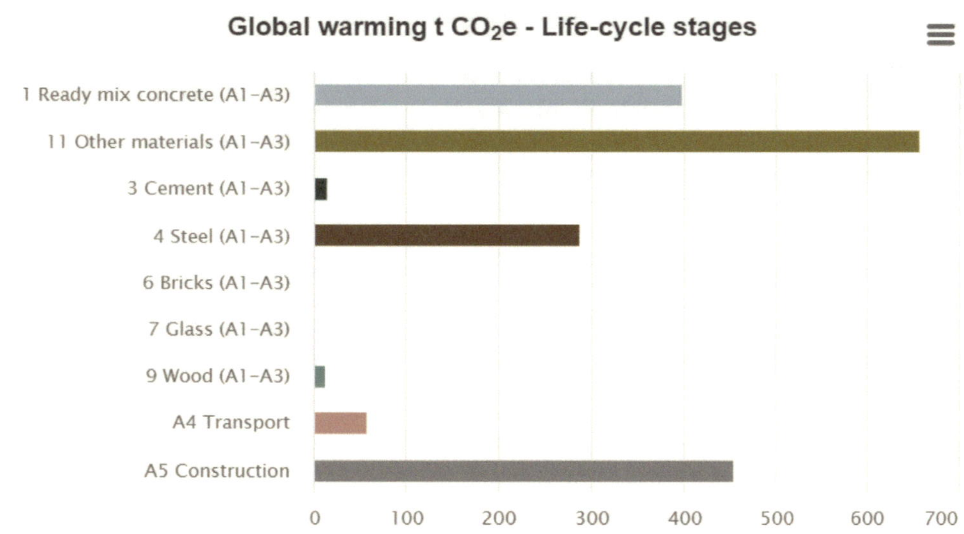

Figure 2.
Global warming potential t CO_2e—life cycle stages.

Result category	Global warming (tCO_2e)	Global warming ($kgCO_2e/m^2$)	Percentage (%)	Mass of raw materials (t)	Mass of raw materials (kg/m^2)
Ready mix concrete (A1–A3)	397.48	39.75	21.11	4951.2	495.12
Cement (A1–A3)	13.58	1.36	0.72	14	1.4
Steel (A1–A3)	286.79	28.68	15.23	71.45	7.15
Bricks (A1–A3)	0.15	0.01	0.01	0.63	0.06
Glass (A1–A3)	1	0.1	0.05	0.8	0.08
Wood (A1–A3)	13.15	1.31	0.7	27.92	2.79
Other materials (A1–A3)	658.86	65.89	34.99	462.32	46.23
Construction materials (A1–A3)	1371,01	137.1	—	5528.32	552.83
Transport to building site (A4)	58.18	5.82	3.09	—	—
Construction/ installation process (A5)	453.96	45.4	24.11	241.14	24.11
Total (A1–A3) + A4 + A5	1883	188	—	—	—

Table 10.
Embodied carbon and materials use by material type.

Figure 3 and **Table 11** display the proportion of embodied carbon in the building components of our case study. The major contributor to embodied carbon emissions is external walls and facade, which accounts for 39.83% of the total carbon, and emits 750 tonCO_2e.

To determine the cause of their high embodied carbon level, these components were analyzed to identify which materials contain the most embodied carbon. Information regarding embodied carbon and global warming potential in terms of tCO2e for resource types is presented in **Figure 4** and summarized in **Table 12**.

The major contributors of resource type to embodied carbon emissions are paints, coatings, and lacquers and ready-mix concrete for external walls and floors that account for 32.73% and 24.08% of the total carbon, respectively, and emit 620 and 450 tonCO_2e.

The global warming potential of construction materials (A1–A3), transport to the building site (A4), and the construction/installation process (A5) are quantified as 1371.01 tCO_2e, 58.18 tCO_2e, and 453.96 tCO_2e, respectively. Cumulatively, this accounts for a total of 1883 tCO_2e. The global warming potential per unit area ($kgCO_2e/m^2$) is computed as 188.3.

5. Conclusions

This chapter is anticipated to provide a substantial addition to the existing literature, addressing the scarcity of information regarding greenhouse gas (GHG)

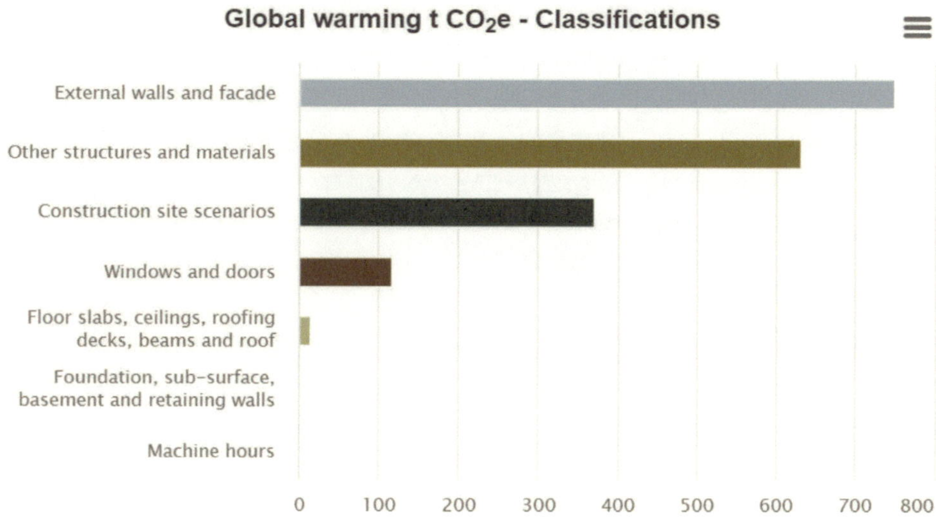

Figure 3.
Global warming potential tCO_2e—classifications.

Item	Value	Unit	Percentage (%)
External walls and facade	750	tCO_2e	39.83
Other structures and materials	630	tCO_2e	33.54
Construction site scenarios	370	tCO_2e	19.64
Windows and doors	120	tCO_2e	6.23
Floor slabs, ceilings, roofing decks, beams, and roof	14	tCO_2e	0.73
Foundation, sub-surface, basement, and retaining walls	0,33	tCO_2e	0.02
Machine hours	0,25	tCO_2e	0.01

Table 11.
Global warming tCO_2e- classifications.

emissions and embodied carbon emissions associated with wastewater treatment plants, and their implications for climate change. It aims to illustrate the potential climate change ramifications of an advanced wastewater treatment facility engaged in treating domestic wastewater, with specific attention to nitrogen and phosphorus treatment, alongside carbon treatment.

When assessing the impact of wastewater treatment plants on climate change, it becomes evident that both embodied carbon and GHG emissions significantly contribute. Embodied carbon encompasses the carbon emissions generated during the construction, material production, transportation, and commissioning phases of wastewater treatment plants. These emissions typically have enduring effects, persisting throughout the facility's lifespan. Conversely, operational carbon refers to emissions resulting from daily plant activities. Such as energy consumption, heating, and lighting.

Both types of carbon emissions influence the overall carbon footprint of wastewater treatment plants. Embodied carbon tends to have a pronounced impact during the construction and installation stages, while operational carbon's effect extends

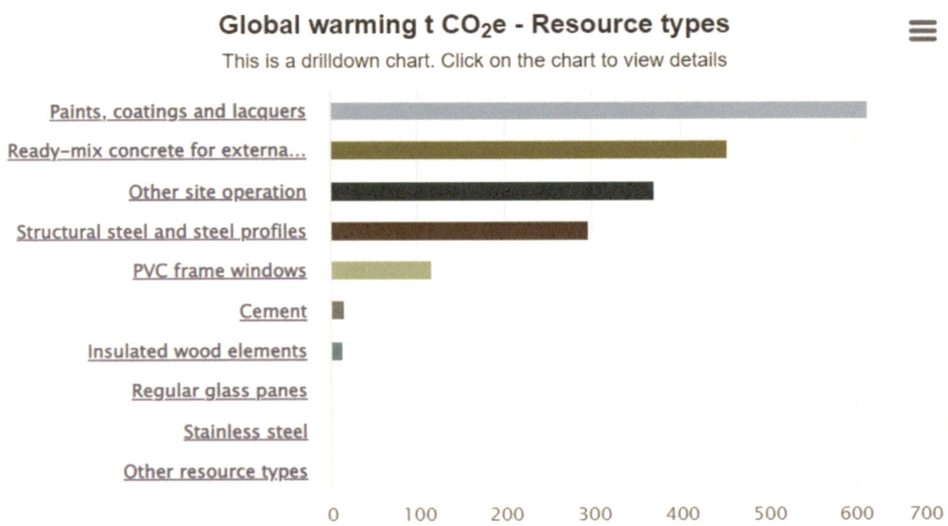

Figure 4.
Global warming potential tCO_2e—resource types.

Items	Value	Unit	Percentage (%)
Paints, coatings, and lacquers	620	tCO_2e	32.73
Ready-mix concrete for external walls and floors	450	tCO_2e	24.08
Other site operation	370	tCO_2e	19.64
Structural steel and steel profiles	300	tCO_2e	15.74
PVC frame windows	120	tCO_2e	6.16
Cement	14	tCO_2e	0.76
Insulated wood elements	13	tCO_2e	0.7
Regular glass panes	1	tCO_2e	0.05
Stainless steel	0.95	tCO_2e	0.05
Other resource types	1.6	tCO_2e	0.09

Table 12.
Global warming tCO_2e—resource types.

throughout the plant's operational life. Thus, both types of emissions play crucial roles in affecting climate change.

The carbon footprint serves as a vital sustainability indicator, facilitating the assessment of wastewater treatment plants' environmental impact on climate change. The widespread adoption of carbon footprint analysis underscores the urgency of addressing GHG emissions from these plants in pursuit of global carbon neutrality objectives.

Based on the findings of one of the investigation, the carbon footprint stemming from various sources was determined. Specifically, from domestic wastewater, 77,100 tons of CO_2 equivalent were emitted, along with 122.83 tons from fuel consumption, 6807 tons from electricity usage, and 95.71 tons from chemical usage. Additionally, the disposal of sludge contributed to 74.5 tons of CO_2 equivalent to the overall carbon footprint. Notably, electricity production from biogas accounted for 1654 tons of CO_2 equivalent. In comparison, the carbon footprint data from 2016 indicated 81,600 tons of CO_2 equivalent from wastewater, 40.44 tons from fuel consumption, 8406 tons from electricity consumption, and 49.5 tons from chemical consumption. The carbon footprint associated with sludge disposal in 2016 was calculated at 68.4 tons of CO_2 equivalent [27].

Our research measured greenhouse gas emissions and found that methane (CH_4) emissions were equivalent to 91.05 tons of carbon dioxide, while nitrous oxide (N_2O) emissions were equivalent to 0.343 tons of carbon dioxide. Methane is commonly produced by microorganisms in anaerobic environments, such as those found in certain types of tanks. Nitrous oxide, on the other hand, is a byproduct of the nitrogen cycle and is produced by bacteria during the processes of nitrification and denitrification, which can also occur in tanks.

Indirect GHG emissions predominantly stem from electricity usage, comprising 15.20 tCO_2eq. The wastewater treatment plant exhibited relatively lower energy consumption, primarily due to two reasons: firstly, gravity-based wastewater collection obviates the need for pumping energy, and secondly, the absence of wet-well pumping in treatment plants owing to embedded kinetic energy in wastewater. Wet-well pumping typically accounts for approximately 40% of total electricity consumption in wastewater treatment plants. The pronounced GHG emissions from electricity are linked to Türkiye's energy grid mix, which heavily relies on coal power plants. The cumulative GHG emissions (combining direct and indirect emissions) for the plant were estimated at 106.59tCO_2eq.

The examination of the structure of WWTP revealed that ready-mix concrete and steel contribute the highest amount of embodied carbon, approximately 397.48 and 286.79 $tonCO_2e$, respectively. The construction material stage (A1-A3) constitutes the highest proportion of embodied carbon (1371.01 $tonCO_2e$), followed by the construction/installation stage A5 (453.96 $tonCO_2e$) and transportation to the building site A4 stage (58.18 $tonCO_2e$). The major contributor to embodied carbon emissions is external walls % facade, which accounts for 39.83% of the total carbon, and emits 750 $tonCO_2e$. The major contributors of resource type to embodied carbon emissions are paints, coatings and lacquers, and ready-mix concrete for external walls & floors which account for 32.73% and 24.08% of the total carbon, respectively, and emit 620 and 450 $tonCO_2e$.

Greenhouse gas emissions are assessed for three key stages of a building's lifecycle: the production of construction, the transportation of these materials to the building site, and the on-site construction and installation. The material production stage (A1–A3) is responsible for 1371.01 tonnes of carbon dioxide equivalent (tCO_2e). Transporting the materials (A4) add a further 58.18 tCO_2e, while the construction and installation process (A5) contributes 453.96 tCO_2e. Overall, these three stages results in a total emission of 1883 tCO_2e. This translates to 188.3 kilograms of carbon dioxide equivalent per square meter ($kgCO_2e/m^2$) of building area.

Given the significant GHG emissions from wastewater treatment plants, their sustainability is an urgent concern in the context of global carbon neutrality. Mitigation strategies targeting GHG emissions focus on regulating and controlling operational parameters such as aeration intensity, dissolved oxygen (DO) concentration, chemical oxygen demand (COD)/nitrogen ratio (COD/N), sludge retention time (SRT), temperature, and pH. Additionally, mitigating nitrous oxide (N_2O) emissions from wastewater treatment plants is crucial for reducing the carbon footprint, as N_2O is a potent greenhouse gas.

Effective N_2O mitigation strategies involve site-specific monitoring, identifying generation pathways, regulating operational parameters, validating models at pilot scale, and verifying results in the field. Notably, even a modest 1% reduction in N_2O production could yield a substantial 30% reduction in the carbon footprint of wastewater treatment processes. Studies have shown that N_2O emissions can surpass half of the total carbon footprint in certain wastewater treatment plants employing processes such as A/O (anaerobic/oxic), A_2O (anaerobic-anoxic-oxic), and sequencing batch reactor (SBR). Therefore, addressing N_2O emissions is critical for achieving significant reductions in the carbon footprint of wastewater treatment processes.

Operational carbon could be reduced by following techniques:

1. Energy efficiency: Increasing the energy efficiency of equipment and processes within the plants through innovative technologies and practices can help reduce operational carbon emissions by lowering electricity and heat consumption.

2. Use of renewable energy: Utilizing renewable energy sources such as solar power, wind energy, or biogas to meet the energy needs of the plants can reduce operational carbon emissions.

3. Process optimization: Optimizing treatment processes to enhance efficiency can lead to reduced operational carbon emissions. This may involve using less energy and chemicals in wastewater treatment.

On the other hand, embodied carbon could be reduced by following techniques:

1. Sustainable material use: Opting for sustainable materials during the construction and expansion of treatment plants can reduce embodied carbon emissions. These may include low-carbon concrete, recycled materials, or locally sourced materials.

2. Carbon neutral methods: Employing carbon-neutral or carbon-negative technologies to offset or absorb carbon emissions generated during construction and operation processes can help reduce embodied carbon. For example, tree planting programs or promoting biological processes that mitigate carbon emissions.

3. Waste management and recycling: Reducing, recycling, and reusing waste generated during construction and operation can lower embodied carbon emissions. This can be achieved through the development of waste management plans and adopting best practices for waste reduction.

An experimental study was conducted on the energy consumption dependent on the aeration periods of the activated sludge process in a wastewater treatment plant in Portugal. Throughout the experimental process, data on biochemical oxygen demand (BOD), chemical oxygen demand (COD), total nitrogen (TN), and total phosphorus (TP) were monitored to maintain wastewater quality standards. Aeration periods for denitrification and nitrification were optimized, both with and without aeration. The aeration control system showed a 13% decrease in specific energy consumption and a reduction of 1.2 tons of CO_2 emissions compared to previous years with the same standard optimization data [28].

Research suggests that wastewater reuse can effectively offset both embodied energy and greenhouse gas (GHG) emissions. Achieving sustainability in treatment plants greatly depends on efficient resource recovery operations, which requires skilled personnel capable of optimizing parameters based on *in situ* operating conditions. The establishment of wastewater infrastructure demands proficient technicians and labourers, as the operation of treatment plants heavily relies on these human resources. Therefore, it is crucial to provide adequate training programs to equip personnel with a thorough understanding of the biological processes and to cultivate their intuitive ability to operate treatment plants and resource recovery facilities effectively.

Declarations

The authors have no relevant financial or non-financial interests to disclose. The authors have no competing interests to declare that are relevant to the content of this chapter.

Conflict of interest

The authors declare no conflict of interest.

Author details

Büşra Çayir[1] and Başak Kılıç Taşeli[2*]

1 ASCE Kanyon 1, Ankara, Türkiye

2 Department of Environmental Engineering, Giresun University, Giresun, Türkiye

*Address all correspondence to: basak.taseli@giresun.edu.tr

References

[1] Friedlingstein P, O'Sullivan M, Jones MW, Andrew RM, Bakker DCE, Hauck J, et al. Global carbon budget 2023. Earth System Science Data;**2023** (15):5301-5369. DOI: 10.5194/essd-15-5301-2023

[2] IPCC: TCW. In: Pachauri RK, Meyer L, editors. AR5 Synthesis Report: Climate Change. Copenhagen, Denmark: IPCC; 2014

[3] Kumar A. Global warming, climate change and greenhouse gas mitigation. In: Kumar A, Ogita S, Yau YY, editors. Biofuels: Greenhouse Gas Mitigation and Global Warming. New Delhi: Springer; 2018. pp. 1-16. DOI: 10.1007/978-81-322-3763-1_1

[4] IPCC: Climate Change 2014: Mitigation of Climate Change. Contribution of Working Group III to the Fifth Assessment Report of the Intergovernmental Panel on Climate Change [Edenhofer O, Pichs-Madruga R, Sokona Y, Farahani E, Kadner S, Seyboth K, Adler A, Baum I, Brunner S, Eickemeier P, Kriemann B, Savolainen J, Schlömer S, von Stechow C, Zwickel T and Minx JC (eds.)]. Cambridge, United Kingdom and New York, NY, USA: Cambridge University Press; 2014

[5] Ravishankara AR, Daniel JS, Portmann RW. Nitrous oxide (N2O): The dominant ozone-depleting substance emitted in the 21st century. Science. 2009;**326**:123-125. DOI: 10.1126/science.1176985 %J Science

[6] Hong Y, Xinyu G, Jingbo G, Hui W, Liang Z, Liru F, et al. Contribution of nitrous oxide to the carbon footprint of full-scale wastewater treatment plants and mitigation strategies—A critical review. Environmental Pollution. 2022; **314**:120295

[7] Wanjiru E, Xia X. Optimal energy-water management in urban residential buildings through grey water recycling. Sustainable Cities and Society. 2017;**32**: 654-668

[8] Zhang Q, Liu S, Wang T, Dai X, Baninla Y, Nakatani J, et al. Urbanisation impacts on greenhouse gas (GHG) emissions of the water infrastructure in China: Trade-offs among sustainable development goals (SDGs). Journal of Cleaner Production. 2019;**232**:474-486

[9] Dixit MK. Embodied energy analysis of building materials: An improved IO-based hybrid method using sectoral disaggregation. Energy. 2017;**124**:46-58

[10] Niu K, Wu J, Qi L, Niu Q. Energy intensity of wastewater treatment plants and influencing factors in China. Science of the Total Environment. 2019;**670**: 961-970

[11] Xu J, Luo P, Lu B, Wang H, Wang X, Wu J, et al. Energy-water nexus analysis of wastewater treatment plants (WWTPs) in China based on statistical methodologies. Energy Proced. 2018;**15**:259-264

[12] Gallego-Schmid A, Tarpani RRZ. Life cycle assessment of wastewater treatment in developing countries: A review. Water Research. 2019;**153**:63-79

[13] Jeong H, Minne E, Crittenden JC. Life cycle assessment of the City of Atlanta, Georgia's centralised water system. International Journal of Life Cycle Assessment. 2015;**20**(6):880-891

[14] Li Y, Luo X, Huang X, Wang D, Zhang W. Life cycle assessment of a municipal wastewater treatment plant: A case study in Suzhou, China. Journal of Cleaner Production. 2013;**57**:221-227

[15] Singh P, Kansal A, Carliell-Marquet C. Energy and carbon footprints of sewage treatment methods. Journal of Environmental Management. 2016;**165**: 22-30

[16] Chen W, Yang S, Zhang X, Jordan ND, Huang J. Embodied energy and carbon emissions of building materials in China. Building and Environment. 2022;**207**:108434

[17] Pai V, Elzarka H. Whole building life cycle assessment for buildings: A case study ON HOW to achieve the LEED credit. Journal of Cleaner Production. 2021;**297**:126501

[18] Hellweg S, Milà i Canals L. Emerging approaches, challenges and opportunities in life cycle assessment. Science (American Association for the Advancement of Science). 2014;**344**: 1109-1113

[19] Teng Y, Li K, Pan W, Ng T. Reducing building life cycle carbon emissions through prefabrication: Evidence from and gaps in empirical studies. Building and Environment. 2018;**132**:125-136

[20] Jungclaus M, Esau R, Olgyay V, Rempher A. Reducing Embodied Carbon in Buildings: Low-Cost, High-Value Opportunities. California, USA: RMI; 2021. Available from: http://www.rmi.org/insight/reducing-embodied-carbon-in-buildings

[21] BS EN ISO 14040:2006+A1:2020.: Environmental Management -Life Cycle Assessment -Principles and Framework. London, UK: British Standards Institute; 2020

[22] Waldman B, Huang M, Simonen K. Embodied carbon in construction materials: A framework for quantifying data quality in EPDs. Buildings & Cities. 2020;**1**(1):625-636

[23] Li L. Integrating climate change impact in new building design process: A review of building life cycle carbon emission assessment methodologies. Cleaner Engineering and Technology. 2021;**5**:100286

[24] IPCC: 2006 IPCC Guidelines for National Greenhouse Gas Inventories – A Primer, Prepared by the National Greenhouse Gas Inventories Programme. In: Eggleston HS, Miwa K, Srivastava N, Tanabe K, editors. Global Environmental Strategies (IGES). Hayama, Japan: IPCC IGES; 2006

[25] Taşeli BK. Sustainability assessment of wastewater treatment plants. In: Water Chemistry. London, UK: IntechOpen; 2020. DOI: 10.5772/intechopen.88338

[26] Gibbons OP, Orr JJ, Archer-Jones C, Arnold W, Green D. How to Calculate Embodied Carbon. 2nd ed. London, UK: The Institution of Structural Engineers; 2022

[27] Güller S, Determination of Muğla waste water treatment plant carbon footprint [thesis]. Muğla, Türkiye: Muğla Sıtkı Koçman University; 2018

[28] Silva MM, Cristovão L, Marinho D, Esteves E, Fraqueza G, Martins A. Reduction of carbon emissions in a mediterranean urban wastewater treatment plant. WIT Transactions on Ecology and the Environment. 2021;**251**: 87-94. DOI: 10.2495/WS210091

Chapter 3

Continuous Tubular Photoreactor with TiO_2 Nanoparticles Catalyst for Batik Industrial Wastewater Treatment

Dessy Ariyanti, Dina Lesdantina, Ariana Aisa, Farida Diyah Hapsari, Filicia Wicaksana and Wei Gao

Abstract

Like many other textile industries, batik requires a lot of water, and its use of different dyes and other chemicals accounts for most of the water pollution (17–20%). It can reduce light penetration, which in turn may decrease photosynthetic activity, and it may also have an impact on the quantity of oxygen available for the biodegradation of aquatic microbes. Many dyes, including diazo (-N=N-), direct, and basic dyes, are challenging to work with using standard techniques, and the byproducts may occasionally be quite hazardous. Photocatalysis with various semiconductor materials has been the focus of investigations in recent years for the wastewater treatment application. Among many semiconductors, because of its strong photocatalytic activity, outstanding stability, lack of toxicity, and affordability, TiO_2 is still the most extensively used photocatalyst. When a catalyst (TiO_2) is exposed to ultraviolet (UV) light, hydroxyl radicals (•OH) are produced in the UV/TiO_2 photocatalytic oxidation process. Organic materials therefore mineralize into CO_2, H_2O, and inorganic components. In this chapter the application of continuous tubular photoreactor with TiO_2 nanoparticles as a catalyst for Batik wastewater treatment is reviewed.

Keywords: photocatalysis, TiO_2, wastewater treatment, dyes degradation, photoreactor

1. Introduction

Synthetic dyes are organic compounds that are widely used in various industries such as the textile industry. One form of environmental contamination that produces a lot of wastewater from the dyeing process is heavy metals and organic pollutants. If this wastewater is not adequately treated, it could pose a hazard to the ecosystem [1]. Dye wastewater tends to cause eutrophication by lowering the amount of dissolved oxygen in the water body and is difficult for nature to break down [2, 3].

Various treatment methods have been developed such as physical, biological, and chemical treatments. Physical methods such as adsorption and sedimentation have widely been used, as well as chemical methods such as coagulation and flocculation.

The current method of treating wastewater has not been able to fully degrade the pollutants into less harmful molecules; instead, it only transfers the toxins to another medium. In the meantime, total breakdown can be achieved through biological treatment using microorganisms, but this requires a longer process and specific handling due to their extreme sensitivity to environmental factors [4]. Advanced oxidation process (AOP) such as photocatalysis is one of the technologies developed to degrade organic pollutants such as dyes. The process involves semiconductor material as catalyst and photon to excite an electron and holes that are responsible for the photodegradation. TiO_2 is a semiconductor that can be used as a catalyst in the photocatalysis process [5] with advantages of low price, non-toxicity, high chemical stability, and good response to light [6].

Research on photocatalysis in degrading dyes organic pollutants has been massively conducted globally with various type of dyes. The batik business in Indonesia is a sort of textile sector that has grown rapidly at the household or small- to medium-sized enterprise (SME) level, producing vast amounts of wastewater that contains dyes and is ill-equipped with appropriate wastewater treatment technologies. In order to meet the demand for wastewater treatment methods that are both sustainable and environmentally benign in order to attain zero waste in the batik industry, this chapter reports on the development of a continuous tube photoreactor using TiO_2 nanoparticles as a catalyst.

2. Batik wastewater characteristics

A historical artifact created by the Indonesian people, batik is a blend of artistic expression with modern technology. It comes in a variety of colors and patterns. Before creating a gorgeous batik fabric, the pattern is drawn with wax onto the mori cloth. Subsequently, the fabric is submerged in a dye solution. Finally, the fabric is washed multiple times to achieve an appealing hue. Massive-scale batik production generates a substantial volume and quality of wastewater during the processing period. Synthetic dyes are utilized in the production of widely appreciated batik. The wastewater that is generated conceals the risks inherent in batik fabric. Batik wastewater contains dyes, wax, paraffin, resin, and other chemical components [5, 6]. Synthetic dyes are divided into three categories: cationic, anionic, and non-ionic dyes.

Charge is the primary distinction between cationic and anionic dyes. The molecules of cationic dyes are positively charged, whereas those of anionic dyes are negatively charged. Both cationic and anionic dyes are water-soluble. Cationic dyes include Crystal violet, Methylene blue, Basic blue, and Basic red [7, 8]. Methyl orange, Remazol, Alizarin Yellow R, and Eriochrome black T (EBT) are anionic dyes [9]. Remazol dyes are categorized as reactive azo dyes because it forms an ester bond with the fiber when they bind to it [10]. The structures of various dyes can be found in **Table 1**. Chromophores and auxochromes are the basic components in dye molecules. The colors that are visible are the result of chromophores. Auxochromes, nevertheless, act to enhance the affinity for the fibers and facilitate the solubility of the chromophore in water [11]. Batik wastewater contains a high concentration of electrolytes, which makes it difficult for light to penetrate [10], particularly solar

Dye	IUPAC name	Structure	MW	Ref
Methyl orange (MO)	Sodium; 4-[[4(dimethylamino) phenyl]diazenyl] benzenesulfonate	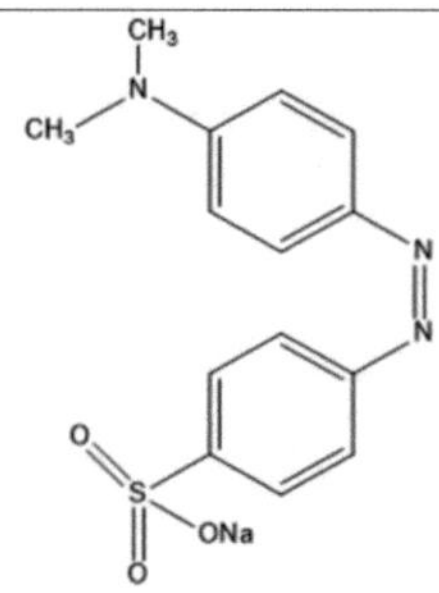	327.33	[12]
Eriochrome black T (EBT)	Naphthalenesulfonic acid, 3-hydroxy-4-[(1-hydroxy-2-naphthyl) azo]-7-nitro-, sodium salt	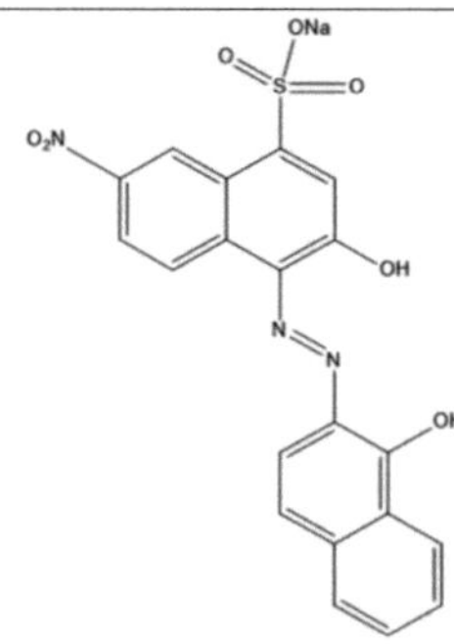	461.37	[12]
Methylene blue (MB)	[7-(Dimethylamino) phenothiazine-3-ylidene] dimethylazanium; chloride	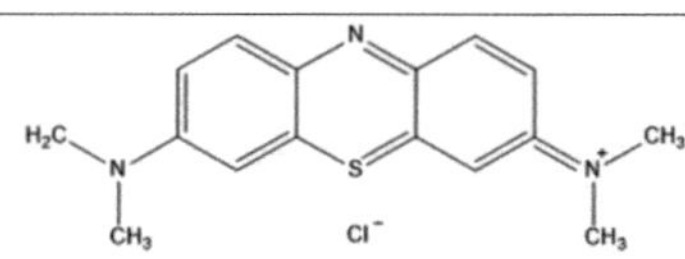	319.85	[12]
Crystal violet (CV)	Tris(4-(dimethylamino) phenyl) methylium chloride	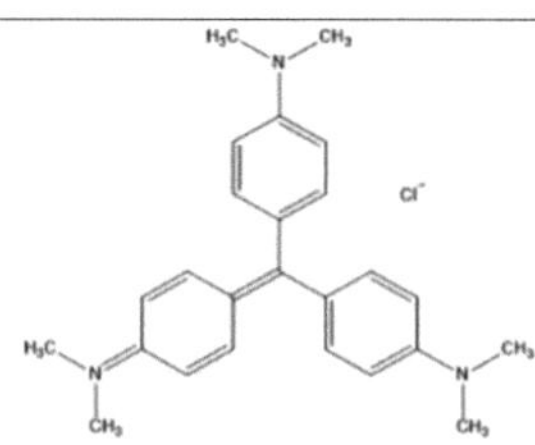	407.21	[12]
Remazol dye Red Rb/Reactive red 198	Tetrasodium;(3Z)-5-[[4-chloro-6-(3-sulfonatoanilino)-1,3,5-triazin-2-yl] amino]-4-4-oxo-3-[[4-(2-sulfonatooxyethyl-sulfonyl)phenyl] hydrazinylidene] naphthalene-2,7-disulfonate	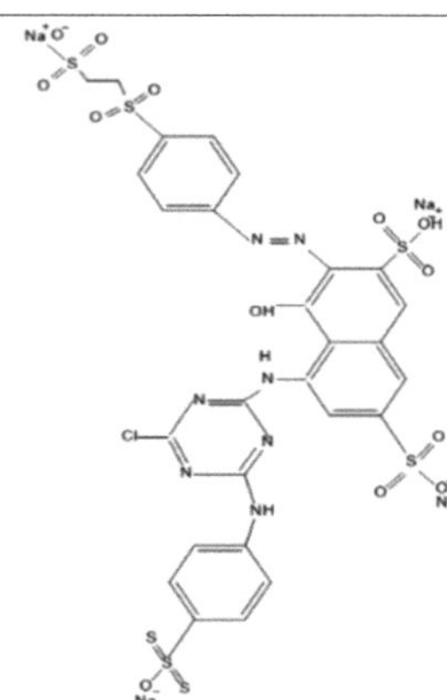	984.2	[10]

Dye	IUPAC name	Structure	MW	Ref
Remazol yellow G/Reactive yellow 14	disodium; 3-chloro-4-[4-[[2-methoxy-5-(2-sulfonatooxyethyl sulfonyl) phenyl] diazenyl]-3-methyl-5-oxo-4H-pyrazol-1-yl]-5-methyl benzene sulfonate	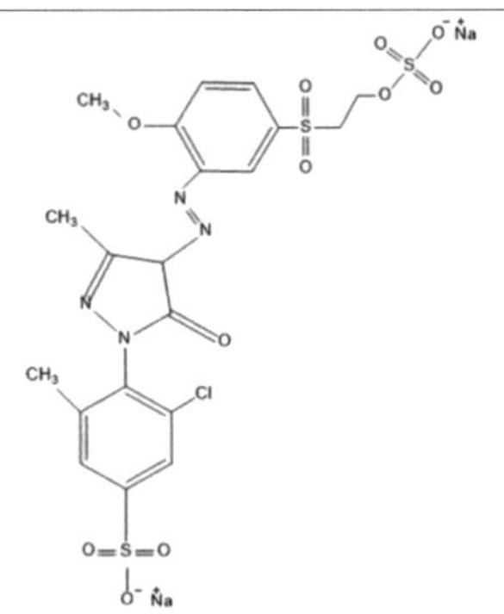	669	[13]
Remazol brilliant blue R	disodium 1-amino-9,10-dioxo-4-({3-[2-(sulfonatooxy) ethanesulfonyl] phenyl}amino)-9,10-dihydroanthracene-2-sulfonate	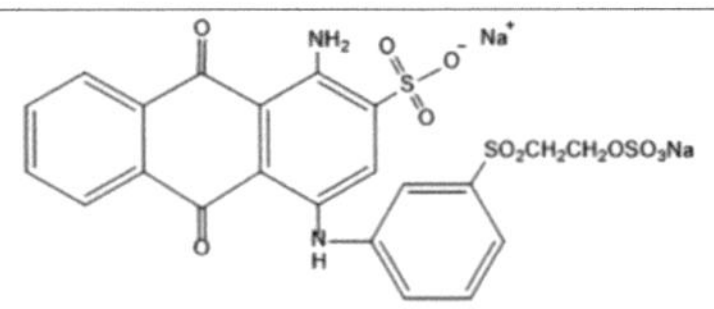	626.53	[14]
Remazol brilliant violet 5R	trisodium;5-acetamido-4-hydroxy-3-[[2-hydroxy-5-(2-sulfonatooxyethyl sulfonyl)phenyl] diazenyl]naphthalene-2,7-disulfonate	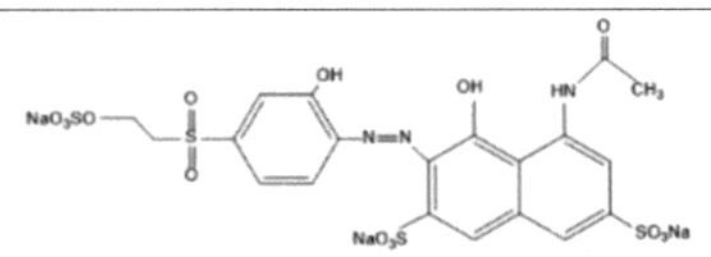	735.6	[12]

Table 1.
Chemical structure of dyes.

irradiation, a natural degrading agent. Moreover, the presence of batik dyes in aquatic environments can give rise to a multitude of adverse effects, such as respiratory issues, cancer, allergies, and ecological imbalances.

3. TiO_2 Photocatalysis for organic degradation

The term "photocatalysis" was used for the first time in 1911 to explain the effect of the illumination of ZnO on the bleaching of Prussian blue. The ZnO acts like a catalyst, which increases chemical reaction rate without getting consumed during the reaction process. Under ultraviolet (UV) light, Fujishima and Honda observed the phenomena of photocatalytic water splitting on a TiO_2 electrode [15], opening a new world of possibilities to produce clean hydrogen.

The term "photocatalysis" refers to the use of a catalyst to speed up a photoreaction. Among these is photosensitization, a process in which an initial absorption of radiation by a different molecular entity known as the photosensitizer causes a photochemical modification in one molecular entity [16]. Depending on the particular photoreaction, the catalyst may interact with a principal photoproduct, the substrate in either its excited or ground state, or both to speed up the reaction. The schematic of photocatalytic processes is displayed in **Figure 1**. After the catalyst C absorbs the light, the sensitized photoreaction can happen in two ways: (i) through

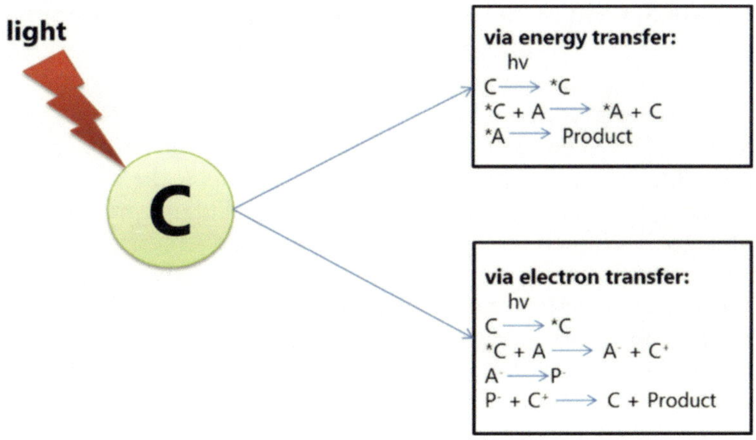

Figure 1.
The schematic of different photocatalytic pathway, C-catalyst, A-reactant (adapted from [17]).

energy transfer by converting the reactant of interest, A, into an activated state that can be oxidized more readily than its ground state; and (ii) via electron transfer, by acting either as an electron donor or acceptor [17].

The advantage of photocatalysis over traditional or current technologies [18] are as follows: (1) with the ability to use renewable and pollution-free solar energy, photocatalysis offers a good substitute for the energy-intensive conventional treatment procedures (adsorption, ultrafiltration, reverse osmosis, coagulation by chemical agents, and ion exchange on synthetic adsorbent resins); (2) unlike traditional treatments that move contaminants from one phase to another, photocatalysis produces harmless products; (3) a range of dangerous substances in various wastewater streams can be destroyed via the photocatalytic method; (4) photocatalysis requires little chemical input, mild reaction conditions, and a short reaction time; and (5) it can be used with hydrogen and produces little secondary waste.

As it offers exceptional advantages, for the past 40 years, the development of photocatalysis technology has progressed rapidly in term of materials and application. Furthermore, it will become very attractive as the process can be widely used in various applications such as degradation of organic contaminants in water, removal of metallic and other inorganic pollutants, recovery of valuable metals and chemicals, water disinfections and bacterial removal, purification and disinfection of air, hydrogen generation, functional materials including self-cleaning and anti-reflective material, and new synthetic routes for chemical processes [19].

4. Photocatalysis for degradation of organic pollutant

Nowadays almost all human activity discharges a wide variety of contaminants into the environment. There are approximately 30,000 chemicals commercially available with less knowledge of their occurrence and fate in the environment. Not to mention newly emerging micro-pollutants (e.g., pharmaceuticals, antibiotics, and pathogens) that are now easily being detected in the environment. The biggest challenge is the industrial dyes and textile dyes, which have been a long-term environmental threat. During the production process, 10–20% of all dye products produced worldwide are lost in textile wastes and discharged into aquatic bodies as effluents. Organic dyes are one of the leading groups of pollutants released into wastewaters from textile and other industrial processes [18].

Advanced oxidation processes (AOPs) such as ozonolysis, ozone/H_2O_2 peroxide process, UV/ozone, UV/H_2O_2, Fe^{2+}/H_2O_2 Fenton process, irradiation with electrons, vacuum ultraviolet and ultrasonic, photocatalysis, and combinations of the stated methods have been extensively explored to mitigate a great variety of pollutants present in various environmental media. Numerous studies have focused on photocatalysis using different semiconductor materials in recent years. Hydroxyl radicals (•OH) are produced during the photocatalytic oxidation process when the catalyst is exposed to light. Organic materials therefore mineralize into CO_2, H_2O, and inorganic components [20].

Photocatalytic reaction primarily depends on photon energy and the response of catalysts. Semiconducting materials are characterized by a filled valence band, a vacant conduction band, and a band gap between two bands.

If a semiconducting material acts as a catalyst, it will perform as charge sensitizers by the irradiation of light and stimulate the redox process. The fundamental steps in the process of semiconductor photocatalysis are shown in **Figure 2**. An electron from the valence band is excited to the conduction band when light strikes the photocatalyst surface and the energy of the incident light is equal to or greater than the photocatalyst bandgap energy. This results in an excess of negative charge in the conduction band, e−, and an electron deficiency or hole, h+, in the valence band. They can participate in redox reactions and are equivalents of oxidizing and reducing agents, respectively. Any organic species that is adsorbed on the semiconductor's surface could undergo sequential oxidation and reduction processes with these holes and electrons to produce the required products. [21]. Photocatalysts such as TiO_2, ZnO [22], Fe_2O_3 [23], CdS [24], WO_3 [25], Cu_2O [26], and ZnS [27] have been used to effectively break down a variety of organic contaminants. TiO_2 is the most often utilized photocatalyst among those semiconductors because of its strong photocatalytic

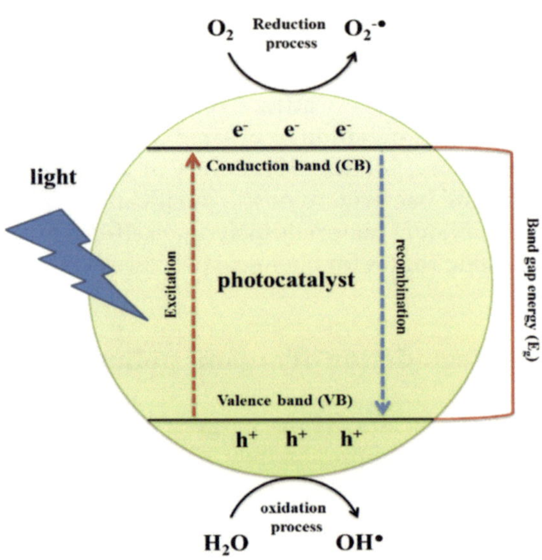

Figure 2.
The schematic of semiconductor photocatalytic mechanism.

activity, high stability, non-toxicity, and affordability when employed in the wastewater treatment industry [28].

In order to achieve greater activity and outperform TiO_2-based photocatalysts, new photocatalytic materials with unique compositions and structures are being thoroughly investigated. But there aren't many examples of successful practical applications. From the technical and economical point of view, alternative photocatalytic materials that are as "good" as TiO_2 are hard to find. One of the most viable and practical approaches in developing better photocatalysts remains the modification of TiO_2 in various methods (e.g., impurity doping, sensitization, surface modification, integration with other nanostructured materials) [29].

5. Basic structures and properties of TiO_2 semiconductor

Since the water splitting breakthrough using semiconductor reported by Fujishima and Honda in 1972 [15], titanium dioxide (TiO_2) has been thoroughly studied as a potentially useful material in a variety of study domains, particularly those pertaining to the environment and energy [30]. TiO_2 is a transition metal oxide semiconductor with three distinct polymorph crystalline structures, namely rutile, anatase, and brookite. Each structure has distinctive parameters, which contribute to the different physical and chemical properties. The size and crystal structure of the particles determine how stable a certain phase of TiO_2 is. The most stable phase for particles larger than 35 nm is rutile. For nanoparticles smaller than 11 nm, the most stable phase is anatase. When it comes to microparticles in the 1135 nm range, brookite is most stable. Bulk rutile, anatase, and brookite have band gaps of 3.0, 3.2, and 3.14 eV, which implied its photon absorption capability in the wavelength smaller than 431, 387, and 395 nm, respectively [31]. It has been discovered that the anatase form offers the best properties for photocatalytic oxidation. When compared with rutile TiO_2, the anatase has greater photocatalytic activity. One argument that could apply is the fact that anatase shows a smaller indirect band gap than direct band gap. Indirect band gap semiconductors typically have longer charge carrier lifetimes than do direct gap semiconductors [32].

6. Photocatalytic oxidation using TiO_2

In order to initiate the inter-band transition of electrons between the lowest unoccupied molecule orbital (LUMO) and the highest occupied molecule orbital (HOMO), the photocatalytic reaction needs a minimum photon energy that is greater than TiO_2's band gap. The photo-excited electrons are de-excited via multiple pathways following irradiation: (i) they recombine with holes on the semiconductor surfaces; (ii) they recombine with holes in the bulk volume; (iii) they are transferred to the absorbed organic/trapped or inorganic redox species (typically oxygen molecules in the aerated solution); and (iv) the holes migrate to the surfaces and receive electrons from the absorbed redox species via oxidation reactions. TiO_2 is a semiconductor with a wide bandgap (3.0–3.2 eV) that requires the excitation light wavelength range shorter than ca. 400 nm ($Eg = hc/\lambda \approx 1240/\lambda$) [29].

The overall photocatalytic process including the reduction-oxidation with its surrounding component can be divided into the following steps and described further in Eqs. (1)–(12) [29]:

$$TiO_2 + hv \rightarrow TiO_2\left(e_{CB}^{-} + h_{VB}^{+}\right) \quad (1)$$

$$TiO_2\left(h_{VB}^{+}\right) + H_2O \rightarrow TiO_2 + H^{+} + OH\cdot \quad (2)$$

$$TiO_2\left(h_{VB}^{+}\right) + OH^{-} \rightarrow TiO_2 + OH\cdot \quad (3)$$

$$TiO_2\left(e_{CB}^{-}\right) + O_2 \rightarrow TiO_2 + O_2\cdot^{-} \quad (4)$$

$$O_2\cdot^{-} + H^{+} \rightarrow HO_2\cdot \quad (5)$$

$$HO_2\cdot + HO_2\cdot \rightarrow H_2O_2 + O_2 \quad (6)$$

$$TiO_2\left(e_{CB}^{-}\right) + H_2O_2 \rightarrow OH\cdot + OH^{-} \quad (7)$$

$$H_2O_2 + O_2\cdot^{-} \rightarrow OH\cdot + OH^{-} + O_2 \quad (8)$$

$$H_2O_2 + hv \rightarrow 2OH\cdot \quad (9)$$

$$\text{Organic compound} + OH\bullet \rightarrow \text{degradation products} \quad (10)$$

$$\text{Organic compound} + TiO_2\left(h_{VB}^{+}\right) \rightarrow \text{oxidation products} \quad (11)$$

$$\text{Organic compound} + TiO_2\left(e_{CB}^{-}\right) \rightarrow \text{reduction products} \quad (12)$$

A semiconductor's surface undergoes photocatalytic processes when photons are absorbed, which can lead to the breakdown of organic molecules or the splitting of water. Because of its ability to gather photons, it can be used to degrade organic molecules, remove contaminants from a variety of media, and oxidize or reduce materials to generate hydrogen and hydrocarbons.

7. Photoreactor for TiO_2 photocatalytic process

The performance of TiO_2 photocatalytic activity is commonly measured by the reaction rate of the degradation process. In general, the degradation of photocatalysis proceeds according to pseudo first-order reaction kinetics. Prefixing the order of a reaction with "pseudo-" has become commonplace since, in actuality, one of the reactants in catalytic systems functions as a catalyst while maintaining its concentration constant. In a similar vein, when the substrate is present in considerable excess, it completely covers the catalyst surface, and its disappearance becomes mostly concentration dependent. A pseudo zero order photocatalytic process will be seen with such saturation coverage [33]. Because the concentration of adsorption in the dark is so small, the Langmuir-Hinshelwood model is frequently employed to describe and analyze reaction kinetics.

Both organic and oxygen molecules must participate in the photocatalytic destruction of organic contaminants. Since oxygen is usually in high concentration, its concentration during photocatalysis can be thought of as constant. The photocatalytic rate, as per the law of mass-action, can be represented as [34]:

$$r = -\frac{dC(t)}{dt} = kC(t) \tag{13}$$

where k is the rate constant, and $C(t)$ is the concentration of reactants that adsorb on the semiconductor surface at time t. Photocatalysis requires the adsorption of organic materials on the semiconductor surface. It is believed that the adsorption-desorption (A–D) equilibrium complies with a Langmuir isotherm. However, in typical photocatalytic circumstances, the organic concentration is typically quite low, therefore the Eq. (13) can be simplified to:

$$\ln\frac{C_0}{C_t} = kt \tag{14}$$

where C_0 is initial concentration of organic substances before photocatalysis, and C_t is the reactive concentration at time t during the photocatalytic reaction. Eq. (14) can be easily used to calculate the k of different photocatalytic reactions.

Numerous factors and operational parameters affect the efficiency of photocatalytic processes, particularly when it comes to organic degradation. After a thorough analysis of the key factors, it was determined that the initial reactant concentration, light wavelength, feed flow rate, irradiation time, and pH had decreasing effects on process performance, while the increase in dissolved oxygen, light intensity distribution, photocatalyst loading to some extent, and air flow rate had increasing effects [32]. The execution of TiO_2 photocatalysis technology in the wastewater treatment application requires more research and development especially in the process and kinetics aspects of the photocatalysis itself. Several types of photocatalytic reactors have been proposed and investigated for the last 20 years, which can mainly be divided into two categories according to the catalyst form: (i) TiO_2 immobilized on support materials, e.g., quartz sand, glass, glass wool matrix, ceramic or polymer membranes, noble metal etc., and (ii) TiO_2 suspended in aqueous medium [32]. Even though it has a high operating cost and necessitates additional treatment after the degradation process, the TiO_2 suspended reactor has a higher photocatalytic area to reactor volume ratio, higher mass transfer and degradation efficiency, fairly uniform catalyst distribution, and an adjustable amount of nanoparticle suspension in the reactor to deal with different compositions of treated solution. In the meantime, TiO_2 immobilized reactors do not require catalyst separation or recycling, and their degrading efficiency is lower than that of suspension mode. Additionally, they cannot have their catalyst loading adjusted to address [35].

Several different reactors were designed such as novel tube light reactor, multiple tube light reactor, fluidized bed photoreactor, fixed bed photocatalytic reactor, batch photoreactor, slurry continuous flow reactor, hexagonal annular photoreactor; spinning disk photoreactors; solar photocatalytic reactor; annular photocatalytic reactor, and tubular photocatalytic reactor, etc. Some of them are designed with the suspended type of catalyst and the other using immobilized catalyst within the reactor [32]. From the technical application point of view the immobilized TiO_2 is preferred because, following the photocatalytic process, it does not require any additional post-treatment for the recovery of catalyst particles. On the other hand, films containing immobilized TiO_2 powders had issues with the catalyst particles' detachment. Furthermore, a portion of the porous structure is lost through the sintering process during the heating procedure that fixes the photocatalyst. They result in inefficient mass transfer by decreasing the photocatalyst's area-to-volume ratio. As a result, there appears to be a factor of 10 difference in the efficiency of reactors with immobilized photocatalysts compared to those with dispersed TiO_2 particles (slurry) [32].

The overview of the photoreactor with the optimized parameters is summarized in **Table 2**. The photocatalytic activity of TiO_2 is a multifaceted function that relies on the physical characteristics of the particles. Types of dye's component to degrade are more profound to be the driving force of the optimum degradation rather than the design of the photocatalytic reactor.

Photoreactor System	Optimal Condition	Photodegradation Performance	Ref.
• **Characteristics:** a pilot-scale falling film photoreactor (FFR) in slurry mode has a surface area of 60x120 cm and is tilted at 35° (local latitude) • **Light Source:** genuine solar light (35 W) • **Scale:** 25 L	• **Pollutant:** 10 ppm • **Catalyst:** 100 ppm • **Light Intensity:** 1.1 mW/cm^2 • **pH:** 2 (2–10) • **Oxidating Agent:** 10 mM H_2O_2 • **Stirring/Flow rate:** 30 L/h	• **Pollutant:** Methylene Blue • **Catalyst:** Pure TiO_2, and 1 $^{wt.\% Ag/TiO}{}_2$ (by the simple chemical reduction) • **Efficiency:** 99%	[36]
• **Characteristics:** double-walled horizontal glass reactor (DHGR) • **Scale:** 1–6 L • **Light Source:** • UV lamps (400 W)	• **Pollutant:** 560 mg/L • **Catalyst:** n.a • **pH:** 3	• **Pollutant:** Remazol Red RGB • **Catalyst:** Pure TiO_2 and 3 mol% Ag–TiO_2 (by the modified sol–gel method) • **Efficiency:** 94.29% (COD) and 98.44% (Color), for the immobilized catalyst	[37]
• **Characteristics:** continuous type of reactor which is named CS4PR and built using a glass vessel • **Scale:** 500 mL • **Light Source:** UV Lamp (Philips TUV 36 W/G 36 T8)	• **The ratio of catalyst, C sulphate ion & V sample** 9: 500 ppm: 500 mL/5 mg.L of RhB • **pH:** 6.5–7.5 • **Stirring/Flow rate:** 2–10 mL/s	• **Pollutant:** Rhodamine B (RhB) • **Catalyst:** TiO_2/NO_3- film coated ceramic • **Efficiency:** 100%	[38]
• **Characteristics:** homemade reactor • **Scale:** 4 L • **Light Source:** sunlight irradiation	• **Pollutant:** 10 ppm • **Catalyst:** 1 g/L	• **Pollutant:** Methylene Blue • Methyl Orange • Rhodamine B • **Catalyst:** TiO_2 Degussa P25 (anatase: rutile = 3:1) • **Efficiency:** 85%; 49% and 96%	[39]

Table 2.
Previous research on the development of photocatalytic reactor.

8. Continuous tubular photoreactor with TiO_2 nanoparticles catalyst for batik industrial wastewater treatment

8.1 Continuous tubular photoreactor prototype

A continuous tubular photoreactor prototype has been developed as shown in **Figure 3**. The system consists of five photoreactor tubes that are designed to allow fluid to flow either in parallel or in series depending on the ball valve that is mounted on the pipe. The box panel is a place where all the buttons to turn on or turn off system parts are stored. The frame serves as a structural support for all the equipment and is made from hollow pipe of ST 42 steel. The frame's primary dimensions are 1200 × 365 × 1700 mm (l × w × h), and it has four roller wheels at the bottom to facilitate system mobility. The feed tank with dimensions of 600 × 300 × 300 mm (w × w × h) is made of 304 stainless steel plate with a thickness of 1 mm. It also has a glass cover on the front to make it easier to watch during the process. Polyvinyl chloride (PVC) pipes Sch 40 with a 1/2-inch diameter are used in the fluid flow system. These pipes are connected using elbow pipes at the ends and tee pipes at each intersection with the same diameter and thickness. Two primary valves are positioned at the end of the flow output and close to the pump. There are also sixteen ball valves to control the flow whether operating in series or parallel. A flowmeter tube with an 18 L/min capacity is mounted on the system's top side.

The tubular photoreactor is made of 304 stainless steel, 90 cm in length and 3/4 inch in diameter with capacity 12 gallons per minute. Inside it is fitted with a UV light with a power of 40 Watt, which is shielded by a glass sleeve of 85 cm. When the main valve and pump are turned on, the solution will flow into the photoreactor. The solution flow rate is controlled with a flowmeter.

The lower the flow rate, the longer the flow will stay in the photoreactor, and the higher the flow rate, the faster the flow will exit the photoreactor. If the system is run in series, the solution will enter photoreactor 1 and photocatalytic activity will occur in the tube. Once the solution exits photoreactor 1, it proceeds to photoreactor 2 for the same procedure, and so on, until the solution reaches photoreactor 5. This system also can run in parallel. The solution will enter simultaneously into photoreactors 1, 2, 3, 4, and 5. Photocatalytic activity also occurs simultaneously in each tube photoreactor.

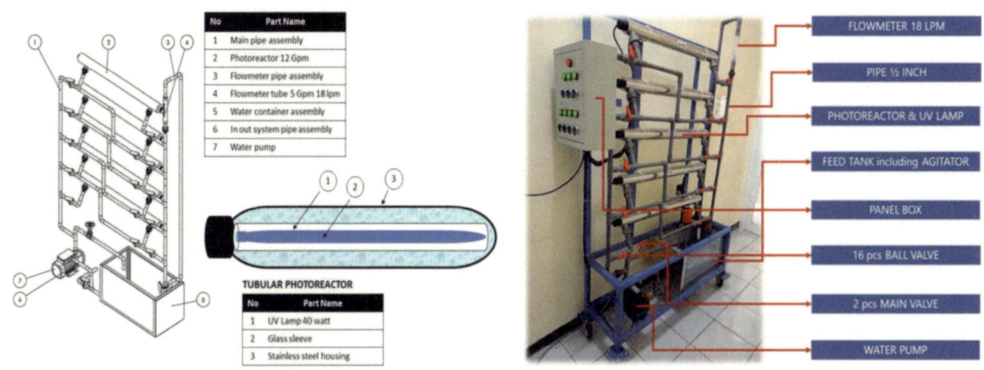

Figure 3.
Continuous tubular photoreactor system.

8.2 Continuous tubular photoreactor performance

Methylene blue solution as a representation of batik waste was run in series on a continuous tubular photoreactor with flow rate and catalyst loading variation. The initial concentration was 10 ppm, and to make sure that photocatalysis was the only factor contributing to the degradation process, the adsorption was performed for 30 minutes before the photocatalytic process began. Photocatalysis is carried out until a color change occurs in the solution, and every 30 minutes the solution is taken to analyze the remaining methylene blue concentration. The catalyst loading used is 0.25; 0.50; 0.75; 1 g/L and from that the best performance of the photodegradation was carried out by catalyst loadings of 0.75 and 1.00 g/L with almost complete degradation up to 97% (**Figure 4**). The degradation process was completely achieved in 30–60 minutes with the highest kinetics rate 0.059/min calculated using pseudo-first-order kinetics from the Langmuir-Hinshelwood model, which can be used to explain the reaction kinetics of methylene blue photodegradation (Eq. 14) as mentioned in previous section.

The optimum catalyst loading in every system varies depending on the reactor and position of the photon source. Thus, in every system it has optimum catalyst as adding too much catalyst dose can cause dispersion, thereby reducing the effectiveness of the UV lamp from entering the solution. TiO_2 has super hydrophilic properties when irradiated by UV light [40–42]. In addition, hydrophilic substances function effectively with alcohol and water, whereas hydrophobic substances function optimally with ester and alkane substances [43].

In this study, methylene blue was dissolved into distilled water, so it can maximize the performance of TiO_2 by photo induction. Nevertheless, the TiO_2 surface can exhibit hydrophobic characteristics under certain conditions when exposed to visible or ultraviolet light; this transformation is reversible [44]. The same as Shondo et al. who conducted research on modification of the Au-TiO_2 surface with octadecylphosphonic acid (ODP) self-assembly monolayer produce extraordinary water repellent, they found the contact angle (CA) of the catalyst was $< 5°$ and it became CA $> 163°$ by means the substance changes from hydrophobic to hydrophilic during

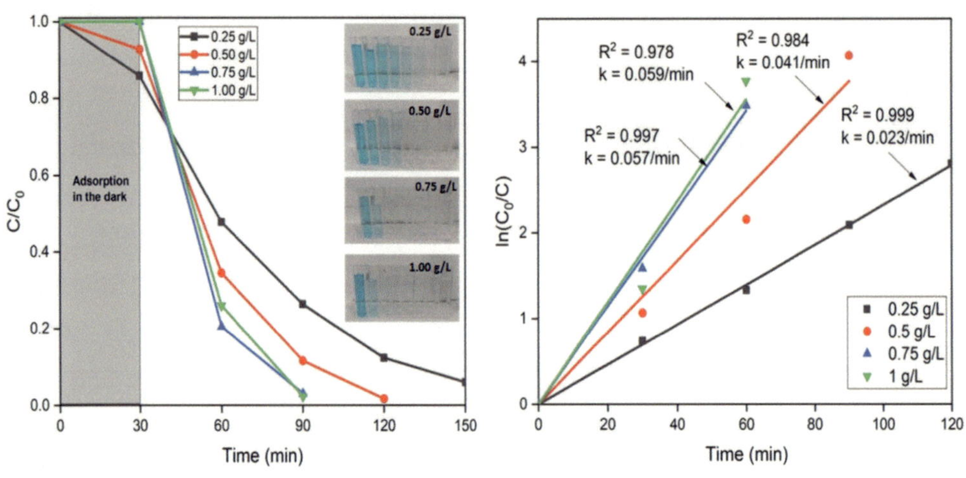

Figure 4.
The methylene blue degradation profile and kinetics constant of photocatalysis using TiO_2 nanoparticles at different catalyst loading.

photocatalysis and it is strongly influenced by light intensity [41, 45]. In addition, the diffusion of products and reactants is influenced by the wettability of the catalyst surface. In addition to catalyst loading, different flow rates ranging from 1 to 9 L/min were used to study the degradation profile of methylene blue, which serves as a proxy for Batik dye wastewater. The result demonstrates that the fastest degradation was attained at a flow rate of 3 L/min, with a degradation kinetics constant of k = 0.01762/min (**Figure 5**).

Several factors do influence photocatalysis such as light intensity, temperature, pH, pollutant concentration, and catalyst loading. Besides that, flow rate is one of the driven parameters in the photodegradation process on tubular photoreactor at a certain level. Previous experiments show that at low flow rate (less than 0.6 L/min) the TiO_2 nanoparticle tend to sediment in the lower side of the photoreactor [46]. While higher flow rate not only gives good suspension of TiO_2 nanoparticles in the methylene blue solution but also led to maximum contact as well as optimum opacity for the UV light to penetrate and activate the TiO_2 surface for photodegradation. When the number of active sites on the catalyst increases, decolorization of MB occurs. The formation of hydroxyl and superoxide as active radicals is the key point driving the degradation reaction [47]. In this system, flow rate of 1 L/min is considered a low flow rate. Sedimentation of TiO_2 nanoparticles occurs at the bottom side of the tubular photoreactor, so that the amount of TiO_2 that is active as a catalyst in photocatalysis is reduced. While the optimal TiO_2 nanoparticle for methylene blue degradation is achieved at a flow rate of 3 L/min, this also leads to the maximum contact time and optimum opacity for the UV light to penetrate and activate the TiO_2 surface for photodegradation. Given that the photodegradation rate, or *k* value, is lower than the rate attained by the 3 L/min, high flow rates (6–9 L/min) may have different impacts.

High flow rates have the potential to cause the suspension to turbulence, which can disrupt both the reaction and penetration of UV radiation. Additional research on the effects of flow rate (2–10 L/min) in disinfection-based photocatalysis reveals that the process is more effective at lower flow rates (2 L/min) because applying a high flow rate could cause mechanical stress and disrupt the process's light flow and sequence [48].

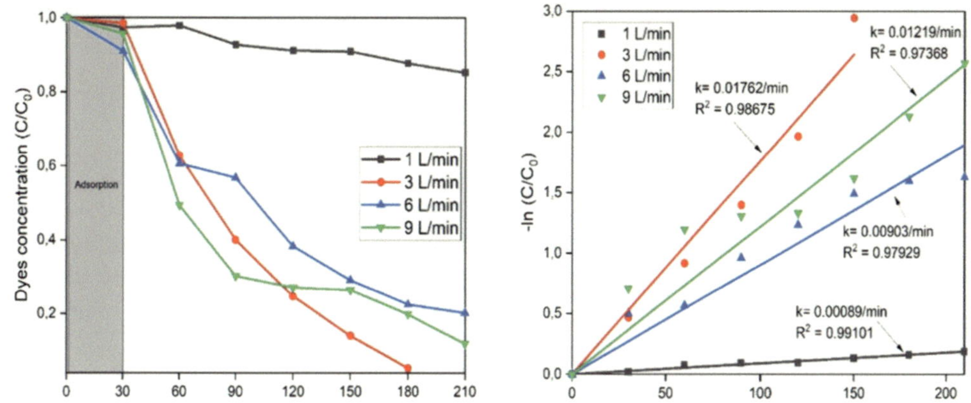

Figure 5.
The methylene blue degradation profile and kinetics constant of photocatalysis using TiO_2 nanoparticles at different flow rates.

8.3 Degradation of batik wastewater

The developed continuous tubular photoreactor is also used to treat batik wastewater from the batik industry in Semarang, Indonesia. The wastewater consists of Remazol Brilliant Blue, Remazol Yellow, Remazol Brilliant Violet, and Remazol Red. Like Remazol, reactive azo dyes are frequently used to color cloth fibers. Remazol yellow's presence of peaks at 259 and 420 nm, whose wavelengths match the color of the substance's UV–VIS spectrum, serves as evidence [49]. Then, Remazol Brilliant Violet or Remazol Brilliant Blue most likely owns the peak at 564 nm [50]. Over time, the absorbance of batik wastewater decreases. Batik wastewater's initial absorbance values were 0.844, 0.537, and 0.203 at wavelengths of 259, 420, and 564 nm, respectively.

The result shows that dyes contained in the solution were no longer detectable within 4 hours. The Remazol Brilliant Violet or Remazol Brilliant Blue degraded within 60 min at a kinetics constant of 0.0065/min, followed by Remazol yellow degraded within 3.5 hours at 0.0037/min of kinetics constant. Smaller organic contaminants containing water and oxygen molecules are produced by this photocatalytic breakdown [51], making them safer for the environment. The most efficient batik waste degradation was observed 30 minutes after the adsorption process. This was also seen by Ariyanti et al. [48] who demonstrated that practically all RhB could be degraded at a catalyst dosage of 0.5 g/L once adsorption conditions were met for a maximum of 15 minutes (**Figure 6**).

In photocatalytic reactions, electrons from the valence band are excited to the conduction band, producing electron hole pairs (e−/h+) when photons with energy equal to or greater than the energy gap of semiconducting materials are absorbed [52]. When water molecules or hydroxyl ions (OH^-) interact with photogeneration gaps (h+) in the valence band, free radicals are produced.

Conversely, excited electrons (e) in the conduction band combine with the oxygen in the solution to form superoxide radical anions ($\bullet O_2^-$) [52]. Then, superoxide radicals react with H^+ ions to produce hydrogen peroxide, which breaks down into OH radicals. The production of reactive oxygen species during photocatalysis can trigger

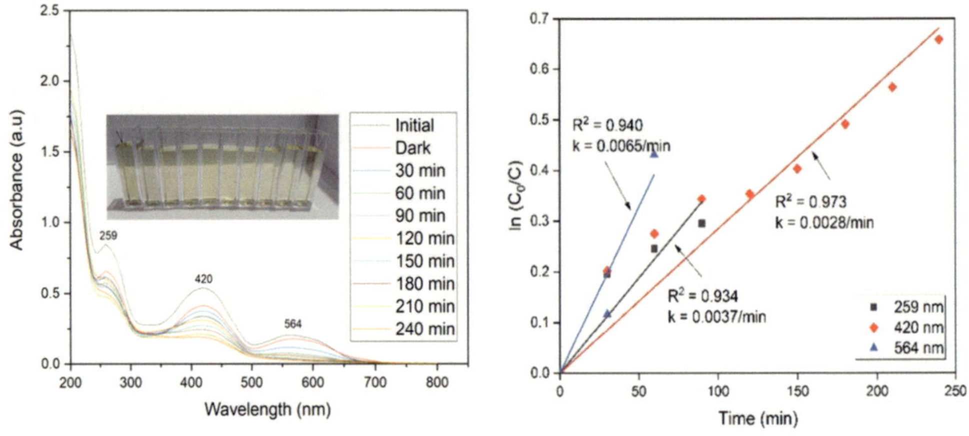

Figure 6.
UV–visible spectrum and visualization of batik wastewater after photocatalysis.

oxidation and reduction, which may lead to the chromophores in the azo-reactive dye being cleaved. Based on Figure 5.5(a), the chemical structure of azo dyes consists of the backbone, auxochrome groups, chromophores groups, and solubilizing groups [53, 54]. Chromophore (N=N) cleavage will result in the formation of intermediate amine compounds and sulfonic acid [55]. The batik wastewater becomes decolored as a result of the chain breaking caused by the insertion of protons and electrons in the diazo N=N bond [56]. The azo linkages (N=N) in the molecules are broken down by photocatalysis, which is involved in the breakdown of wastewater pollution. The process starts with the azo bond cleavage (N=N) and continues with the cleavage of single and double bonds.

Breaking bonds with the lowest energy come first for initiators, then those with the highest energy. The degradation pathways for Remazol Yellow, Remazol Brilliant Blue, Remazol Red, and Remazol Brilliant Violet are shown in Figures 5.5(b-e). At the point where the bond energy is lowest, which is C−S (259 kJ/mol), the bond breaks. Remazol undergoes C−S bond cleavage throughout. All things considered, Remazol Yellow is the azo dye that degrades the slowest due to the occurrence of the bonds C=N (615 kJ/mol) and C=O (745 kJ/mol). Aside from that, other types of Remazol have more phi (π, electron transition) compared to Remazol Yellow. The decomposition of Remazol Brilliant Blue R and Remazol Yellow FG using doped TiO_2 was also found in the research. The results show that the Remazol Brilliant Blue R decomposed more than Remazol Yellow [14].

9. Conclusions

The Implementation of TiO_2 photocatalysis as one of advanced oxidation technology (AOP) for complete removal of organic pollutants such as dyes in wastewater is very promising. Many types and designs of photocatalytic reactors are developed; among them the tubular photocatalytic reactors have more technical visibility and economic viability to be implemented for Batik wastewater industry application.

Continuous tubular photoreactor with TiO_2 nanoparticles catalyst developed for batik industrial wastewater treatment consists of five reactors arranged in parallel connected by a 1/2-inch pipe and equipped with a circulation pump, flowmeter, feed tank, panel box, and valve. This system has been tested for photocatalysis in series with nanostructured TiO_2 of 0.75–1.5 g/L catalyst loading. The degradation of methylene blue reached more than 97% and the reduction of Remazol dye in batik wastewater achieved 61%. Another processing parameter that strongly influences the performance of photoreactor is the flow rate, and the flow rate of 3 L/min is the most effective condition for the degradation of dyes in wastewater with the kinetic constant of 0.01762/min. Degradation of dyes in wastewater is a challenge and improvement still needs to be address in terms of increasing the degradation rate as well as catalyst recovery.

Acknowledgements

Authors acknowledge Universitas Diponegoro, Indonesia for the research financial support.

Author details

Dessy Ariyanti[1*], Dina Lesdantina[1], Ariana Aisa[1], Farida Diyah Hapsari[1], Filicia Wicaksana[2] and Wei Gao[2]

1 Faculty of Engineering, Department of Chemical Engineering, Universitas Diponegoro, Semarang, Indonesia

2 Department of Chemical and Materials Engineering, The University of Auckland, Auckland, New Zealand

*Address all correspondence to: dessy.ariyanti@che.undip.ac.id

References

[1] Darmawanti T, Suhartana S, Widodo DS. Pengolahan Limbah Cair Industri Batik dengan Metoda Elektrokoagulasi Menggunakan Besi Bekas Sebagai Elektroda. Jurnal Kimia Sains dan Aplikasi. 2010;**13**(1):18-24

[2] Agustina TE, Bustomi A. Pada Proses Pengolahan Limbah Pewarna Sintetik. Jurnal Teknik Kimia Universitas Sriwijaya. 2016;**22**(1):65-72

[3] Indonesia SNR. Peraturan Pemerintah Nomor 22 Tahun 2021 tentang Pedoman Perlindungan dan Pengelolaan Lingkungan Hidup. Vol. 1. Jakarta: Sekretariat Negara Republik Indonesia; 2021

[4] Jannah IN, Muhimmatin I. Pengelolaan Limbah Cair Industri Batik menggunakan Mikroorganisme di Kecamatan Cluring Kabupaten Banyuwangi. Warta Pengabdian. 2019;**13**(3):106-115

[5] Rashidi H, Sulaiman NMN, Hashim NA, Bradford L, Asgharnejad H, Madani LM. Wax removal from textile wastewater using an innovative hybrid baffle tank. Journal of the Textile Institute. 2021;**112**(2):223-232

[6] Almazán-Sánchez PT, Linares-Hernández I, Solache-Río MJ, Martínez-Miranda V. Textile wastewater treatment using iron-modified clay and copper-modified carbon in batch and column systems. Water, Air, and Soil Pollution. 2016;**227**(4):100

[7] Salleh MAM, Mahmoud DK, Karim WAWA, Idris A. Cationic and anionic dye adsorption by agricultural solid wastes: A comprehensive review. Desalination. 2011;**280**(1-3):1-13

[8] Demirbas A. Agricultural based activated carbons for the removal of dyes from aqueous solutions: A review. Journal of Hazardous Materials. 2009;**167**(1-3):1-9

[9] Atas MS, Dursun S, Akyildiz H, Citir M, Yavuz CT, Yavuz MS. Selective removal of cationic micro-pollutants using disulfide-linked network structures. RSC Advances. 2017;**7**(42):25969-25977

[10] Aryanti N, Nafiunisa A, Kusworo TD, Wardhani DH. Separation of reactive dyes using natural surfactant and micellar-enhanced ultrafiltration membrane. Journal of Membrane Science and Research. 2021;**7**(1):20-28

[11] Solayman HM, Hossen MA, Abd Aziz A, Yahya NY, Leong KH, Sim LC, et al. Performance evaluation of dye wastewater treatment technologies: A review. Journal of Environmental Chemical Engineering. 2023;**11**(3):109610

[12] Attallah OA, Al-Ghobashy MA, Nebsen M, Salem MY. Removal of cationic and anionic dyes from aqueous solution with magnetite/pectin and magnetite/silica/pectin hybrid nanocomposites: Kinetic, isotherm and mechanism analysis. RSC Advances. 2016;**6**(14):11461-11480

[13] Ketut SD, Oktofa RD, Ketut S. Color removal of textile wastewater using indirect electrochemical oxidation with multi carbon electrodes. Environment Asia. 2018;**11**(3):170-181

[14] Purnawan C, Wahyuningsih S, Aniza ON, Sari OP. Photocatalytic degradation of Remazol brilliant blue R and Remazol yellow FG using TiO_2

doped Cd, Co, Mn. Bulletin of Chemical Reaction Engineering and Catalysis. 2021;**16**(4):804-815

[15] Fujishima A, Honda K. Electrochemical photolysis of water at a semiconductor electrode. Nature. 1972;**238**:37-38

[16] Balzani V, Bergamini G, Ceroni P. Photochemistry and photocatalysis. Rendiconti Lincei. 2017;**28**:125-142

[17] Castellote M, Bengtsson N. Applications of Titanium Dioxide Photocatalysis to Construction Materials. Switzerland AG: Springer Nature; 2011. pp. 5-10

[18] Saravanan R, Gracia F, Stephen A. Basic principles, mechanism, and challenges of Photocatalysis. In: Nanocomposites for Visible Light-Induced Photocatalysis. Switzerland AG: Springer Nature; 2017. pp. 19-40

[19] Li M, Lin HY, Huang CP. Nanotechnostructured catalysts TiO2 nanoparticles for water purification. In: Nanotechnologies for Water Environment Applications. VA, USA: American Society of Civil Engineers (ASCE); 2009. pp. 43-92

[20] Ajmal A, Majeed I, Malik RN, Idriss H, Nadeem MA. Principles and mechanisms of photocatalytic dye degradation on TiO_2 based photocatalysts: A comparative overview. RSC Advances. 2014;**4**(70):37003-37026

[21] Nakata K, Fujishima A. TiO_2 photocatalysis: Design and applications. Journal of Photochemistry and Photobiology C: Photochemistry Reviews. 2012;**13**(3):169-189

[22] Hamdam Momen M, Amadeh A, Heydarzadeh Sohi M, Moghanlou Y. Photocatalytic properties of ZnO nanostructures grown via a novel atmospheric pressure solution evaporation method. Materials Science and Engineering: B. 2014;**190**:66-74

[23] Ai Z, Gao Z, Zhang L, He W, Yin JJ. Core-shell structure dependent reactivity of $Fe@Fe_2O_3$ nanowires on aerobic degradation of 4-chlorophenol. Environmental Science & Technology. 2013;**47**(10):5344-5352

[24] Fu X, Zhang Y, Cao P, Ma H, Liu P, He L, et al. Radiation synthesis of CdS/reduced graphene oxide nanocomposites for visible-light-driven photocatalytic degradation of organic contaminant. Radiation Physics and Chemistry. 2016;**123**:79-86

[25] Mohite SV, Ganbavle VV, Rajpure KY. Solar photoelectrocatalytic activities of rhodamine-B using sprayed WO3 photoelectrode. Journal of Alloys and Compounds. 2016;**655**:106-113

[26] Feng L, Zhang C, Gao G, Cui D. Facile synthesis of hollow Cu_2O octahedral and spherical nanocrystals and their morphology-dependent photocatalytic properties. Nanoscale Research Letters. 2012;**7**:1-10

[27] Sharma M, Jain T, Singh S, Pandey OP. Photocatalytic degradation of organic dyes under UV-visible light using capped ZnS nanoparticles. Solar Energy. 2012;**86**(1):626-633

[28] Ola O, Maroto-Valer MM. Review of material design and reactor engineering on TiO_2 photocatalysis for CO_2 reduction. Journal of Photochemistry and Photobiology C: Photochemistry Reviews. 2015;**24**:16-42

[29] Park H, Park Y, Kim W, Choi W. Surface modification of TiO_2 photocatalyst for environmental applications. Journal of Photochemistry

and Photobiology C: Photochemistry Reviews. 2013;**15**(1):1-20

[30] Chen X, Liu L, Huang F. Black titanium dioxide (TiO_2) nanomaterials. Chemical Society Reviews. 2015;**44**(7):1861-1885

[31] Fujishima A, Zhang X, Tryk DA. TiO_2 photocatalysis and related surface phenomena. Surface Science Reports. 2008;**63**(12):515-582

[32] Zangeneh H, Zinatizadeh AAL, Habibi M, Akia M, Hasnain IM. Photocatalytic oxidation of organic dyes and pollutants in wastewater using different modified titanium dioxides: A comparative review. Journal of Industrial and Engineering Chemistry. 2015;**26**(December):1-36

[33] Gaya UI. Heterogeneous photocatalysis using inorganic semiconductor solids. In: Heterogeneous Photocatalysis Using Inorganic Semiconductor Solids. Dordrecht: Springer Science + Business Media; 2014. pp. 43-71

[34] Liu B, Zhao X, Terashima C, Fujishima A, Nakata K. Thermodynamic and kinetic analysis of heterogeneous photocatalysis for semiconductor systems. Physical Chemistry Chemical Physics. 2014;**16**(19):8751-8760

[35] Ong CS, Lau WJ, Goh PS, Ng BC, Ismail AF, Choo CM. The impacts of various operating conditions on submerged membrane photocatalytic reactors (SMPR) for organic pollutant separation and degradation: A review. RSC Advances. 2015;**5**(118):97335-97348

[36] Wetchakun K, Wetchakun N, Sakulsermsuk S. An overview of solar/visible light-driven heterogeneous photocatalysis for water purification: TiO_2- and ZnO-based photocatalysts used in suspension photoreactors. Journal of Industrial and Engineering Chemistry. 2019;**71**:19-49

[37] Akram TM, Ahmad N, Shaikh IA. Characterization and optimization of photocatalytic activity of sol gel-synthesized TiO_2 and Ag-doped TiO_2 through degradation of synthetic textile effluent by UV lamp-assisted experimental setup. Polish Journal of Environmental Studies. 2019;**28**(4):2571-2583

[38] Damodar RA, You SJ, Ou SH. Coupling of membrane separation with photocatalytic slurry reactor for advanced dye wastewater treatment. Separation and Purification Technology. 2010;**76**(1):64-71

[39] Yap WQ, Chin YH, Leong KH, Saravanan P, Sim LC. Design of photoreactor with high sunlight concentration for improved photocatalytic degradation of dye pollutant. IOP Conference Series: Earth and Environmental Science. 2021;**646**(1)

[40] Kameya Y, Yabe H. Optical and superhydrophilic characteristics of TiO_2 coating with subwavelength surface structure consisting of spherical nanoparticle aggregates. Coatings. 2019;**9**(9):547

[41] Shondo J, Veziroglu S, Stefan D, Mishra YK, Strunskus T, Faupel F, et al. Tuning wettability of TiO_2 thin film by photocatalytic deposition of 3D flower- and hedgehog-like Au nano- and microstructures. Applied Surface Science. 2021;**537**:147795

[42] Zhang X, Jin M, Liu Z, Tryk DA, Nishimoto S, Murakami T, et al. Superhydrophobic TiO_2 surfaces: Preparation, photocatalytic wettability conversion, and superhydrophobic-superhydrophilic patterning. Journal of Physical Chemistry C. 2007;**111**(39):14521-14529

[43] Wei M, Kuang Y, Duan Z, Li H. The crucial role of catalyst wettability for hydrogenation of biomass and carbon dioxide over heterogeneous catalysts. Cell Reports Physical Science. 2023;**4**(5):101340

[44] Stevens N, Priest CI, Sedev R, Ralston J. Wettability of photoresponsive titanium dioxide surfaces. Langmuir. 2003;**19**(8):3272-3275

[45] Maeda H, Kobayashi T, Konishi S. Patterning of wettability using the photocatalytic decomposition of hydrophobic self-assembled monolayer on the TiO_2 pattern. Japanese Journal of Applied Physics. 2017;**56**(6):2-6

[46] Suryaman D. Titanium dioxide suspension photocatalysis in a tubular photoreactor and titanium dioxide reuse on the removal of phenol. Jurnal Air Indonesia. 2018;**4**(2):160-166

[47] Abdellah MH, Nosier SA, El-Shazly AH, Mubarak AA. Photocatalyticdecolorizationofmethylene blue using TiO_2/UV system enhanced by air sparging. Alexandria Engineering Journal. 2018;**57**(4):3727-3735

[48] Sichel C, Blanco J, Malato S, Fernández-Ibáñez P. Effects of experimental conditions on *E. Coli* survival during solar photocatalytic water disinfection. Journal of Photochemistry and Photobiology A: Chemistry. 2007;**189**(2-3):239-246

[49] Bhuiyan MSH, Miah MY, Paul SC, Das AT, Saha O, Rahaman MM, et al. Green synthesis of iron oxide nanoparticle using Carica papaya leaf extract: Application for photocatalytic degradation of remazol yellow RR dye and antibacterial activity. Heliyon. 2020;**6**(8):e04603

[50] Zainip VJ, Adnan LA, Elshikh MS. Decolorization of Remazol brilliant violet 5R and Procion red MX-5B by Trichoderma species. Tropical Aquatic and Soil Pollution. 2021;**1**(2):108-117

[51] Mohammadi M, Sabbaghi S, Binazadeh M, Ghaedi S, Rajabi H. Type-1 α-Fe_2O_3/TiO_2 photocatalytic degradation of tetracycline from wastewater using CCD-based RSM optimization. Chemosphere. 2023;**336**(May):139311

[52] Shahid-ul-Islam BS, Kamali MR. Review on green biomass-synthesized metallic nanoparticles and composites and their photocatalytic water purification applications: Progress and perspectives. Chemical Engineering Journal Advances. 2023;**14**(January):100460

[53] Benkhaya S, Cherkaoui O, Assouag M, Mrabet S, Rafik M, El HA. Synthesis of a new asymmetric composite membrane with bi-component collodion: Application in the ultrafiltration of baths of reagent dyes of fabric rinsing/padding. Journal of Materials and Environmental Science. 2016;**7**(12):4556-4569

[54] Benkhaya S, M'rabet S, El Harfi A. Classifications, properties, recent synthesis and applications of azo dyes. Heliyon. 2020;**6**(1):e03271

[55] Mazzanti M, Caramori S, Fogagnolo M, Cristino V, Molinari A. Turning waste into useful products by photocatalysis with nanocrystalline TiO_2 thin films: Reductive cleavage of azo bond in the presence of aqueous formate. Nanomaterials. 2020;**10**(11):1-17

[56] Rauf MA, Meetani MA, Hisaindee S. An overview on the photocatalytic degradation of azo dyes in the presence of TiO_2 doped with selective transition metals. Desalination. 2011;**276**(1-3):13-27

Chapter 4

Application of Advanced Oxidation Processes for Domestic and Industrial Wastewater Treatment

Alexis Rubén Bracamontes-Ruelas,
José Rafael Irigoyen-Campuzano,
Luis Arturo Torres-Castañon and Liliana Reynoso-Cuevas

Abstract

Domestic and industrial wastewaters are complex matrices that contain a large variety of pollutants. Consequently, the conventional processes at wastewater treatment plants (WWTPs) cannot remove these. These pollutants remain in the effluent and are discharged into different environmental compartments worldwide, generating a range of negative impacts on the environment and human health. In this chapter, general features and the application of the most common advanced oxidation processes (AOPs) for the treatment of domestic and industrial wastewater are described. Also, the feasibility of scaling up advanced oxidation processes for pollutants removal (emerging and conventional) and the advantages and complications of each type of advanced oxidation process when applied to wastewater treatment (domestic and industrial) are shown.

Keywords: emerging pollutants, wastewater, advanced oxidation processes, tertiary process, conventional pollutants, removal

1. Introduction

Most of the human activities (daily household activities, urban runoff, manufacturing, and service processes) generate wastewater [1], which is a very complex matrix due to the mixture of pathogens, heavy metals, organic matter, particulate solids, and other toxic substances. Due to this complexity of municipal and industrial wastewater, conventional wastewater treatment processes usually cannot provide high-quality treatment, so advanced oxidation processes are used [2].

Advanced oxidation processes are powerful techniques that can remove (a general term used in this chapter to define the mineralization, removal, transformation, or degradation of the conventional or emerging pollutants mentioned in the text) pathogens and harmful organic compounds in a liquid medium by generating hydroxyl radicals ($HO^{\bullet}$) through various chemical pathways [3, 4]. Hydroxyl radicals ($HO^{\bullet}$) have a high oxidative power amounting to 2.8 eV [5].

Among the various techniques of advanced oxidation processes employed to generate hydroxyl radicals ($HO^{\bullet}$), photo-peroxidation (UV-H_2O_2), Fenton, photo-Fenton, indirect ozonation, photocatalysis, and sonolysis are the most common [6].

The main advantages of advanced oxidation processes over other methods are that they are entirely environmentally friendly, are carried out *in situ*, and do not produce large quantities of toxic wastes [7]. In addition, it has been shown that some advanced oxidation processes tend to increase the biodegradability of certain pollutants in the treated wastewater [8].

On the other hand, advanced oxidation processes nowadays have been used for the removal of persistent organic pollutants in wastewater, such as emerging pollutants [9–11], because the persistent or emerging pollutants are those that remain in the environment and can cause problems to human health [12].

Generally, advanced oxidation processes are coupled with tertiary processes to treat domestic wastewater to remove emerging pollutants in WWTPs that handle conventional processes since they cannot degrade or remove them [12, 13]. In contrast, in the industrial sector, advanced oxidation processes are used as the principal wastewater treatment [14, 15] since the industry generates more toxic waters [16] than the commonly called dump, which is the domestic wastewater found in dwellings.

However, the purpose of using advanced oxidation processes, either as tertiary or primary processes for domestic and industrial wastewater treatment, is to produce a high-quality effluent that can be reintroduced into the environment without adversely affecting flora, fauna, or human health.

Therefore, this chapter aims to outline, in general terms, the most common advanced oxidation processes for removing conventional and emerging pollutants from domestic (as a tertiary or coupled process) and industrial wastewater. It also discusses the advantages and disadvantages of each process and their potential for scale up.

2. Domestic and industrial wastewater

As water is used for most human activities, it becomes polluted, making it unsafe for human consumption, which must be treated to be returned to the environment in the best possible way [17]. In general, wastewater can be classified into domestic wastewater and industrial wastewater.

2.1 Domestic wastewater

Domestic wastewater is the water generated by human activities at home, including washing clothes, dishes, hands, body, vehicles, and household cleaning. Domestic wastewater can be categorized (**Figure 1**) as yellow (contains urine), brown (contains feces as well as cleaning water), black (contains urine, feces, and bacterial activity), and gray wastewater (contains water from cooking, laundry, and human cleaning). Domestic wastewater contains millions of human enteric bacteria and other microorganisms [18]. Like other types of wastewater (industrial wastewater), domestic wastewater may contain conventional pollutants or substances that affect its quality, such as organic carbon (organic matter), nitrogen-containing organic matter, inorganic matter, suspended and dissolved solids, microorganisms, and heavy metals [19].

The above-mentioned conventional pollutants, referring to the Mexican standard NOM-001-SEMARNAT-2021 (which establishes limits for conventional pollutants in

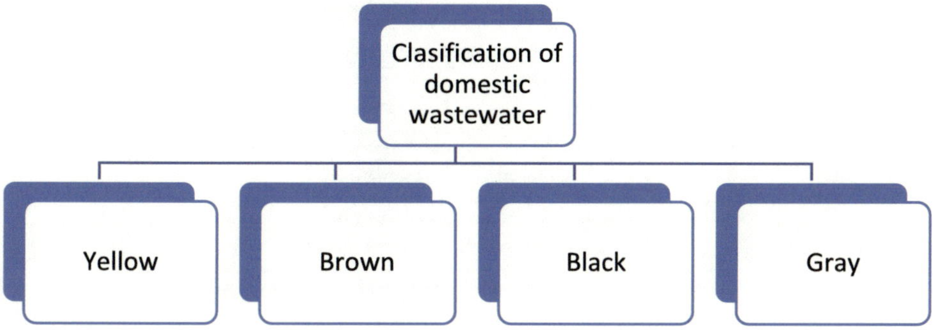

Figure 1.
General classification of domestic wastewater.

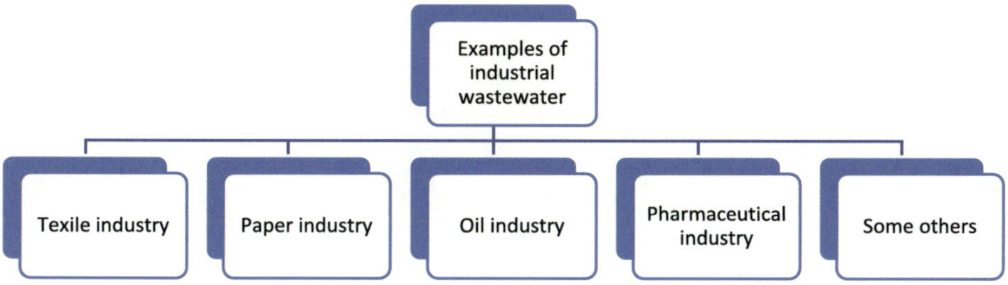

Figure 2.
Examples of types of industries that generate wastewater in their processes.

wastewater), can be evaluated in domestic and industrial wastewater, for example, as chemical oxygen demand (COD), total suspended solids (TS), total organic carbon (TOC), and others [20].

2.2 Industrial wastewater

On the other hand, regarding industrial wastewater, different industries rely on water for their processes, producing wastewater. Some industries that generate wastewater from their processes include textile, paper, oil, pharmaceutical, and others (**Figure 2**) [21].

Industrial wastewater, unlike domestic, is very different in its pollutant load depending on the sector (based on the process being handled) that generates it (e.g., textile or paper industry). It is known that industrial wastewater usually contains a mixture of several pollutants, such as persistent organic pollutants, endocrine-disrupting chemicals, and many others [22].

Furthermore, another characteristic of industrial wastewater is that it is difficult to treat, and an individual analysis has to be made beforehand to determine a specific treatment based on what is required [23].

2.3 General characteristics of domestic and industrial wastewater

Domestic and industrial wastewater characteristics (conventional pollutants) can be divided into physical, chemical, organic, and biological [24], as seen in **Figure 3**.

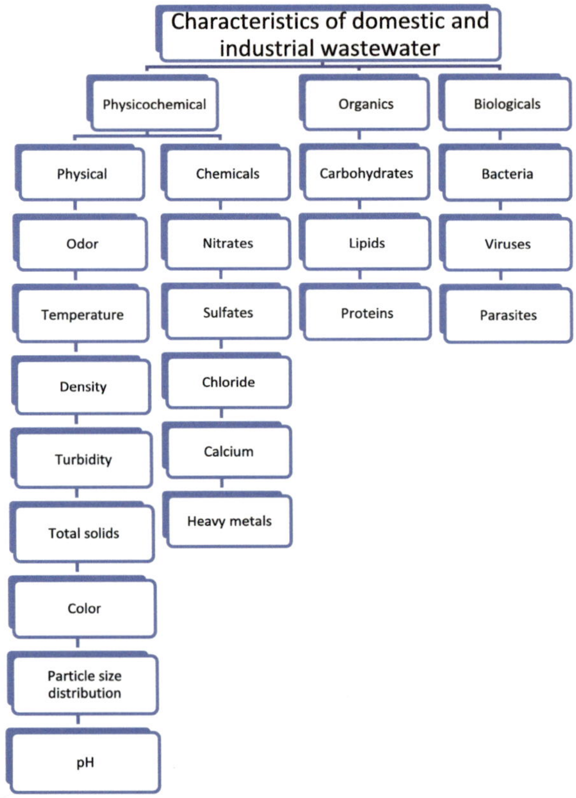

Figure 3.
General characteristics of domestic and industrial wastewater.

The physical, chemical, organic, and biological characteristics (**Figure 3**) that may be present in domestic and industrial wastewater, depending on the regulations used in a given country, may be considered as conventional pollutants.

However, in recent decades, in domestic and industrial wastewater, not only the presence of conventional pollutants (general characteristics of wastewater) has been reported. Other types of pollutants, called emerging pollutants, have also been detected [25].

2.4 Emerging pollutants in wastewater

Emerging pollutants are natural or synthetic compounds that are not subject to legislation. Thus, they are not monitored in the environment and have the potential to cause damage to the environment and human health [26]. One of the characteristics of emerging pollutants is that they are commonly found and quantified in the environment in concentrations varying from ng L^{-1} to μg L^{-1} [27]. Emerging pollutants can be classified as personal care products, hormones, flame retardants, industrial additives, endocrine-disrupting chemicals, pharmaceuticals, nanomaterials, and pesticides. Due to their presence, the emerging pollutants can also convincingly diminish the water quality [28].

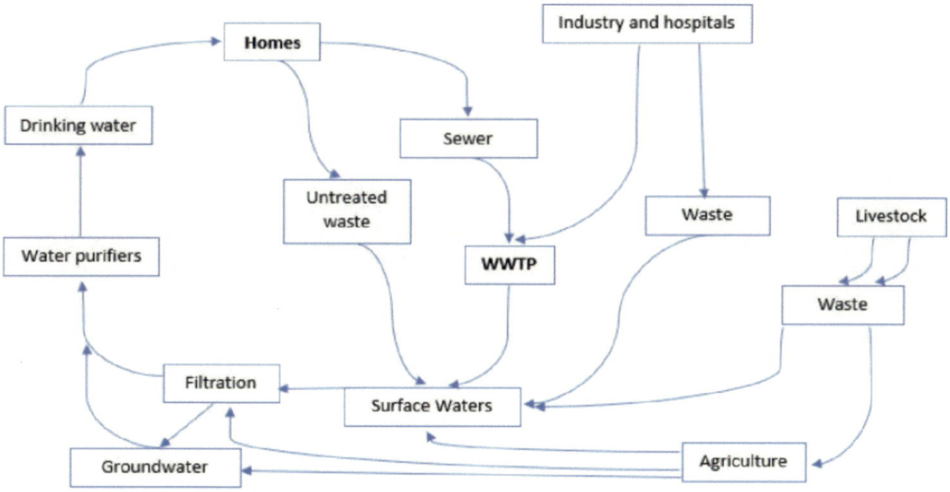

Figure 4.
Distribution of emerging pollutants through the water cycle in the environment [31].

In addition, current research has shown that certain emerging pollutants can cause harm to humans, such as thyroid disruption, DNA damage, and liver metabolism impairment [29].

It should be noted that conventional wastewater treatment plants, which cannot remove emerging pollutants [30], through raw sewage and effluent generate the main entry of emerging pollutants into the water cycle. **Figure 4** shows how emerging pollutants are widely distributed through the water cycle in different environmental compartments until they reach human consumption (e.g., drinking water). Emerging pollutants can reach human consumption through contaminated drinking water or the trophic chain. Many animals or crops can interact with emerging pollutants as they are distributed, as shown in **Figure 4** [31].

Therefore, it is essential to use methods such as advanced oxidation processes to eliminate emerging and conventional pollutants (general characteristics of wastewater) so that they cannot be distributed in the environment through the water cycle.

3. Advanced oxidation processes for the removal of conventional and emerging pollutants in domestic and industrial wastewater

Advanced oxidation processes were first proposed for drinking water treatment in the 1980s. Subsequently, they have been adopted by the scientific community for the treatment of various wastewater due to their oxidative capacity for removing organic matter [15].

Moreover, the advanced oxidation processes are defined as those that can be carried out at an ambient temperature and pressure and generate hydroxyl radicals ($HO^{\bullet}$) as oxidative species for pollutant removal [15, 32, 33].

It is important to note that advanced oxidation processes require chemicals as raw materials and sometimes energy (e.g., UV radiation) to generate hydroxyl radicals ($HO^{\bullet}$) [34], for example, Fenton (Fe^{2+}/H_2O_2) and photo-Fenton ($Fe^{2+}/H_2O_2/$

UV radiation). In general, a catalyst (e.g., Fe^{2+}) and an oxidant (e.g., H_2O_2) are the chemicals used to perform advanced oxidation processes to generate hydroxyl radicals ($HO^\bullet$) [35].

Hydroxyl radicals ($HO^\bullet$) are non-selective radicals that can remove pollutants through mechanisms such as electron transfer, hydrogen abstraction, and radical addition [34].

Consequently, various research projects have continuously proposed advanced oxidation processes to remove difficult-to-remove organic pollutants (emerging and conventional) from domestic and industrial wastewater [32]. It has been shown that advanced oxidation processes entirely or partially remove the target pollutants (emerging and conventional pollutants) and do not only concentrate them in one phase or transfer them to another phase compared to other treatment processes (e.g., coagulation, flocculation, and membranes) of domestic and industrial wastewater [34].

Therefore, the following subsections describe the most common advanced oxidation processes for wastewater treatment. These are photo-peroxidation (UV-H_2O_2), Fenton, photo-Fenton, indirect ozonation, photocatalysis, and sonolysis [6]. The advantages and disadvantages of their application, as well as their potential scale up, are also discussed.

3.1 Photo-peroxidation (UV-H_2O_2) process

To effect the photo-peroxidation (UV-H_2O_2) process, hydrogen peroxide (H_2O_2) must be photolyzed by UV radiation at wavelengths between 200 and 300 nm, causing homolytic fission of the O-O bond belonging to the hydrogen peroxide (H_2O_2) molecule and hydroxyl radicals ($HO^\bullet$) are produced as shown in Eq. (1), as well as other oxidative species such as perhydroxyl radicals ($HO_2^\bullet$), as shown in Eq. (2) [36, 37].

$$H_2O_2 + hv \rightarrow 2HO^\bullet \quad (1)$$

$$HO^\bullet + H_2O_2 \rightarrow H_2O + HO_2^\bullet \quad (2)$$

The reaction rate for the generation of hydroxyl radicals ($HO^\bullet$) is faster at $pH > 10$ (alkaline medium), so the process must be carried out in alkaline wastewater for the removal of the target pollutants (conventional and emerging pollutants). This, in turn, can be a disadvantage since chemicals for pH adjustment in certain types of wastewaters must be used as inputs before performing this advanced oxidation process. Another drawback of photo-peroxidation (UV-H_2O_2) for wastewater treatment is that the molar absorption of hydrogen peroxide (H_2O_2) in the UV region is insufficient, so high concentrations of hydrogen peroxide (H_2O_2) must be used for the removal of the target pollutants (conventional and emerging pollutants) [36].

In summary, the photo-peroxidation (UV-H_2O_2) process is dependent on the initial concentration of hydrogen peroxide (H_2O_2), the intensity of the UV light, the pH of the wastewater, and the constituents contained in the wastewater [38].

Regarding the removal of conventional pollutants, one example is the research of Philippopoulos and Poulopoulos [39] where the photo-peroxidation (UV-H_2O_2) process was applied for the oil industry wastewater treatment. This work shows that low percentages of chemical oxygen demand (COD) removal are obtained in the treated wastewater using this oxidation process, since the maximum removal percentage obtained was 45%, at a hydraulic retention time of 150 minutes, with an initial hydrogen peroxide (H_2O_2) concentration of 6660 mg L^{-1}. However, the removal of some pollutants present in the treated wastewater, such as isobutane, n-pentanol, phenol, o-cresol, and m-cresol amounted to almost 100%, although there was a low removal of ethylene glycol of almost 20%.

Similarly, research by Della Roca et al. [40] showed that the percentages of total organic carbon (TOC) removal provided by the photo-peroxidation (UV-H_2O_2) method are very low since only approximately 5% of the total organic carbon (conventional pollutant) was removed.

Also, it has been noted in several investigations that the scaling of reactors to carry out photocatalytic processes belonging to advanced oxidation processes (e.g., photo-peroxidation (UV-H_2O_2)) is still a challenge, due to complications in the light source, as a linear model is usually used and has many drawbacks despite its widespread use in several reactors, which makes it a challenge at present [41].

Then, based on the information gathered, the photo-peroxidation (UV-H_2O_2) process tends to have diverse complications, making it a problematic advanced oxidation process to scale up. Furthermore, the process has low removal yields for conventional pollutants such as chemical oxygen demand (COD) and total organic carbon (TOC).

3.2 Fenton process

Meanwhile, the advanced oxidation process, called Fenton, was reported in 1894 by Henry John Horstman Fenton [42], when one of his students found that a mixture of hydrogen peroxide (H_2O_2), iron salts, and tartaric acid dyeing an orange color to the water or solution containing them. This led to the discovery of the Fenton reaction shown in Eq. (3) [43].

Eq. (3) shows how the catalyst reaction (Fe^{2+}) with the oxidizing agent (H_2O_2) generates hydroxyl radicals ($HO^{\bullet}$) as the main product, and ferric ions (Fe^{3+}), among other species, as by-products [44, 45]. The orange color of the water, when Fenton's student performed the above experiment, is now attributed to the ferric ions (Fe^{3+}) generated as by-products in Eq. (3) [46].

$$Fe^{2+} + H_2O_2 \rightarrow Fe^{3+} + OH^- + HO^{\bullet} \quad (3)$$

Currently, it has been reported utilizing Eq. (4) that the ferric ions (Fe^{3+}), if there is any remaining hydrogen peroxide (H_2O_2), can react with the hydrogen peroxide (H_2O_2) and subsequently generate other oxidative species such as perhydroxyl radicals ($HO_2^{\bullet}$) [44, 45].

$$Fe^{3+} + H_2O_2 \rightarrow Fe^{2+} + H^+ + HO_2^{\bullet} \quad (4)$$

Nevertheless, it is necessary to say that to carry out the Fenton reaction (Eq. (3)) and generate the hydroxyl radicals ($HO^{\bullet}$) (oxidative species of interest) for the treatment of domestic and industrial wastewater, an acidic pH in the aqueous medium to be treated of about pH 3 must be maintained [44, 46].

The compound commonly used in several studies as a supplier of ferrous ions (Fe^{2+}) for the Fenton process is ferrous sulfate heptahydrate ($FeSO_4 \bullet 7H_2O$) [46–48].

Eq. (3), Eq. (5), and Eq. (6) have been used for the dosing of both the oxidant (H_2O_2) and the supplier catalyst ($FeSO_4 \cdot 7H_2O$) to perform the Fenton process and to treat wastewater, either industrial or domestic [31, 46].

$$\text{Quantity}\left(mg\right)\text{H}_2\text{O}_2 = \frac{17}{8}\left[COD, mg\,\text{L}^{-1}\right]\left(\text{V}\left(\text{L}\right)\right) \quad (5)$$

$$\text{Quantity}\left(mg\right)FeSO_4 \bullet 7\text{H}_2\text{O} = \left(278{,}010\right)\left(\text{nmol}\,Fe^{2+}\right) \quad (6)$$

The dosages of hydrogen peroxide (H_2O_2) and ferrous sulfate heptahydrate (Fe^{2+} supplier) for domestic and industrial wastewater treatment have been made according to the initial value of the wastewater's total Chemical Oxygen Demand (COD) because hydroxyl radicals ($HO^{\bullet}$) are non-selective oxidants, and if, for example, they were calculated according to the amount of conventional or emerging pollutants of interest to be eliminated, these quantities would be underestimated, since there is a large number of organic pollutants in the wastewater that could interfere with its removal due to the nature of the hydroxyl radicals ($HO^{\bullet}$). Therefore, they are generally included in the total chemical oxygen demand (COD) determined for the wastewater.

In Eq. (5), the total chemical oxygen demand (COD) of the wastewater to be treated is calculated in mg L^{-1} and the volume of wastewater to be treated in L. In Eq. (6), the moles of ferrous ion (Fe^{2+}) are calculated by means of the (H_2O_2) 1:1 (Fe^{2+}) molar ratio shown for the Fenton reaction in Eq. (3).

On the other hand, the Fenton process has shown good performance in treating domestic and industrial wastewater, as several studies have demonstrated, it removes conventional and emerging pollutants such as total suspended solids (TSS), chemical oxygen demand (COD), pathogens (e.g., fecal coliforms), tartrazine, tetrachloromethane, azo dyes, orange II, acetylsalicylic acid, tetracycline, 2,4-dichlorophene, bisphenol A, estrogens, pesticides, and so on [46, 49]. In addition, the Fenton process has been used and demonstrated its feasibility as a tertiary process at the laboratory level to treat and remove emerging pollutants from a conventional WWTP effluent from a secondary activated sludge process [46]. This demonstrates the viability of using this process for domestic and industrial wastewater treatment.

However, the Fenton process drawbacks are related to the scale up associated with the implementation costs [50] and to the fact that the process must be divided into two stages: one to adjust the wastewater to be treated to the acidic pH (pH 3) and add the Fenton reagents (Fe^{2+} and H_2O_2); and the other to increase the pH to neutral (pH 7) and separate the ferric ions (Fe^{3+} and orange color of wastewater) generated as by-products process from the effluent produced by sedimentation processes [46]. It is noteworthy to say that Fenton's advanced oxidation process may involve additional costs, due to the amount of acid required to be added to the wastewater to be treated to adjust the pH 3 (ideal pH level to manage the removal of pollutants), since if the pH of the wastewater to be treated is not adjusted to 3 beforehand, the removal effectiveness of the Fenton process of the conventional and emerging pollutants of interest may be affected.

Considering what was mentioned above, Fenton's advanced oxidation process shows the capacity to remove several conventional and emerging pollutants. Nevertheless, there are still areas of opportunity to improve and scale it up to treat domestic and industrial wastewater.

3.2.1 Photo-Fenton process

The photo-Fenton process is a modified version of the traditional Fenton process where an additional energy source, such as UV radiation, is added. The Fenton reaction is identified as a powerful reaction when UV radiation is incorporated into Fenton's advanced oxidation process for the removal of organic pollutants. Several studies have shown that the UV radiation used in the Fenton process increases the removal of organic pollutants commonly achieved by the traditional Fenton process [51].

The UV radiation added to the traditional Fenton process generates a photoreduction of ferric ions (Fe^{3+}) to ferrous ions (Fe^{2+}) as shown in Eq. (7), where hydroxyl radicals ($HO^{\bullet}$) are generated as compared to the Fenton process which generates perhydroxyl radicals ($HO_2^{\bullet}$) (Eq. (4)) [52]. The reaction shown in Eq. (7), which is generated by UV radiation in the photo-Fenton process, significantly increases the removal efficiency of organic pollutants because the reaction of Eq. (7) is added to the traditional Fenton reaction (Eq. (3)), which is also carried out at the same time in this process.

$$Fe^{3+} + H_2O + hv \rightarrow Fe^{2+} + H^+ + HO^{\bullet} \tag{7}$$

The pH handled in the advanced oxidation process, photo-Fenton, for the treatment of domestic and industrial wastewater, is the same pH 3 commonly used in the traditional Fenton process, and also the catalyst (Fe^{2+}) supplier used is ferrous sulfate heptahydrate ($FeSO_4 \bullet 7H_2O$) [53, 54].

For the dosing of hydrogen peroxide (H_2O_2) and ferrous sulfate heptahydrate ($FeSO_4 \bullet 7H_2O$), Eq. (3), Eq. (5), and Eq. (6) can also be used in the same way in the photo-Fenton process.

The photo-Fenton process has been proposed as a tertiary treatment of domestic wastewater to remove persistent organic pollutants, such as emerging pollutants [55].

The photo-Fenton process has demonstrated its ability in research to remove various recalcitrant pollutants (e.g., emerging pollutants) in domestic and industrial wastewater of different types (e.g., oil refinery wastewater, tea industry wastewater, livestock wastewater, among others). Examples of good removal of emerging pollutants by the photo-Fenton process have included polymers, pesticides, reactive dyes, penicillin, 2-chlorophenol, and others [56]. This shows, as an advantage, the remarkable capacity of the photo-Fenton process to treat different types of wastewater and pollutants, as well as the traditional Fenton process.

Furthermore, another advantage that can be included is that the photo-Fenton process can be carried out with sunlight to treat wastewater (domestic and industrial) [57, 58], compared to other advanced oxidation processes that employ lamps that work at specific wavelengths.

Nonetheless, the complications presented by the photo-Fenton advanced oxidation process are the same as those presented by the Fenton traditional advanced oxidation process shown in subsection 3.2 of this chapter.

Finally, another problem that arises, and which is a great challenge, when wanting to scale up the advanced oxidation process, photo-Fenton, if it is desired to use sunlight, is that the reactor must be made of a translucent material so that it captures as much sunlight as possible to maximize the performance of the photo-Fenton reaction. So, despite the excellent development of the photo-Fenton process and its good performance in wastewater treatment, some challenges remain.

This chapter emphasizes only the process of homogeneous Fenton and photo-Fenton.

3.3 Indirect ozonation

Indirect ozonation, for its part, is usually better than direct ozonation because direct ozonation is very selective. Indirect ozonation produces hydroxyl radicals ($HO^{\bullet}$) by decomposition of the ozone molecule (O_3) [59]. To carry out indirect ozonation in the wastewater to be treated, $pH > 8$ must be managed [59–61].

In this type of advanced oxidation process, an ozone (O_3) generator has to be considered [62], which can complicate the applicability and scalability of the process.

Another complication of indirect ozonation is that the ozone (O_3) dosages to treat domestic and industrial wastewater must be defined in the laboratory beforehand. Moreover, the number of ozone (O_3) generators used to treat the wastewater of interest is defined according to the volume to be treated, the ozone (O_3) generation capacity of the equipment used (generator), and the ozone (O_3) dosages previously defined in the tests carried out at laboratory level to treat the wastewater of interest.

Nevertheless, the advanced oxidation process, indirect ozonation, according to the information gathered, is inefficient for removing persistent pollutants (e.g., some emerging pollutants) [63].

Given the above information, the advanced oxidation process, indirect ozonation, presents problems such as applicability, scalability, and difficulty in removing persistent pollutants (e.g., emergency pollutants) from the wastewater (domestic and industrial) to be treated, which makes it an ineffective method.

3.4 Photocatalysis

Instead, photocatalysis is a series of chemical reactions that require light to carry out and it is difficult in most cases to perform them in the dark [64]. It should be noted that the most important energy source in advanced oxidation processes, photocatalytic, to which it owes part of its name, is light [65].

Although the reactions in photocatalysis vary per process for wastewater treatment, the general process that occurs in photocatalysis in the different processes can be specified globally in four steps [66]:

1. Absorption of light to generate electron–hole pairs.

2. Separation of excited charges.

3. Transfer of holes and electrons to the surface of the photocatalyst.

4. And finally, the use of charges to carry out redox reactions to remove pollutants (conventional and emerging) from interest.

The photocatalysts commonly used in these advanced oxidation processes are usually solids [67].

The dosages of the photocatalyst, light radiation, and pH to be handled in the advanced photocatalytic oxidation processes are previously established based on laboratory or pilot tests with the wastewater (domestic or industrial) to be treated. It is worth mentioning that radiation, pH, and photocatalyst dosages vary according to the nature and composition of the wastewater to be treated.

Nonetheless, although advanced photocatalysis oxidation processes have demonstrated good capacity to remove emerging pollutants, there are still many challenges to improve the processes, such as the separation of the photocatalyst from the actual treated wastewater for reuse and disposal, the development of photocatalytic materials, and the lack of real-world application of these photocatalysis processes for wastewater treatment [67].

To conclude, advanced photocatalysis oxidation processes still need to be further investigated and scaled up to real-world applications despite providing good pollutants removal in treated wastewater.

Note that due to the large number of photocatalysis processes, only a general description is given in this subsection of the present document.

3.5 Sonolysis

Sonolysis is the advanced oxidation process that uses ultrasound as the main tool to produce hydroxyl radicals ($HO^{\bullet}$) in an aqueous medium (wastewater) to remove different types of conventional and emerging pollutants [68]. To be considered ultrasound, the waves generated must have a frequency greater than 20 kHz [69].

The high acoustic intensity generated by the ultrasound radiates to the liquids (e.g., domestic and industrial wastewater) and produces cavitation [70].

The general principle of producing hydroxyl radicals ($HO^{\bullet}$) in sonolysis to treat wastewater of interest (domestic and industrial) is divided into the following steps: first, sound energy and dissolved bubbles lead to the growth of bubbles near adiabatic collapse, better known as cavitation, then cavitation generates extreme temperature and pressure conditions without bubbles collapsing, and finally, they reach a violent collapse.

In the wastewater to be treated, cavitation is the process that produces the hydroxyl radicals ($HO^{\bullet}$), which are the oxidative species of interest that are used to remove the conventional and emerging pollutants of interest [68].

A drawback of the sonolysis process for the treatment of domestic and industrial wastewater is that if the wastewater to be treated contains a high load of solid particles (e.g., total suspended solids), these can dissipate the ultrasonic waves and weaken the dissipation energy of the process, decreasing the removal efficiency of conventional or emerging pollutants of importance. This indicates that wastewater with a high solids load can only be treated directly by sonolysis after first undergoing a solids removal process. Other parameters or factors that affect the removal provided by the sonolysis process are the frequency with which the ultrasonic waves are generated, the temperature of the wastewater to be treated, and the properties of the pollutants, whether conventional or emerging, to be removed [71].

In addition, the concentration of volatile compounds in the wastewater to be treated affects the concentration of cavitation bubbles and thus decreases the removal of pollutants provided by the process [72].

However, given all the complications presented by the sonolysis, which interferes with the production of hydroxyl radicals ($HO^{\bullet}$) for the removal of conventional and emerging pollutants in domestic and industrial wastewater, it is suggested to use it as a tertiary or polishing process wastewater treatment because if the raw wastewater presents high loads of solids or volatile compounds, the removal capacity can be affected. It should be mentioned that the removal of conventional and emerging pollutants of interest by sonolysis is usually slow [73].

4. Overview and suggestions to carry out the advanced oxidation processes

Certain constituents present in the wastewater (domestic and industrial) to be treated can also inhibit advanced oxidation processes. Chloride (Cl^-) and bromide (Br^-) can act as scavengers of hydroxyl radicals ($HO^\bullet$) [44] and, to some extent, inhibit the removal of pollutants in wastewater (conventional or emerging) provided such processes. So, for example, depending on the type of chlorination used in a conventional WWTP for water treatment and disinfection, an advanced oxidation process cannot be used as a tertiary or quaternary process because chlorine can act as a scavenger of the hydroxyl radicals ($HO^\bullet$) and inhibit the advanced oxidation process to some extent. It should be emphasized that it is necessary to identify in which form the chlorine (e.g., chloride) is added for treating wastewater (domestic and industrial).

As a suggestion, if the wastewater (domestic and industrial) to be treated is highly loaded with the aforementioned anions (Cl^- and Br^-), additional processes for their removal can be contemplated and combined with the advanced oxidation processes, but this would increase the cost of treating the wastewater (domestic and industrial) to be treated considerably.

Additionally, it is not advisable to use hydrochloric acid (HCl) in advanced oxidation processes that require pH adjustment to acidic levels of the wastewater (domestic and industrial) to be treated, since its nature and its dissociation into chlorides (Cl^-) and protons (H^+) in the wastewater could affect the removal of pollutants (conventional and emerging) of interest, because indirectly through the addition of hydrochloric acid (HCl), chlorides (Cl^-) would be added to the wastewater (domestic and industrial).

As a final part, to define the amounts of oxidant to be added to the wastewater (domestic and industrial) to be treated by the advanced oxidation processes that employ hydrogen peroxide (H_2O_2) as the oxidant, as well as the Fenton and photo-Fenton processes, Eq. (5) reported in the present chapter can be used for that purpose. The amounts of catalyst can be established stoichiometrically using the reaction of the catalyst with the oxidant (H_2O_2), which it presents in the advanced oxidation process to be employed, in particular, for the generation of the hydroxyl radicals ($HO^\bullet$) for the treatment of wastewater (domestic and industrial). An example of this is in subsection 3.2 relating to the Fenton process.

5. Conclusions

Because poorly treated wastewater is the main entry point for conventional and emerging pollutants into the water cycle and the environment, and it can have both indirect and direct effects on flora, fauna, and human health, processes such as advanced oxidation must be used to treat domestic and industrial wastewater, either as tertiary or primary processes.

On the other hand, among the advanced oxidation processes discussed in this chapter, Fenton's advanced oxidation process and its variant (photo-Fenton) are, according to the information gathered, the best option for the treatment of domestic and industrial wastewater and the removal of conventional and emerging pollutants from wastewater. In addition, compared to the other advanced oxidation processes mentioned earlier in this chapter, the Fenton processes have the least difficulty scaling up. Photocatalysis, indirect ozonation, and cavitation are the advanced oxidation processes that lack scalability and should be further developed.

Finally, there is a need for further improvement of advanced oxidation processes to increase their applicability to the treatment of domestic and industrial wastewater at the macro level and to improve the quality of treated domestic and industrial wastewater by reducing the presence of conventional and emerging pollutants.

Acknowledgements

This research was funded by the National Council of Humanities, Science and Technology (CONAHCYT), with scholarship grant number 805457 and project number CF-2019/102967. Acknowledges to Dagoberto Rodríguez-Ortiz from CIMAV-DGO for his valuable technical support and collaboration.

Conflict of interest

The authors declare no conflict of interest.

Notes/thanks/other declarations

Firstly, thanks to my God and my family for their support and understanding. To Dr. Liliana Reynoso Cuevas (thesis director and mentor) for her support and advice to carry out my research project and different publications at national and international levels. This work is dedicated to your grandmother, who was like a mother to me (Beatriz Ruelas).

Author details

Alexis Rubén Bracamontes-Ruelas[1], José Rafael Irigoyen-Campuzano[1], Luis Arturo Torres-Castañon[1] and Liliana Reynoso-Cuevas[2*]

1 Department of Sustainable Engineering, Center for Research in Advanced Materials, Durango, Mexico

2 CONAHCYT-Center for Research in Advanced Materials, Durango, Mexico

*Address all correspondence to: liliana.reynoso@cimav.edu.mx

References

[1] Gopalakrishnan G, Jeyakumar RB, Somanathan A. Challenges and emerging trends in advanced oxidation technologies and integration of advanced oxidation processes with biological processes for wastewater treatment. Sustainability. 2023;**15**:4235. DOI: 10.3390/su15054235

[2] Nikbeen T, Nayab AK. Transformation of traditional wastewater treatment methods into advanced oxidation processes and the role of ozonation. Journal of Ecological Engineering. 2023;**24**:173-189. DOI: 10.12911/22998993/162777

[3] Mahbub P, Duke M. Scalability of advanced oxidation processes (AOPs) in industrial applications: A review. Journal of Environmental Management. 2023;**345**:118861. DOI: 10.1016/j.jenvman.2023.118861

[4] Saviano L, Brouziotis AA, Padilla Suarez EG, Siciliano A, Spampinato M, Guida M, et al. Catalytic activity of rare earth elements (REEs) in advanced oxidation processes of wastewater pollutants: A review. Molecules. 2023;**28**:6185. DOI: 10.3390/molecules28176185

[5] Ma D, Yi H, Lai C, Liu X, Huo X, An Z, et al. Critical review of advanced oxidation processes in organic wastewater treatment. Chemosphere. 2021;**275**:130104. DOI: 10.1016/j.chemosphere.2021.130104

[6] Pandis PK, Kalogirou C, Kanellou E, Vaitsis C, Savvidou MG, Sourkouni G, et al. Key points of advanced oxidation processes (AOPs) for wastewater, organic pollutants and pharmaceutical waste treatment: A mini review. ChemEngineering. 2022;**6**:8. DOI: 10.3390/chemengineering6010008

[7] Umair M, Kanwal T, Loddo V, Palmisano L, Bellardita M. Review on recent advances in the removal of organic drugs by advanced oxidation processes. Catalysts. 2023;**13**:1440. DOI: 10.3390/catal13111440

[8] Lima JPP, Tabelini CHB, Aguiar A. A review of Gallic acid-mediated Fenton processes for degrading emerging pollutants and dyes. Molecules. 2023;**28**:1166. DOI: 10.3390/molecules28031166

[9] Wang H, Wang Y, Dionysiou DD. Advanced oxidation processes for removal of emerging contaminants in water. Water. 2023;**15**:398. DOI: 10.3390/w15030398

[10] Manassero A, Alfano OM, Satuf ML. Degradation of emerging pollutants by photocatalysis: Radiation modeling and kinetics in packed-bed reactors. Water. 2022;**14**:3608. DOI: 10.3390/w14223608

[11] Olea-Mejia O, Brewer S, Donkor K, Amado-Piña D, Natividad R. Photo-Fenton catalyzed by Cu_2O/Al_2O_3: Bisphenol (BPA) mineralization driven by UV and visible light. Water. 2022;**14**:3626. DOI: 10.3390/w14223626

[12] Bermúdez LA, Pascual JM, Martínez MMM, Poyatos Capilla JM. Effectiveness of advanced oxidation processes in wastewater treatment: State of the art. Water. 2021;**13**:2094. DOI: 10.3390/w13152094

[13] Solanki VS, Pare B, Gupta P, Jonnalagadda SB, Shrivastava R. A review on advanced oxidation processes (AOPs) for wastewater remediation. Asian Journal of Chemistry. 2020;**32**:2677-2684. DOI: 10.14233/ajchem.2020.22806

[14] Marcinowski P, Bury D, Krupa M, Ścieżyńska D, Prabhu P, Bogacki J. Magnetite and hematite in advanced oxidation processes application for cosmetic wastewater treatment. Processes. 2020;**8**:1343. DOI: 10.3390/pr8111343

[15] Deng Y, Zhao R. Advanced oxidation processes (AOPs) in wastewater treatment. Current Pollution Reports. 2015;**1**:167-176. DOI: 0.1007/s40726-015-0015-z

[16] Dutta D, Arya S, Kumar S. Industrial wastewater treatment: Current trends, bottlenecks, and best practices. Chemosphere. 2021;**285**:131245. DOI: 10.1016/j.chemosphere.2021.131245

[17] Khan SAR, Ponce P, Yu Z, Golpîra H, Mathew M. Environmental technology and wastewater treatment: Strategies to achieve environmental sustainability. Chemosphere. 2022;**286**:131532. DOI: 10.1016/j.chemosphere.2021.131532

[18] Koul B, Yadav D, Singh S, Kumar M, Song M. Insights into the domestic wastewater treatment (DWWT) regimes: A review. Water. 2022;**14**:3542. DOI: doi.org/10.3390/w14213542

[19] Obaideen K, Shehata N, Sayed ET, Abdelkareem MA, Mahmoud MS, Olabi AG. The role of wastewater treatment in achieving sustainable development goals (SDGs) and sustainability guideline. Energy Nexus. 2022;**7**:100112. DOI: 10.1016/j.nexus.2022.100112

[20] Mexican Official Standard NOM-001-SEMARNAT-2021 [Internet]. 2021. Available from: https://www.dof.gob.mx/nota_detalle.php?codigo=5645374&fecha=11/03/2022#gsc.tab=0

[21] Singh BJ, Chakraborty A, Sehgal R. A systematic review of industrial wastewater management: Evaluating challenges and enablers. Journal of Environmental Management. 2023;**348**:119230. DOI: 10.1016/j.jenvman.2023.119230

[22] Yu Y, Wu B, Jiang L, Zhang XX, Ren HO, Li M. Comparative analysis of toxicity reduction of wastewater in twelve industrial park wastewater treatment plants based on battery of toxicity assays. Scientific Reports. 2019;**9**:3751. DOI: 10.1038/s41598-019-40154-z

[23] Ahmed J, Thakur A, Goyal A. CHAPTER 1: Industrial wastewater and its toxic effects. In: Shah MP, editor. Biological Treatment of Industrial Wastewater. 1st ed. London: The Royal Society of Chemistry; 2021. pp. 1-14. DOI: 10.1039/9781839165399

[24] Osorio-Rivera MA, Carrillo-Barahona WE, Negrete-Costales JH, Loor-Lalvay XA, Riera-Guachichullca EJ. La calidad de las aguas residuales domésticas. Polo del Conocimiento. 2021;**6**:228-245. DOI: 10.23857/pc.v6i3.2360

[25] Parida VK, Saidulu D, Majumder A, Srivastava A, Gupta B, Gupta AK. Emerging contaminants in wastewater: A critical review on occurrence, existing legislations, risk assessment, and sustainable treatment alternatives. Journal of Environmental Chemical Engineering. 2021;**9**:105966. DOI: 10.1016/j.jece.2021.105966

[26] Kristanti RA, Tirtalistyani R, Tang YY, Thao NTT, Kasongo J, Wijayanti Y. Phytoremediation mechanism for emerging pollutants: A review. Tropical Aquatic and Soil Pollution. 2023;**3**:88-108. DOI: 10.53623/tasp.v3i1.222

[27] Kumar R, Qureshi M, Vishwakarma DK, Al-Ansari N, Kuriqi A,

Elbeltagi A, et al. A review on emerging water contaminants and the application of sustainable removal technologies. Case Studies in Chemical and Environmental Engineering. 2022;**6**:100219. DOI: 10.1016/j.cscee.2022.100219

[28] Arman NZ, Salmiati S, Aris A, Salim MR, Nazifa TH, Muhamad MS, et al. A review on emerging pollutants in the water environment: Existences, health effects and treatment processes. Water. 2021;**13**:3258. DOI: 10.3390/w13223258

[29] Zhang M, Sun Y, Xun B, Liu B. Analysis of the spatial distribution characteristics of emerging pollutants in China. Water. 2023;**15**:3782. DOI: 10.3390/w15213782

[30] Gavrilescu M, Demnerová K, Aamand J, Agathos S, Fava F. Emerging pollutants in the environment: Present and future challenges in biomonitoring, ecological risks and bioremediation. New Biotechnology. 2015;**32**:147-156. DOI: 10.1016/j.nbt.2014.01.001

[31] Bracamontes-Ruelas AR, Ordaz-Díaz LA, Bailón-Salas AM, Ríos-Saucedo JC, Reyes-Vidal Y, Reynoso-Cuevas L. Emerging pollutants in wastewater, advanced oxidation processes as an alternative treatment and perspectives. Processes. 2022;**10**:1041. DOI: 10.3390/pr10051041

[32] Amor C, Marchão L, Lucas MS, Peres JA. Application of advanced oxidation processes for the treatment of recalcitrant agro-industrial wastewater: A review. Water. 2019;**11**:205. DOI: 10.3390/w11020205

[33] Rayaroth MP, Boczkaj G, Aubry O, Aravind UK, Aravindakumar CT. Advanced oxidation processes for degradation of water pollutants—Ambivalent impact of carbonate species—A review. Water. 2023;**15**:1615. DOI: 10.3390/w15081615

[34] Navidpour AH, Abbasi S, Li D, Mojiri A, Zhou JL. Investigation of advanced oxidation process in the presence of TiO_2 semiconductor as photocatalyst: Property, principle, kinetic analysis, and photocatalytic activity. Catalysts. 2023;**13**:232. DOI: 10.3390/catal13020232

[35] Ameta R, Chohadia AK, Jain A, Punjabi PB. In: Ameta SC, Ameta R, editors. Chapter 3 – Fenton and Photo-Fenton Processes. 1st ed. Amsterdam: Elsevier; 2018. pp. 49-87. DOI: 10.1016/B978-0-12-810499-6.00003-6

[36] Lupu G-I, Orbeci C, Bobirică L, Bobirică C, Pascu LF. Key principles of advanced oxidation processes: A systematic analysis of current and future perspectives of the removal of antibiotics from wastewater. Catalysts. 2023;**13**:1280. DOI: 10.3390/catal13091280

[37] Mandavgane SA. Study of degradation of p-toluic acid by photo-oxidation, peroxidation, photo-peroxidation and photo-Fenton processes. Materialstoday: Proceedings. 2020;**29**:1213-1216. DOI: 10.1016/j.matpr.2020.05.478

[38] Kanakaraju D, Glass BD, Oelgemöller M. Advanced oxidation process-mediated removal of pharmaceuticals from water: A review. Journal of Environmental Management. 2018;**219**:189-207. DOI: 10.1016/j.jenvman.2018.04.103

[39] Philippopoulos CJ, Poulopoulos SG. Photo-assisted oxidation of an oily wastewater using hydrogen peroxide. Journal of Hazardous Materials. 2003;**98**:201-210. DOI: 10.1016/S0304-3894(02)00357-6

[40] Della Rocca DG, Victória HFV, Moura-Nickel CD, Scaratti G, Krambrock K, De Noni A, et al. Peroxidation and photo-peroxidation of pantoprazole in aqueous solution using silver molybdate as catalyst. Chemosphere. 2021;**262**:127671. DOI: 10.1016/j.chemosphere.2020.127671

[41] Iyyappan J, Gaddala B, Gnanasekaran R, Gopinath M, Yuvaraj D, Kumar V. Critical review on wastewater treatment using photo catalytic advanced oxidation process: Role of photocatalytic materials, reactor design and kinetics. Case Studies in Chemical and Environmental Engineering. 2024;**9**:100599. DOI: 10.1016/j.cscee.2023.100599

[42] Fenton HJH. LXXIII.—Oxidation of tartaric acid in presence of Iron. Journal of the Chemical Society, Transactions. 1894;**65**:899-910. DOI: 10.1039/ct8946500899

[43] Abe C, Miyazawa T, Miyazawa T. Current use of Fenton reactions in drugs and food. Molecules. 2022;**27**:5451. DOI: 10.3390/molecules27175451

[44] Pignatello JJ, Oliveros E, MacKay A. Advanced oxidation processes for organic contaminant destruction based on the Fenton reaction and related chemistry. Critical Reviews in Environmental Science and Technology. 2006;**36**:1-84. DOI: 10.1080/10643380500326564

[45] Kremer ML. Initial steps in the reaction of H_2O_2 with Fe^{2+} and Fe^{3+} ions: Inconsistency in the free radical theory. Reactions. 2023;**4**:171-175. DOI: 10.3390/reactions4010010

[46] Bracamontes-Ruelas AR, Reyes-Vidal Y, Irigoyen-Campuzano JR, Reynoso-Cuevas L. Simultaneous oxidation of emerging pollutants in real wastewater by the advanced fenton oxidation process. Catalysts. 2023;**13**:748. DOI: 10.3390/catal13040748

[47] Cheng S, Ran X, Ren G, Wei Z, Wang Z, Rao T, et al. Comparison of Fenton and ozone oxidation for Pretreatment of petrochemical wastewater: COD removal and biodegradability improvement mechanism. Separations. 2022;**9**:179. DOI: 10.3390/separations9070179

[48] Santin-Gusman M, Moreno-Andrés J, Cisneros-Abad M, Aguilar-Ramírez S. Optimization for Fenton Process in removal of COD for landfill leachate treatment. International Journal of Environmental Science and Development. 2015;**6**:920-924. DOI: 10.7763/IJESD.2015.V6.722

[49] Kastanek F, Spacilova M, Krystynik P, Dlaskova M, Solcova O. Fenton reaction–unique but still mysterious. Processes. 2023;**12**:432. DOI: 10.3390/pr11020432

[50] Pimentel Prates M et al. Fenton: A systematic review of its application in wastewater treatment. Process. 2023;**11**:2466. DOI: 10.3390/pr11082466

[51] Elmobarak WF, Hameed BH, Almomani F, Abdullah AZ. A review on the treatment of petroleum refinery wastewater using advanced oxidation processes. Catalysts. 2021;**11**:782. DOI: 10.3390/catal11070782

[52] Xavier S, Gandhimathi R, Nidheesh PV, Ramesh ST. Comparative removal of Magenta MB from aqueous solution by homogeneous and heterogeneous photo-Fenton processes. Desalination and Water Treatment. 2015;**57**:12832-12841. DOI: 10.1080/19443994.2015.1054887

[53] Klamerth N, Malato S, Maldonado MI, Agüera A, Fernández-Alba AR. Application of photo-Fenton as a tertiary treatment of

emerging contaminants in municipal wastewater. Environmental Science & Technology. 2010;**44**:1792-1798. DOI: 10.1021/es903455p

[54] Jorge N, Teixeira AR, Silva S, Pirra A, Peres JA, Lucas MS. Homogeneous vs. heterogeneous Photo-Fenton processes in the treatment of winery wastewater. Engineering Proceedings. 2023;**56**:288. DOI: 10.3390/ASEC2023-15405

[55] Garrido-Cardenas JA, Esteban-García B, Agüera A, Sánchez-Pérez JA, Manzano-Agugliaro F. Wastewater treatment by advanced oxidation Process and their worldwide research trends. International Journal of Environmental Research and Public Health. 2020;**17**:170. DOI: 10.3390/ijerph17010170

[56] Buthiyappan A, Abdul Aziz AR, Wan Daud WMA. Degradation performance and cost implication of UV-integrated advanced oxidation processes for wastewater treatments. Reviews in Chemical Engineering. 2015;**31**:263-302. DOI: 10.1515/revce-2014-0039

[57] Sanabria P, Wilde ML, Ruiz-Padillo A, Sirtori C. Trends in Fenton and photo-Fenton processes for degradation of antineoplastic agents in water matrices: Current knowledge and future challenges evaluation using a bibliometric and systematic analysis. Environmental Science and Pollution Research. 2022;**29**:42168-42184. DOI: 10.1007/s11356-021-15938-4

[58] Zhang Y, Shaad K, Vollmer D, Ma C. Treatment of textile wastewater using advanced oxidation processes—A critical review. Water. 2021;**13**:3515. DOI: 10.3390/w13243515

[59] Chiang YP, Liang YY, Chang CN, Chao AC. Differentiating ozone direct and indirect reactions on decomposition of humic substances. Chemosphere. 2006;**65**:2395-2400. DOI: 10.1016/j.chemosphere.2006.04.080

[60] Miklos DB, Remy C, Jekel M, Linden KG, Drewes JE, Hübner U. Evaluation of advanced oxidation processes for water and wastewater treatment – A critical review. Water Research. 2018;**139**:118-131. DOI: 10.1016/j.watres.2018.03.042

[61] Joseph CG, Farm YY, Taufiq-Yap YH, Pang CK, Nga JLH, Li PG. Ozonation treatment processes for the remediation of detergent wastewater: A comprehensive review. Journal of Environmental Chemical Engineering. 2021;**9**:106099. DOI: 10.1016/j.jece.2021.106099

[62] Jesus F, Domingues E, Bernardo C, Pereira JL, Martins RC, Gomes J. Ozonation of selected pharmaceutical and personal care products in secondary effluent— Degradation kinetics and environmental assessment. Toxics. 2022;**10**:765. DOI: doi.org/10.3390/toxics10120765

[63] Zilberman A, Gozlan I, Avisar D. Pharmaceutical transformation products formed by ozonation—Does degradation occur? Molecules. 2023;**28**:1227. DOI: 10.3390/molecules28031227

[64] Yang X, Wang D. Photocatalysis: From fundamental principles to materials and applications. ACS Applied Energy Materials. 2018;**1**:6657-6693. DOI: 10.1021/acsaem.8b01345

[65] Fang Y, Zheng Y, Fang T, Chen Y, Zhu Y, Liang Q, et al. Photocatalysis: An overview of recent developments and technological advancements. Science China Chemistry. 2020;**63**:149-181. DOI: 10.1007/s11426-019-9655-0

[66] Zhu S, Wang D. Photocatalysis: Basic principles, diverse forms of implementations and emerging scientific opportunities. Advances Energy Materials. 2017;**7**:1-24. DOI: 10.1002/aenm.201700841

[67] Rueda-Marquez JJ, Levchuk I, Fernández Ibañez P, Sillanpää M. A critical review on application of photocatalysis for toxicity reduction of real wastewaters. Journal of Cleaner Production. 2020;**258**:120694. DOI: 10.1016/j.jclepro.2020.120694

[68] Madhavan J, Theerthagiri J, Balaji D, Sunitha S, Choi MY, Ashokkumar M. Hybrid advanced oxidation processes involving ultrasound: An overview. Molecules. 2019;**24**:3341. DOI: 10.3390/molecules24183341

[69] Wang N, Li L, Wang K, Huang X, Han Y, Ma X, et al. Study and application status of ultrasound in organic wastewater treatment. Sustainability. 2023;**15**:15524. DOI: 10.3390/su152115524

[70] Gadipelly C, Pérez-González A, Yadav GD, Ortiz I, Ibañez R, Rathod VK, et al. Pharmaceutical industry wastewater: Review of the Technologies for Water Treatment and Reuse. Industrial & Engineering Chemistry Research. 2014;**53**:11571-11592. DOI: 10.1021/ie501210j

[71] Savun-Hekimoğlu B. A review on Sonochemistry and its environmental applications. Acoustics. 2020;**2**:766-775. DOI: 10.3390/acoustics2040042

[72] González-García J, Sáez V, Tudela I, Díez-Garcia MI, Deseada EM, Louisnard O. Sonochemical treatment of water polluted by chlorinated Organocompounds. A review. Water. 2010;**2**:28-74. DOI: 10.3390/w2010028

[73] Wang G, Cheng H. Application of photocatalysis and sonocatalysis for treatment of organic dye wastewater and the synergistic effect of ultrasound and light. Molecules. 2023;**28**:3706. DOI: 10.3390/molecules28093706

Chapter 5

Novel Approaches for Sustainable Management of Chromium Contaminated Wastewater

Manoj Kumar and Harvinder Singh Saini

Abstract

The release of high volumes of untreated effluents containing different forms of chromium into waterbodies and further use of this wastewater for ferti-irrigation purposes pose a direct threat to health of human populations consuming produces from such agricultural fields. The higher concentration of chromium above permissible limits at these sites may pose harm to flora and fauna. The conventional processes used for treatment of chromium-containing effluents have low treatment efficiency, high operational costs, and produce toxic sludge requiring safe disposal. In contrast, the approaches exploiting use of living systems, such as microbes/microbial products and microbes, may provide sustainable treatment options. The emerging advanced/novel treatment technologies based on harnessing metabolic potential of microbiome of the polluted sites have potential to achieve the efficient removal of heavy metals from polluted sites. The success of protocols developed and tested at lab scale needs to be replicated at pilot/industrial to handle high volumes with varying levels of organic co-contaminants and harsh physiological conditions. The presented chapter provides an overview of impact of high chromium levels on ecosystem and various treatment processes with advanced aspect of management of heavy metals to prevent harmful effects on the environment.

Keywords: chromium treatment, heavy metals, metal-containing wastewater treatment, toxicity, Cr (VI), bioremediation

1. Introduction

Due to rapid globalization and industrialization, there is an extensive use of chromium (Cr) in various industries involved in the production of steel and chromium alloys, leather tanning, textile industries, and wood preservation for meeting the need of increasing population. Various types of Cr-containing waste are being generated *viz.* tannery effluent, generation of chromite mining waste leading to soil and groundwater contamination. In light the overall cost including environmental cost of chromium mining it is imperative to develop technologies for recovery/recycling of

IntechOpen

Cr from such wastes. The release of partially/untreated wastewater containing heavy metals, such as chromium (Cr), from various industrial units causes the pollution in the water and sediments of aquifers/waterbodies [1]. Industrially, Cr is mostly used in steel manufacturing, electroplating and chrome plating, leather tanning, textile dyeing, and wood manufacturing. The higher/extensive use of these heavy metals has considerably exacerbated the problems related with water pollution, which has negative effects on economy, and public health-related issues worldwide. Additionally, the use of water from these contaminated waterbodies for agriculture purpose led to accumulation of heavy metals and other pollutants in agricultural crops affecting food safety, human health, and deterioration of ecological niches [2]. Chromium (Cr) is the tenth most common element in the Earth's mantle, with an atomic number of 24 and a relative atomic mass of 51.9. Most of the Cr in soil/water environment is found in the two most common valence states, that is. Cr (VI) and Cr (III). Among them, Cr (VI) is having more aqueous solubility than as compared to Cr (III) state and resultantly more mutagenic and carcinogenic than later [3]. The pollution related to the presence of Cr (VI) in underground aquifers can pose risk to flora and fauna of the receiving waterbodies. Additionally, the consumption of Cr (VI) polluted water can cause adverse health effects among the human population such as skin infections, respiratory diseases, kidney disorder, liver-related disorder, and cardiovascular diseases [4]. Further, the use of wastewater for ferti-irrigation caused accumulation of these pollutants in agricultural soil and direct exposure of human population consuming produces from such sites. Cr (VI) can be toxic to both prokaryotic and eukaryotic cells as it can be easily transported into the cells by the process of facilitated diffusion, using nonspecific anion channels. Inside the cell, Cr (VI) can be reduced to pentavalent, tetravalent, and trivalent forms by various enzymatic systems. Cr (VI) at higher concentrations can form oxyanions/hydroxyl radicals causing strand break in DNA structure [5].

The commonly used methods for lowering/removal of Cr (VI) from effluents/wastewater are based on physical/chemical methods such as adsorption (activated carbon and renewable agricultural waste), ion exchange, membrane filtration, electrochemical methods, hydroxide precipitation, sulfide precipitation, and chelating precipitation, etc. The use of physical/chemical processes has numerous disadvantages such as recurring cost of consumables, high operation cost, and formation of toxic sludge requiring proper disposal in sanitary landfills. In light of this, there is an urgent need to develop sustainable treatment strategies, which might be less expensive than existing treatment options and do not produce toxic sludge/byproducts, which may need cost intensive handling processes. Additionally, a treatment option with possibility of recovery of metals for their reuse in process will make the treatment options more attractive for end users and will be more environmentally friendly. The treatment methods based on use of biological tools, such as plants and microorganisms, are attractive options being used. However, the accumulation of metals in plant and microbial biomass poses a problem in both extraction of metals of their safe disposal. The more appropriate options being conversion of the toxic form to less toxic form by reduction or precipitating out the metals from effluents or *in situ* metal precipitation (ISMP) on sludge or soil matrix, resulting in lowering of their toxic potential. Further, the recent reports on use of microbial products *viz* metallothioneins, siderophore, exo-polymers, etc. to bind or sequester out metals have provided other options, which may contribute to economic growth in the metal use and recycling [6, 7]. Biological system requires metabolically versatile inocula, which may be able to maintain their potential under harsh conditions such as temperature, pH, and nutrients for successful treatment [8]. Recently, with the advent of novel/advanced treatment technologies, including nanotechnology, combined treatment technologies, and synthetic biology tools, such as micro-biome engineering and systems biology, have provided a wide field to develop more sustainable options than other treatments for the restoration of ecosystems affected by presence of high level of heavy metals. The application of bioengineering and molecular biology tools can facilitate the transfer of treatment processes developed at laboratory or pilot scale to field scale level. The establishment of multiple levels of validation of bioremediation will open a new window for easier audit of bioremediation processes [1].

This chapter outlines application of Cr (VI) in various industries, its environmental, and health impact as well as management of chromium contaminated wastewater using conventional, biological, and advanced methods. This chapter also serve as valuable source for scientist and researchers working in remediation of Cr (VI) present in industrial wastewater to protect our environment and ecosystem.

2. Chromium-based industries and pollution

Chromium finds its applications in many industries, especially in steel manufacturing, textile and dyes, leather, and chemical manufacture, such as chromated copper arsenate (CCA), used in wood preservation. Many of these industrial facilities use low-grade chromite ores for a variety of purposes such as producing pigments, polishing metal, preventing corrosion, synthesizing organic materials, tanning leather, and preserving wood as shown in **Figure 1** [9]. The leather industry alone accounts for 40% of the worldwide usage. Chromium is used as a tanning agent for making leather durable and less susceptible to decomposition. Chromium is also used in the production of steel and chromium alloys to improve their hardening and corrosion resistance [10]. Chromium is used in wood preservation in the form of chromated copper arsenate (CCA) to impregnate wood products so that it can be protected from insects and termites [11]. The tetrahalides, CrF_4, $CrCl_4$, and $CrBr_4$ of chromium are used as mordant and catalyst in dyeing processes of chrome dyes

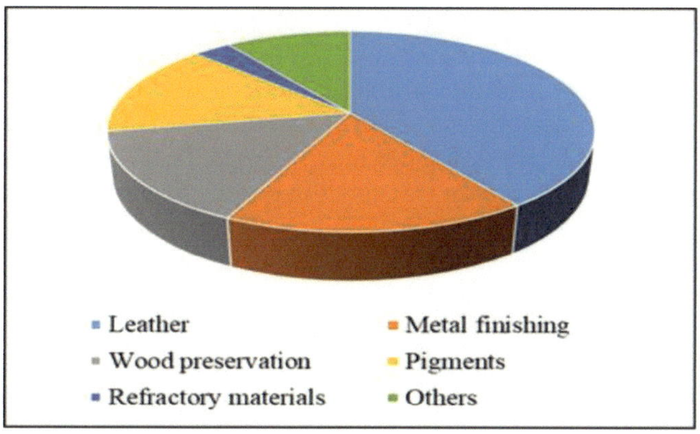

Figure 1.
Consumption of chromium in different industries.

in textile industries. The manufacture/synthesis processes from abovementioned products may produce a large amount of solid wastes and effluents rich in different forms of chromium with other components with pollution potential. Large number of heavy metal contaminants, including Cr (VI),are produced from industries such as tannery processing, chromate preparation, metal preparation, production of the colors chrome and ferrochrome, and temper steel welding [12].

The most important cause of chromium contamination in the environment is the release of untreated wastewater from various industries involved in either the production or utilization of chromium. Additionally, the disposal of toxic sludge from various industries and treatment plants can cause soil pollution, ultimately percolating to groundwater sources. Many of industries release varying amount of heavy metals found in their wastewater, which are released into the environment either directly or indirectly [13]. Due to the wide variations in the nature of these wastewaters depending on the type of industry, treatment of these wastewaters is a major topic of concern for achieving removal of heavy metals. In addition to this, leaching and weathering of chromite ores reserves can lead to major contamination of soil and water, causing environmental concern [14].

3. Impact of high level of chromium on environment and human health

In recent times, due to unrestricted discharge by point and nonpoint sources of varied industries, heavy metal, such as Cr (VI), caused irreparable damage to environment and animal/human health. The persistent release of Cr (VI) contaminated wastewater has led to its buildup in water and soil environment, causing adverse effects on flora and fauna. Some of these effects/impacts of Cr (VI) are discussed in the section below.

3.1 Cr (VI) impact on environment

Due to extensive use of Cr (VI) in many applications causes unprecedented damage to the environment. Recent studies have shown the accumulation of Cr (VI) beyond permissible limits in groundwater sources, wastewater effluents, and soil environments. In plants, Cr (VI) can be used for development, productivity, and metabolism [15]. However, plants exposed to higher concentrations of Cr (VI) had been reported for adverse effects on their physiology, morphology, and structural makeup. The nutrient, as well as water uptake mechanisms, may be affected, resulting in slowing down the growth and photosynthetic activity [16]. In literature, Wang et al., [17] studied that the exposure of maize plant to 300 mg/l of Cr (VI) as $K_2Cr_2O_7$ can cause the alterations in leave morphology of maize plant. Pandey et al., [18] reported inhibition of electron transport in photosystem-I and photosystem-II in presence of high concentration of Cr (VI) in which plants system. Also, the higher concentrations of Cr (VI) can cause oxidative damage to DNA and membrane lipids, as well as chlorosis and necrosis in plants [19].

The accumulation of Cr (VI) in soil may affect the overall productivity of agricultural crops as it had been reported that 94% reduction in stem growth was evident in plants of 32 species exposed to 1000 mg/kg plant weight as compared to control plants without exposure to any Cr (VI) [20]. The Cr (VI) bioaccumulation had been reported in fish, aquatic plants, algae, and invertebrates leading to hazardous and toxic effects on overall growth and development [21]. It was reported that accumulation of Cr (VI) in aquatic system cause increase in Cr (VI) level in higher trophic levels of food chains due to the phenomenon of bioaccumulation. This can lead to more harmful consequences in terms of toxicity in higher trophic level of food chain. More importantly, the potential risk of Cr (VI) has to studied in more depth in order to evaluate its concern regarding wildlife and environmental damages [22].

3.2 Cr (VI) impact on human health

Cr (VI) is reported to be highly soluble in water as compared to Cr (III) form, which can lead to its absolute presence in water sources. Lower concentrations of Cr (VI) are necessary for number of bodily functions such as lowering blood cholesterol level by decreasing low-density lipoproteins (LDLs), maintaining blood sugar level and as chromium picolinate used in supplements for gaining muscle mass [23, 24]. However, at higher concentrations, it can lead to cytotoxic and genotoxic effects on animal cells. Accidental breathing of Cr can cause irritation in the lungs and nose. Further, it can lead to lung cancer, kidney dysfunction, asthma, ulcers, and diarrhea [25]. Acute exposure can lead to choking, wheezing, and shortness of breath, whereas chronic exposure can lead to development of ulcers and perforations in respiratory, gastrointestinal, bronchitis, septum parts, etc. In literature, Cr (VI) can cause damage to blood cells, causing kidney and liver failure and also can lead to toxicity in bloodstream due to its oxidative nature [26]. In light of the serious effects of environment and overall health of flora and fauna, including human beings due to high levels of different forms of chromium, there is an urgent need to develop environmentally friendly and sustainable treatment systems for wastes/effluents having their high levels. Hence, due to severe environmental damage and its harmful health effects, we should design protocols for the treatment of effluents containing Cr (VI) as a pollutant.

4. Conventional treatment technologies

Cr (VI) can be removed from wastewater using variety of methods, including physical/chemical and biological methods. Various physical/chemical methods, which can be used for the treatment of Cr (VI) containing wastewater are described below.

4.1 Physical methods

Various physical methods, such as adsorption, ion exchange, membrane filtration, and electrochemical removal, can be used for the removal of Cr (VI) from wastewater.

4.1.1 Adsorption

Adsorption method is a surface deposition based on phase transfer process to remove chromium from contaminated water. It is based on the principle of purification or sorption of metal onto the pores of the sorbent, which reduces the concentration of toxic metal ions from contaminated wastewater [27]. Adsorption process is simple to carry out and cost-effective for the treatment of wastewaters. For adsorption, various types of adsorbent materials can be used, such as activated carbon and agricultural waste, as mentioned below [28].

4.1.1.1 Activated carbon

The most widely used adsorbent materials for the removal of Cr (VI) from wastewater is activated carbon. Different materials, such as coconut shell, wood, and sawdust, can be used for making activated carbon [27]. Activated carbon can be classified into two forms, that is. powdered activated carbon (PAC) and granular activated carbon (GAC) [29]. Commercially available activated carbon includes: Filtrasorb-100, F-400 (commercially activated carbon) and GA-3 [30, 31]. The advantage of using activated carbon includes low price, high surface area (500 to 1500 m^2/g), and easy operation as compared to conventional methods [27]. The high adsorption capacity of activated carbon depends on pH and temperature during the adsorption process. An increase in temperature was reported to decrease the adsorption capacity for the removal of chromium, whereas a low pH was reported to increase the adsorption capacity of activated carbon [32]. Moreover, activated carbon needs to be chemically activated to enhance its adsorption capacity and should be regenerated after each treatment cycle [27].

4.1.1.2 Use of agricultural waste

Agricultural waste or its byproducts can be a good option to remove Cr (VI) from wastewaters, for example, pine needles, cactus [33], timber, grain crops, peanut shells, hazelnut shell, bark, and tea leaves [34]. There are various mechanisms used in bio-sorption, such as chemisorption or complexation, with pores present on its surface due to adsorption and ion exchange [35]. The advantage of this method is that the adsorbent material is derived from low-cost agricultural waste, which can be used for the recovery of metals such, as Cr (VI) [27]. However, the main disadvantage of this technique is that plant material cannot be directly used without chemical pretreatments (using acid or base), for example, treatment of orange peels with calcium chloride ($CaCl_2$) and sodium hydroxide (NaOH), can result in increased adsorption capacity because of the transformation of methyl esters as inhibiting groups to carboxylate ligands. As a result, the metal binding capability was greatly enhanced. The metal uptake can be increased by increasing number of binding sites, better ion-exchange ability, and formation of new functional groups. The application of these chemical pretreatments might increase the cost of treatment hindering its use for long-term treatment [34–36].

4.1.2 Ion exchange-based removal

Ion exchange methods are based on removal of chromium using a support having resins for ion exchange. The support materials used in ion exchange columns can include Dowex 2X4 [37], Ambersep 132 [38], solvent-impregnated resin with quaternary ammonium salt (aliquat 336), polyacrylate anion exchanger having strong base functional groups Amberlite IRA 458, and Amberlite IRA 958 [39]. The Cr (VI) containing wastewater was made to pass through resin bed under pressure, where chromium is removed by ion-exchange mechanism. When resin capacity is exhausted due to binding of chromium, the column is backwashed to remove trapped solids and then it can be regenerated using HCl [40]. Ion exchange method requires low maintenance. However, it is expensive, susceptible to fouling, and readily clogged by organic materials and other compounds found in wastewater. In general, pretreatment is required before using ion exchange resin on wastewater that contains significant quantities of suspended particles and salts, thereby limiting its application for the treatment of industrial wastewater having Cr (VI) as a pollutant [39]. The use of acid for the regeneration of ion exchange columns makes this process more expensive. Additionally, the release of this acid after the regeneration process from ion exchangers can cause secondary pollution [9, 40].

4.1.3 Membrane filtration

The fundamental component of membrane processes is the use of a semipermeable membrane to divide a solution into two distinct streams. The use of different membranes for the treatment of wastewater containing Cr (VI) determines the separation selectivity. There are many types of membranes, which can be used for the removal of Cr (VI) from wastewater such as polymeric, inorganic, and liquid membranes. The inorganic membranes are very costly, but they are chemically and thermally stable [27]. The polymer-based membrane filters made up of polyethersulphone with carboxymethyl sulphonate (CMC) as complexing agent was found to achieve 99.5% efficiency of 10 mg/l of Cr (VI). However, these polymer membranes generally face biofouling problem [41]. Liquid membranes have high diffusion rates, but they instable due membrane expansion and liquid loss during the removal of Cr (VI) [27]. The main disadvantage of using membrane filtration is its high cost and recovery process for the removal of Cr (VI) is not efficient.

4.1.4 Electrochemical methods

It is also referred to as electro-dissolution or electrocoagulation method. The combination of electrochemical and chemical reactions makes this method of choice to be used for the treatment Cr (VI) present in solid, liquid, and gaseous matters [42]. Electrocoagulation process uses two different electrodes namely: anode (sacrificial electrode) that is usually aluminum or iron and a cathode (liberates hydrogen gas) as shown in **Figure 2** [43]. Electrochemical removal process has two stages: [1] Reduction of Cr (VI) to Cr (III) and [2] Precipitation as hydrated chromium complex. The

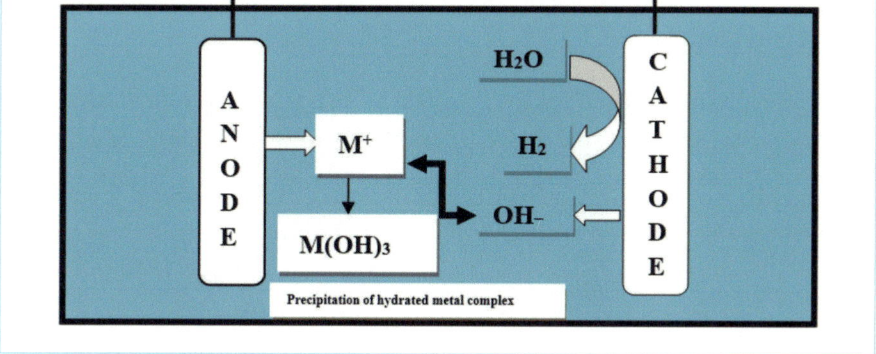

Figure 2.
Electrocoagulation of chromium.

electro-dissolution method involves dissolution of iron (anode) under strong acidic condition followed by reduction of Cr (VI) as shown in reactions of **Figure 2** [44]. It is widely used in electroplating industry to treat industrial wastewater, but its efficiency is affected by the presence of other ion, especially salt ions and change in pH.

Reaction mechanisms of chromium reduction in electrochemical cell are as follows:

$$FeO + 2H_2O \rightarrow Fe^{2+} + 2H_2 + O_2 + 2e^- \quad (1)$$

$$6Fe^{2+} + Cr_2O_7^{2-} + 14H^+ \rightarrow 6Fe^{3+} + 2Cr^{3+} + 7H_2O \quad (2)$$

4.2 Chemical precipitation methods

Chemical precipitation is used as a common method for the removal of chromium from industrial wastewater. Chemical precipitation is a relatively simple process in which different chemicals react with heavy metals to form insoluble precipitates. These precipitates can be separated from the water by sedimentation or filtration, followed by its decantation for appropriate discharge [45]. The main disadvantages of chemical precipitation include high operational and maintenance cost, energy input, manual oversight, and generation of large sludge [34]. Chemical treatment of wastewater containing heavy metals includes different methods as described below.

4.2.1 Hydroxide precipitation

Hydroxide precipitation is the most commonly used method among chemical precipitation. It is based on the principle that metal hydroxide complex, which can be formed by the addition of Ca $(OH)_2$ or NaOH in wastewater containing heavy metals. These metal hydroxide complexes can be removed by the process of sedimentation or flocculation [46]. In literature, it was reported that the addition of 4% Ca $(OH)_2$ achieved complete removal of 30 mg/l of Cr (VI) at a pH of 8.3 [47]. The advantage of using hydroxide precipitation is that it is simple and low-cost method used for the removal of heavy metals from industrial effluents. The disadvantages of hydroxide precipitation include generation of large amount of sludge due to amphoteric nature of hydroxides, and it can create problem in maintaining pH of the wastewater, which can inhibit metal hydroxide precipitation.

4.2.2 Sulfide precipitation

Sulfide precipitation is used for the removal of heavy metals from wastewater by the addition of iron sulfide or pyrite [48]. Sulfide precipitation is non-amphoteric in nature, so it does not require maintenance of pH for the removal of heavy metal. The main disadvantage of this process includes evolution of H_2S fumes and formation of colloidal precipitates that cause separation problems [34].

4.2.3 Chemical precipitation with other methods

Chemical precipitation (i.e., hydroxide and sulfide precipitation) can be used with other methods, such as nano-filtration or with ion exchange method in order to overcome some of the limitations of hydroxide or sulfide precipitation method. Sulfide precipitation can be used with nano-filtration for the removal of heavy metals. Ion exchange method can be used in combination with chemical precipitation for the removal of heavy metals. But the combination of these methods can be expensive as acid used in regeneration of ion exchange columns can cause secondary pollution [41].

4.2.4 Chelating precipitation

This method uses chelating precipitation for the removal of heavy metals from wastewater. N-bis-(dithiocarboxy) piperazine (BDP) and 1,3,5- hexahydrotriazin-edithiocarbamate (HTDC) were used for the removal of complex heavy metal from wastewater containing Ni^{2+}, Cr^{6+} and Cu^{2+} [41]. Both BDP and HTDC can effectively lower the concentration of different heavy metals in wastewater to less than 0.5 mg/l. In literature, it was reported that the use of potassium ethyl xanthate (KEX) for the complete removal of 1000 mg/l of Cu^{2+} from wastewater [49]. All the above mentioned methods are conventionally used for the removal of Cr (VI) as large volumes can be treated in short time; however, they have several disadvantages. The use of physicochemical methods requires specific conditions for proper treatment and can be affected by interferences of other pollutants, which actually reduce their efficiency for proper removal of Cr (VI). Moreover, generation of higher quantities of sludge, high operational costs, and secondary pollution caused due to the release of chemicals/acids used in the treatment process make them environmentally unsustainable [50]. The various advantages and disadvantages of various physical/chemical methods are given in **Table 1**.

Technique	Method	Advantages	Disadvantages
Physical methods	Adsorption	• Porous sorbent hence high capacity. • Suited for highly contaminated sites. • Effective and fast kinetics.	• Laborious and costly. • Performance depends upon type of carbon used. • Complexing agent requires to improve its performance.
	Ion exchange method	• Energy efficient, effective removal.	• Costly as required backwashing and regeneration of column. • Causes secondary pollution.
	Use of agricultural waste	• Low cost. • Convenient raw material. • Little environmental pollution and easy recovery of metals.	• Agricultural waste requires chemical pretreatments may increase the cost of treatment and may result in biomass loss. • Causes secondary pollution.
	Membrane filtration	• Less space required and specific in action. • Low energy demand.	• Membrane fouling expensive process. • Require proper maintenance.
	Electrochemical methods	• Low cost and high selectivity. • No additional chemical required.	• Sludge production. • Costly and require energy consumption and maintenance.

Technique	Method	Advantages	Disadvantages
Chemical precipitation	Hydroxide precipitation	• Rapid, easy, and cost-effective method.	• It generates large amount of sludge. • Maintenance of pH is required for effective removal. • Complexing agent can inhibit metal removal. • Causes secondary pollution.
	Sulfide removal	• Rapid, easy, and more effective than hydroxide precipitation as pH maintenance is not needed.	• Formation of hydrogen sulfide. • Formation of colloidal precipitates that cause separation problems.
	Combination of methods	• Rapid, easy, and more effective	• Secondary pollution, Expensive process
	Chelating precipitation	• Easy and effective at lower concentration of heavy metals.	• Generate large amount of sludge, Secondary pollution.

Table 1.
Advantages and disadvantages of various physical and chemical methods used for the treatment of Cr (VI) from wastewater [34].

5. Bioremediation: an eco-friendly approach

The biological treatment for achieving removal/reduction of chromium from wastewater provides an environmentally sustainable alternative to conventional treatment system. The use of microorganisms, such as bacteria, fungi, and yeast to reduce the toxic hexavalent chromium at polluted site to relatively less-toxic trivalent chromium, is an attractive option. Different microbial species, in mixed or pure cultures, are being reported for wide array of enzymes, which can achieve reduction of Cr (VI) [8]. These microbes belong to both aerobic and anaerobic domains and can work effectively at interface of such environment, thus providing multitude of options for natural bioremediation processes. Different groups of microorganisms can be used for the removal of Cr (VI) from wastewater, and these include the following.

5.1 Bacteria

Bacteria are diverse group of microorganisms, which can grow/divide rapidly and possess array of different enzymes. As per various literature reports, bacterial cells and their enzymes can reduce Cr (VI) to Cr (III), which is a significantly less hazardous metal [51]. Cr (VI) is mutagenic and toxic to bacteria, but it was observed that gram-positive bacteria are more resistant to Cr (VI) than gram-negative bacteria because of their thick cell wall. In literature, it was reported that 10–12 mg/l of Cr (VI), inhibited most soil bacteria in liquid media, such as *Pseudomonas aeruginosa* CCTCC AB93066, whereas Cr (III) at this concentration was found to be nontoxic [52, 53].

The reduction of Cr (VI) can be carried by either pure culture or by mixed bacterial culture. There are various reports for the treatment of Cr (VI) by bacterial isolates but only few of them can perform under both aerobic and anaerobic conditions [54]. Most of the studies for the reduction of Cr (VI) were carried out using pure microbial culture.

In literature, various bacterial isolates were reported for the reduction of Cr (VI), for example, *Cellulosimicrobium* sp. isolated from common effluent treatment plant (CETP) of a tannery plant, which was able achieve 97% reduction of 100 mg/l Cr (VI) after 72 h incubation [9]. Similarly, Banerjee et al. [55] isolated *Bacillus* strain TCL from Tasra coalmine lake (TCL), Jharia (Jharkhand), which was able to achieve complete reduction of 200 mg/l Cr (VI) in Luria-Bertani (LB) broth after 16 h incubation. Zeng et al. [56] isolated *Oceanobacillus* sp. W4 from chromium-contaminated soil, which was able to achieve 74.2% reduction of 200 mg/l Cr (VI) after 72 h incubation. Similarly, *Acinetobacter junii* strain b2w was able to achieve 98.24% reduction of 10 mg/l Cr (VI) to Cr (III) form. This was due to the involvement of various mechanisms bioaccumulation and efflux pumps used for the removal of Cr (VI) by *Acinetobacter junii* strain b2w [57].

However, the efficiency of pure cultures under field-scale applications can be affected by their survival under extreme environmental conditions and establishing themselves in competition with other predominating bacterial species already adapted to such conditions [58]. Hence, mixed bacterial consortium might have a better chance to survive during field-scale studies as compared to pure culture [59]. The metabolic versatility of different groups of microorganisms can complement each other to achieve better cleanup efficiency as compared to that achieved by individual pure cultures [60]. Bacteria can use different mechanisms by which it can achieve removal of heavy metals such as bioaccumulation, bio-sorption, bio-leaching, bio-transformation based detoxification, or *in situ* metal precipitation (ISMP) based on the presence of enzymes, metallothioneins and/or exopolysaccharides. Singh et al. [61] reported an SRB (sulfate reducing bacteria) consortium, which was able to achieve 96% reduction of 50 mg/l Cr (VI) in Postgate growth medium supplemented with 2.58% sodium lactate as carbon source using a batch type small scale bioreactor. Ma et al. [58] isolated mixed bacterial consortium comprising *Aeromonas*, *Pseudogracilibacillus*, and *Acinetobacter*, which was able to achieve complete reduction of 10 mg/l Cr (VI) in acetate minimal medium after an incubation of 72 h.

5.2 Fungi

Fungi, mostly yeasts isolated from different chromium-contaminated sites, had been reported for their resistance to high levels of chromium (VI). This metal resistance was attributed to the presence of cell wall and production of different enzymes, which carry out extracellular and intracellular precipitation, redox reactions, and active intake into the cells [62]. In literature, Xu et al. [63] isolated *Paecilomyces lilacinus* XLA, which was able to achieve 96.86% reduction of 100 mg/l Cr (VI) in minimum mineral (MM) medium after an incubation of 14 days. Long et al. [64] isolated *Penicillium oxalicum* SL2, which was able to achieve 89.6% reduction of 217 mg/l Cr (VI) from electroplating wastewater after an incubation of 96 h. Igeihon and Babalola, [65] isolated *Penicillium* sp., which was able to achieve 50% reduction of 16.1 mg/l Cr (VI) in Norkrans medium after an incubation of 20 days. Kumar and Dwivedi [66] isolated *Trichoderma lexii,* which was able to achieve 99.4% reduction of 50 mg/l Cr (VI) in potato dextrose broth medium after an incubation of 120 h. Hence, the expulsion of Cr (VI) was reported to be prominent in fungi.

5.3 Algae

The photosynthetic organisms, such as cyanobacteria or algae, have a ubiquitous distribution, and there are reports regarding their tolerance or resistance of algae to

Cr (VI) [67]. The resistance to Cr (VI) in algae was attributed to presence of polysaccharides, proteins, or lipids in their cell wall, which can act as binding agent of heavy metals as it contains carboxyl, sulfate, amino, and hydroxyl groups. Some studies suggested sequestration of Cr (VI) by its complexation with organic compounds in algal exudates/biomass [68]. Megharaj et al. [69] reported suppression of algal growth when it was exposed to 40 mg/kg of Cr (VI). Gupta and Rastogi [68] isolated *Oedogonium hatei* from freshwater, which was reported for bio-sorption of 30 mg/g of Cr (VI) from 100 mg/l Cr (VI) aqueous solution. Das et al. [70] reported a salt-tolerant micro-alga *Chlorella vulgaris* NIOCCV for its ability to achieve complete reduction of 3.22 mg/l of Cr (VI) from tannery wastewater after 12 days incubation. Costa et al. [71] isolated *Chlorella vulgaris,* which was able to achieve complete reduction of 10 mg/l Cr (VI) in the presence of photocatalyst experiment carried with titanium dioxide (TiO_2) after an incubation of 16 h.

5.4 Cyanobacteria

Cyanobacteria also known as blue-green algae have been reported for affinity to heavy metals due to their higher extracellular mucilage content [72]. Garnham and Green [73] reported the accumulation of 10 nmoles of Cr (VI)/gm dry weight by unicellular non-nitrogen-fixing cyanobacterium *Synechococcus* sp. PCC 6301. However, filamentous nitrogen-fixing cyanobacterium *Anabaena variabilis* was able to achieve 20% reduction of 10 nmoles Cr (VI) to Cr (III). This reduction of Cr (VI) by *A. variabilis* was due to the presence of heterocysts, and so it was able to tolerate 100 mg/l of Cr (VI), whereas the growth of *Synechococcus* sp. PCC 6301 was completely inhibited at this concentration of Cr (VI) [73]. Parveen et al. [74] reported *Synechocystis* sp. strain PUPCCC 62, which was able to reduce 250 nmoles of Cr (VI) per mg of protein. Sen et al. [72] isolated a cyanobacterial consortium having *Limnococcus limneticus* and *Leptolyngbya subtilis,* which was able to achieve 50% reduction of 15 mg/l Cr (VI) in BG-11 medium after 12 days. These studies indicate that cyanobacteria can also be used for the removal of Cr (VI) (**Table 2**).

Group	Culture name	Site of isolation	Reduction efficiency	Reference
Bacteria	*Pseudochrobactrum saccharolyticum* LY10	Chromium-contaminated soil	95% removal of 130 mg/l Cr (VI)	[75]
	Brevibacterium sp. K1	Soil sample	43.6% reduction of 500 mg/l Cr (VI)	[76]
	Stenotrophomonas sp. D6	Soil sample	87.6% reduction of 500 mg/l Cr (VI)	[76]
	Sporosarcina saromensis M52	Sediment sample	Complete reduction of 100 mg/l Cr (VI)	[77]
	Bacillus cereus (TIB3) strain	Mine tailings	82% reduction of 100 mg/l Cr (VI)	[78]
	Bacillus strain TCL	Coalmine lake	Complete reduction of 200 mg/l Cr (VI)	[55]
	Anoxybacillus flavithermus ABF1	Geothermal region	93.71% reduction of 20 mg/l Cr (VI)	[79]

Group	Culture name	Site of isolation	Reduction efficiency	Reference
	Bacillus paramycoides Cr6	Soil under the chrome slag of a chromate plant	100% reduction of 200 mg/l of Cr (VI)	[80]
Fungi	*Fusarium*	Tannery effluent	Complete removal of 10 mg/l Cr (VI)	[81]
	Penicillium commune	Soil and sludge sample	73% reduction of 9.86 mg/l Cr (VI)	[82]
	Trichoderma asperellum strain PTN7	Soil sample	Able to reduce 2.7 mg/l of Cr (VI)	[83]
	Trichoderma asperellum strain PTN10	Soil sample	Able to reduce 8.35 mg/l of Cr (VI)	[83]
	Penicillium oxalicum SL2	Air contaminated with industrial vapors	Complete reduction of 200 mg/l Cr (VI)	[64]
	Trichoderma lexii	Electroplating water	99.4% reduction of 50 mg/l Cr (VI)	[66]
Algae	*Spirogyra*	National chemical laboratory	98% bio-sorption of 5 mg/l Cr (VI)	[84]
	Chlorella vulgaris	Tannery wastewater	Reduce 3.22 mg/l of Cr (VI)	[70]
Cyanobacteria	*Nostoccalcicola* HH-12	Metal contaminated soil	Complete bio-sorption of 20 mg/l Cr (VI)	[85]
	Chroococcus sp. HH-11	Metal contaminated soil	Complete bio-sorption of 20 mg/l Cr (VI)	[85]
	Limnococcus, *Limneticus* and *Leptolyngbya subtilis* consortium	Wetland	50% removal of 15 mg/l Cr (VI)	[72]

Table 2.
Bio-reduction of chromium by different groups of organisms.

6. Advanced removal methods

Recently, advanced approaches are developed with the goal of enhancing effectiveness and sustainability for the elimination of Cr (VI) from wastewater. These methods are essential for addressing the serious environmental and health issues associated with heavy metal pollution. A few of them are mentioned below.

6.1 Using nanotechnology approach

The field of nanotechnology can be used for the removal of heavy metals from industrial wastewaters. It is a very efficient method for the removal of heavy metals from wastewaters due to their stability and small size, large surface area, and easily accessible pore space [86]. In addition to this, due to their small size, they possess efficient Brownian motion properties, making them suspended in wastewater for removal of heavy metals, such as Cr (VI). The use of nanoparticles in conjunction with other adsorbent materials, such as ion exchange resin and zerovalent ions, can increase the bioremediation potential for the removal of heavy metals, such as Cr (VI), present in wastewaters [87, 88]. Currently, various types of nanomaterials can be used for the uptake of heavy metals, such as Cr (VI), present in wastewater such as metal-organic framework (MOFs), metal oxides, carbon materials, and chitosan. Nithiya et al. [89] reported 55.71 mg/g removal of Cr (VI) based on adsorption phenomenon using chitosan/silicagel-based nano-composites. Recently, carbon nano-tubes (CNTs) can be used for the removal of heavy metals from wastewater due to their unique hollow structure, outside surface area, and high activity due to fast transport of water. Further, the surface modification of CNTs with groups, such as hydroxyl and carboxylic acid, can increase solubility of CNTs in wastewater due to high hydrogen bonding capacity, ion exchange, and increasing functionalization using nanoparticles increasing affinity toward heavy metals [90, 91]. In literature, Lyu et al. [88] reported adsorption of 130 mg/g of Cr (VI) using a biochar-supported nanoscale iron sulfide (FeS) composite (CMC-FeS@biochar) having three components *viz.* biochar, carboxymethyl cellulose (CMC), and FeS acting as adsorbent.

6.2 Combined treatment technologies

Combined treatment methods include the use of two techniques to achieve the efficient removal of heavy metals, such as Cr (VI). The removal of chromium from wastewater various approaches such as chemical precipitation, adsorption, membrane filtration, biological treatment, electrocoagulation, and advanced oxidation processes [86, 92]. These methods can be used alone or in combination to effectively remove chromium depending on the specific characteristics of the wastewater and regulatory requirements. Certain systems are designed not solely for the purpose of chromium removal, but also with the intention of recycle it for reuse. Pilot-scale testing is typically done to determine the most suitable treatment approach for a given wastewater stream [13]. As electrochemical and biological treatment can be used in culmination for the removal of heavy metals. In literature, Suganthi et al. [93] reported the removal of color and chemical oxygen demand (COD) present in tannery wastewater by using hybrid membrane bioreactor having activated sludge and electrocoagulation. Hou et al. [94] reported 51.3% removal of overall Cr and 48.1% of hexavalent chromium using electrochemically operated bio-sorption framework followed by adsorption using *Sargassum* sp. Similarly, Moradi and Moussavi et al. [95] reported 100 and 98.27% removal of COD and absolute Cr and sulfide from tannery wastewater using the combination of UVC/VUV photo-bioreactor and electrocoagulation technique. Hence, combination of two or more techniques can be used for efficient removal of heavy metals present in industrial wastewaters.

6.3 Use of synthetic biology tools for improving microbial potential

Synthetic biology tools use combination of molecular biology tools combining systems and cell biology to design/synthesize genes, enzymes, or transcription factors to redesign metabolic pathways for achieving targeted studies. Some of the methods, that is, microbiome engineering and synthetic biology are used nowadays to design synthetic microbes, specifically for achieving bioremediation are described in detail below.

6.3.1 Microbiome engineering

The interaction between microbial population and heavy metals is the basis of bioremediation of metal contaminated environments. The presence of diverse autochthonous microbial population can help in mitigation of heavy metals [96]. Microbiome includes different groups of micro-organisms capable to achieving bioremediation of heavy metals. The knowledge regarding microbiome of polluted environments and synthetic biology can pave ways for the removal of heavy metals from complex wastewater environments [97]. Nowadays, molecular biological tools (MBTs) can be used to identify contaminant degrading capabilities of microbiome present at polluted sites and further designing bioremediation protocols. The microbiome present in soil or wastewater environment can replenish the microenvironments using different methods such as immobilization, reduction of heavy metals to less toxic forms, and binding to cellular entities [98]. However, these activities depend upon efficiency of cultures present, nutrient sources, and physical and environmental factors. The activity of these autochthonous population can be increased by alteration in any of these factors. Also, manipulations of genes can increase siderophore, metallothionine production, and exo-polymer production increasing mitigation of heavy metals. In addition to this, efficient indigenous microbial culture can be used to achieve the desired bioremediation of wastewaters and soil environment [99].

6.3.2 Synthetic biology

The field of synthetic biology allows scientists to design and construct genetically engineered microorganisms capable of achieving reduction Cr (VI) to less toxic Cr (III). The use of interdisciplinary fields, such as biochemistry, genetic, and systems biology can plays a pivotal role in enhancing the efficiency of bioremediation processes using micro-organisms [100]. The field of biochemistry apprises various mechanisms by which microorganisms bind, sequester, or transform heavy metals and also focuses on the characteristics of various biomolecules, such as peptides or synthesis of metal-chelating compounds to achieve metal-binding proteins [92]. It can unravel various enzymes responsible for achieving reduction of heavy metals, such as Cr (VI). The knowledge regarding metabolic pathways and enzymes can be used to construct engineered microbes with enhanced abilities to reduce heavy metals to less toxic forms [101]. To achieve this, various genetic techniques such as gene knockout, overexpression or introduction of genes coding for metal-binding proteins, transporters proteins, efflux pumps, or enzymes can be employed to modify microorganisms for increased biotransformation capabilities [92, 100–102]. In literature, Xue et al. [103] engineered *Pseudomonas putida* KT2440 with self-controlled genetic circuit sensing variation in the concentration of mercury (Hg^{2+}). Nowadays, various gene editing methods can be used such as ZFN (zinc finger nuclease), TALENs (transcription activator-like effector nucleases), and CRISPRs (clustered regularly interspaced short palindromic repeats) used as transcription activators. Also, CRISPR-Cas9 approach, which can be precisely used to engineer microorganisms for specific heavy metal bioremediation applications [104].

In recent times, systems biology can help develop biosensors providing real-time monitoring to detect the presence of heavy metals in wastewater or polluted environments [105]. In literature, Adekunle et al. [106] described the use of floating microbial fuel cells (MFCs) as biosensors for the detecting the presence of copper (Cu) as low as 35–40 μg/L and other heavy metals present in aquatic environments.

Systems biology can provide holistic understanding of complex biological processes, including microbial communities that can play an important role in bioremediation of heavy metals. It includes the identification of microbial species, their functions, and understanding complex interplay/interactions among themselves and environmental factors to optimize bioremediation processes [107]. Also, mathematical modeling and simulation can predict the behavior of microbial communities in response to changing environmental conditions to design and optimization of bioremediation strategies. By integrating experimental data with mathematical models, researchers can develop efficient strategies to remove chromium from wastewater. Systems biology can also be used to monitor the impact of chromium bioremediation on the surrounding environment [108]. The ongoing development of innovative technologies and interdisciplinary collaboration will continue to drive progress in the field of bioremediation.

7. Conclusion

Chromium is widely used in several industrial processes as an important component, which leads to its pervasive presence in wastewaters of industrial origin. Further, the release of these wastewaters into rivers and drains leads to heavy metal presence in water resources causing harm to inhabiting flora and fauna. This leads to increase in the presence of heavy metals pollution in the environment (water/soil). The environment, human health, and the ecosphere are all at risk when excessive levels of heavy metals are released into the air, water, and soil. This chapter discusses pollution related to Cr (VI), and its harmful effects on environment and human health. We have also highlighted the treatment/removal of Cr (VI) using various conventional techniques, having many drawbacks; in contrast, biological treatment offers better alternative for the treatment of Cr (VI) from industrial wastewaters. In bioremediation aspect, different microorganisms, such as bacteria, fungi, algae, and cyanobacteria, can use a variety of mechanisms, such as metal sequestration or reduction for the removal of Cr (VI). In addition to this, advanced removal methods, including nanotechnology, combined treatment technologies, and synthetic biology tools, such as microbiome engineering and systems biology, can be used as one of the efficient and sustainable method for the removal of heavy metals from wastewater. Hence, the removal of Cr (VI) from wastewater is an important need of time in order to prevent harmful effects on environment and human health.

Author details

Manoj Kumar[1] and Harvinder Singh Saini[2*]

1 School of Allied and Healthcare Sciences, GNA University, Phagwara, Punjab, India

2 Department of Microbiology, Guru Nanak Dev University, Amritsar, Punjab, India

*Address all correspondence to: sainihs.micro@gndu.ac.in

References

[1] Kumar M, Saini HS. Reduction of hexavalent chromium (VI) by indigenous alkaliphilic and halotolerant *microbacterium* sp. M5: Comparative studies under growth and nongrowth conditions. Journal of Applied Microbiology. 2019;**127**:1057-1068. DOI: 10.1111/jam.14366

[2] Irshad MA, Nawaz R, Wojciechowska E, Mohsin M, Nawrot N, Nasim I, et al. Application of nanomaterials for cadmium adsorption for sustainable treatment of wastewater: A review. Water, Air, & Soil Pollution. 2023;**234**:54. DOI: 10.1007/s11270-023-06064-7

[3] Bao Z, Feng H, Tu W, Li L, Li Q. Method and mechanism of chromium removal from soil: A systematic review. Environmental Science and Pollution Research. 2022;**29**:35501-35517. DOI: 10.1007/s11356-022-19452-z

[4] Majumder S, Raghuvanshi S, Gupta S. Application of a hybrid biofilter column for the removal of Cr (VI) from aqueous solution using an indigenous bacterial strain *Pseudomonas taiwanensis*. Bioremediation Journal. 2016;**20**:10-23. DOI: 10.1080/10889868.2015.1113923

[5] Chiu A, Shi XL, Lee WKP, Hill R, Wakeman TP, Katz A, et al. Review of chromium (VI) apoptosis, cell-cycle-arrest, and carcinogenesis. Journal of Environmental Science and Health. 2010;**28**:188-230. DOI: 10.1080/10590501.2010.504980

[6] Vikrant K, Kumar V, Vellingiri K, Kim KH. Nanomaterials for the abatement of cadmium (II) ions from water/wastewater. Nano Research. 2019;**12**:1489-1507. DOI: 10.1007/s12274-019-2309-8

[7] Chang J, Deng S, Liang Y, Chen J. Cr (VI) removal performance from aqueous solution by *pseudomonas* sp. strain DC-B3 isolated from mine soil: Characterization of both Cr (VI) bio-reduction and total Cr biosorption processes. Environmental Science and Pollution Research. 2019;**26**:28135-28145. DOI: 10.1007/s11356-019-06017-w

[8] Pradhan D, Sukla LB, Sawyer M, Rahman PK. Recent bio-reduction of hexavalent chromium in wastewater treatment: A review. Journal of Industrial and Engineering Chemistry. 2017;**55**:1-20. DOI: 10.1016/j.jiec.2017.06.040

[9] Bharagava RN, Mishra S. Hexavalent chromium reduction potential of *Cellulosimicrobium* sp. isolated from common effluent treatment plant of tannery industries. Ecotoxicology and Environmental Safety. 2018;**147**:102-109. DOI: 10.1016/j.ecoenv.2017.08.040

[10] Robins KJ, Hooks DO, Rehm BHA, Ackerley DF. *Escherichia coli* NemA is an efficient chromate reductase that can be biologically immobilized to provide a cell free system for remediation of hexavalent chromium. PLoS One. 2013;**8**:1-8. DOI: 10.1371/journal.pone.0059200

[11] Radivojevic S, Cooper PA. Extraction of hexavalent chromium from chromated copper arsenate treated wood under alkaline conditions. Environmental Science and Technology. 2008;**42**:3739-3744. DOI: 10.1021/es702885f

[12] Kamaludeen SPB, Megharaj M, Juhasz AL, Sethunathan N, Naidu R. Chromium-microorganism interactions in soils: Remediation implications. Reviews of Environmental Contamination and Toxicology.

2003;**178**:93-164. DOI: doi.org/10.1007/0-387-21728-2_4

[13] Nur-E-Alam M, Mia MAS, Ahmad F, Rahman MM. An overview of chromium removal techniques from tannery effluent. Applied Water Science. 2020;**10**:205. DOI: 10.1007/s13201-020-01286-0

[14] Das P, Mishra S. Hexavalent chromium [Cr (VI)]: Yellow water pollution and its remediation. Sarovar Saurabh ENVIS Newsl Wetl Ecosyst. 2009;**5**:1-8

[15] Stambulska UY, Bayliak MM, Lushchak VI. Chromium (VI) toxicity in legume plants: Modulation effects of rhizobial symbiosis. BioMed Research International. 2018;**2018**:8031213. DOI: 10.1155/2018/8031213

[16] Plaza SM, ND L. The safety and efficacy of high-dose chromium. Alternative Medicine Review. 2002;**7**:218-235

[17] Wang ZX, Chen JQ, Chai LY, Yang ZH, Huang SH, Zheng Y. Environmental impact and site-specific human health risks of chromium in the vicinity of a ferro-alloy manufactory, China. Journal of Hazardous Materials. 2011;**190**:980-985. DOI: 10.1016/j.jhazmat.2011.04.039

[18] Pandey V, Dikshit V, Shyam R. Hexavalent chromium induced inhibition of photosynthetic electron transport in isolated spinach chloroplasts. In: Photosynthesis. London: IntechOpen; 2013

[19] Tumolo M, Ancona V, De Paola D, Losacco D, Campanale C, Massarelli C, et al. Chromium pollution in European water, sources, health risk, and remediation strategies: An overview. International Journal of Environmental Research and Public Health. 2020;**17**:5438. DOI: 10.3390/ijerph17155438

[20] Lukina AO, Boutin C, Rowland O, Carpenter DJ. Evaluating trivalent chromium toxicity on wild terrestrial and wetland plants. Chemosphere. 2016;**162**:355-364. DOI: 10.1016/j.chemosphere.2016.07.055

[21] Muthukumaravel K, Rajaraman P. A study on the toxicity of chromium on the histology of gill and liver of freshwater fish *Labeo rohita*. Journal of Pure and Applied Zoology. 2013;**1**:122-126

[22] Speer RM, Wise SS, Croom-Perez TJ, Aboueissa AM, Martin-Bras M, Barandiaran M, et al. A comparison of particulate hexavalent chromium cytotoxicity and genotoxicity in human and leatherback sea turtle lung cells from a one environmental health perspective. Toxicology and Applied Pharmacology. 2019;**376**:70-81. DOI: 10.1016/j.taap.2019.05.013

[23] Cabral-Pinto MMS, Inácio M, Neves O, Almeida AA, Pinto E, Oliveiros B, et al. Human health risk assessment due to agricultural activities and crop consumption in the surroundings of an industrial area. Exposure and Health. 2020;**12**:629-640. DOI: 10.1007/s12403-019-00323-x

[24] Mohan D, Rajput S, Singh VK, Steele PH, Pittman CU Jr. Modeling and evaluation of chromium remediation from water using low cost bio-char, a green adsorbent. Journal of Hazardous Materials. 2011;**188**:319-333. DOI: 10.1016/j.jhazmat.2011.01.127

[25] Mohanty K, Jha M, Meikap BC, Biswas MN. Removal of chromium (VI) from dilute aqueous solutions by activated carbon developed from *Terminalia arjuna* nuts activated with

zinc chloride. Chemical Engineering Science. 2005;**60**:3049-3059. DOI: 10.1016/j.ces.2004.12.049

[26] Lewicki S, Zdanowski R, Krzyzowska M, Lewicka A, Debski B, Niemcewicz M, et al. The role of chromium III in the organism and its possible use in diabetes and obesity treatment. Annals of Agricultural and Environmental Medicine. 2014;**21**:331-335

[27] Owlad M, Aroua MK, Daud WAW, Baroutian S. Removal of hexavalent chromium-contaminated water and wastewater: A review. Water, Air, and Soil Pollution. 2009;**200**:59-77. DOI: 10.1007/s11270-008-9893-7

[28] Sujatha S, Sivarethinamohan R. A critical review of Cr (VI) ion effect on mankind and its amputation through adsorption by activated carbon. Materials Today: Proceedings. 2021;**37**:1158-1162. DOI: 10.1016/j.matpr.2020.06.351

[29] Narayanan NV, Ganesan M. Use of adsorption using granular activated carbon (GAC) for the enhancement of removal of chromium from synthetic wastewater by electrocoagulation. Journal of Hazardous Materials. 2009;**161**:575-580. DOI: 10.1016/j.jhazmat.2008.03.113

[30] Hamadi NK, Chen XD, Farid MM, Lu MGQ. Adsorption kinetics for the removal of chromium (VI) from aqueous solution by adsorbents derived from used tyres and sawdust. Chemical Engineering Journal. 2001;**4**:95-105. DOI: 10.1016/S1385-8947(01)00194-2

[31] Hu MJ, Wei YL, Wang YW, Lee JF. Immobilization of chromium (VI) with debris of aquatic plants. Bulletin of Environmental Contamination and Toxicology. 2003;**71**:840-847. DOI: 10.1007/s00128-003-0212-0

[32] Loghavi MM, Mohammadi-Manesh H, Eqra R. Material for lithium-ion batteries. Journal of Solid State Electrochemistry. 2019;**23**:2569-2578

[33] Dakiky M, Khamis M, Manassra A, Me'reb M. Selective adsorption of Cr (VI) in industrial adsorbents. Advances in Environmental Research. 2002;**6**:533-540

[34] Xia S, Song Z, Jeyakumar P, Shaheen SM, Rinklebe J, Ok YS, et al. A critical review on bioremediation technologies for Cr (VI)-contaminated soils and wastewater. Critical Reviews in Environmental Science and Technology. 2019;**49**:1027-1078. DOI: 10.1080/10643389.2018.1564526

[35] Gardea-Torresdey JL, Peralta-Videa JR, Montes M, Rosa GDL, Corral-Diaz B. Bioaccumulation of cadmium, chromium and copper by Convolvulus arvensis L.: Impact on plant growth and uptake of nutritional elements. Bioresource Technology. 2004;**92**:229-235. DOI: 10.1016/j.biortech.2003.10.002

[36] Nguyen TLT, Hermansen JE, Mogensen L. Environmental performance of crop residues as an energy source for electricity production: The case of wheat straw in Denmark. Applied Energy. 2013;**104**:633-641. DOI: 10.1016/j.apenergy.2012.11.057

[37] Sapari N, Idris A, Hamid NHA. Total removal of heavy metal from mixed plating rinse wastewater. Desalination. 1996;**106**:419-422

[38] Lin SH, Kiang CD. Chromic acid recovery from waste acid solution by an ion exchange process: Equilibrium and column ion exchange modelling. Chemical Engineering Journal. 2003;**92**:193-199. DOI: 10.1016/S1385-8947(02)00140-7

[39] Kabay N, Arda M, Saha B, Streat M. Removal of Cr (VI) by solvent impregnated resins (SIR) containing aliquat 336. Reactive and Functional Polymers. 2003;**54**:103-115. DOI: 10.1016/S1381-5148(02)00186-4

[40] Jachuła J, Hubicki Z. Removal of Cr (VI) and As (V) ions from aqueous solutions by polyacrylate and polystyrene anion exchange resins. Applied Water Science. 2013;**3**:653-664. DOI: 10.1007/s13201-013-0110-5

[41] Fu F, Wang Q. Removal of heavy metal ions from wastewaters: A review. Journal of Environmental Management. 2010;**92**:407-418. DOI: 10.1016/j.jenvman.2010.11.011

[42] Lugo-Lugo V, Barrera-Díaz C, Ureña-Núñez F, Bilyeu B, Linares-Hernández I. Bio-sorption of Cr (III) and Fe (III) in single and binary systems onto pretreated orange peel. Journal of Environmental Management. 2012;**112**:120-127. DOI: 10.1016/j.jenvman.2012.07.009

[43] Emamjomeh MM, Sivakumar M. Review of pollutants removed by electrocoagulation and electrocoagulation/flotation processes. Journal of Environmental Management. 2009;**90**:1663-1679. DOI: 10.1016/j.jenvman.2008.12.011

[44] Moussa DT, El-Naas MH, Nasser M, Al-Marri MJ. A comprehensive review of electrocoagulation for water treatment: Potentials and challenges. Journal of Environmental Management. 2017;**186**:24-41. DOI: 10.1016/j.jenvman.2016.10.032

[45] Ku Y, Jung IL. Photocatalytic reduction of Cr (VI) in aqueous solutions by UV irradiation with the presence of titanium dioxide. Water Research. 2001;**35**:135-142. DOI: 10.1016/S0043-1354(00)00098-1

[46] Chen QY, Tyrer M, Hills CD, Yang XM, Carey P. Immobilisation of heavy metal in cement-based solidification/stabilisation: A review. Waste Management. 2009;**29**:390-403. DOI: 10.1016/j.wasman.2008.01.019

[47] Mirbagheri SA, Hosseini SN. Pilot plant investigation on petrochemical wastewater treatment for the removal of copper and chromium with the objective of reuse. Desalination. 2005;**171**:85-93. DOI: 10.1016/j.desal.2004.03.022

[48] Özverdi A, Erdem M. Cu^{2+}, Cd^{2+} and Pb^{2+} adsorption from aqueous solutions by pyrite and synthetic iron sulphide. Journal of Hazardous Materials. 2006;**137**:626-632. DOI: 10.1016/j.jhazmat.2006.02.051

[49] Chang YK, Chang JE, Lin TT, Hsu YM. Integrated copper-containing wastewater treatment using xanthate process. Journal of Hazardous Materials. 2002;**94**:89-99. DOI: 10.1016/S0304-3894(02)00060-2

[50] Su X, Kushima A, Halliday C, Zhou J, Li J, Alan T. Electrochemically-mediated selective capture of heavy metal chromium and arsenic oxyanions from water. Nature Communications. 2018;**9**:4701. DOI: 10.1038/s41467-018-07159-0

[51] Thatoi H, Das S, Mishra J, Rath BP, Das N. Bacterial chromate reductase, a potential enzyme for bioremediation of hexavalent chromium: A review. Journal of Environmental Management. 2014;**146**:383-399. DOI: 10.1016/j.jenvman.2014.07.014

[52] Kang CH, Kwon YJ, So JS. Bioremediation of heavy metals by using bacterial mixtures. Ecological

Engineering. 2016;**89**:64-69. DOI: 10.1016/j.ecoleng.2016.01.023

[53] Kang C, Wu P, Li L, Yu L, Ruan YLB, Gong B, et al. Cr (VI) reduction and Cr (III) immobilization by resting cells of *Pseudomonas aeruginosa* CCTCC AB93066: Spectroscopic, microscopic, and mass balance analysis. Environmental Science and Pollution Research. 2017;**24**:5949-5963. DOI: 10.1007/s11356-016-8356-8

[54] Xu L, Luo M, Jiang C, Wei X, Kong P, Liang X, et al. In vitro reduction of hexavalent chromium by cytoplasmic fractions of *Pannonibacter phragmitetus* LSSE-09 under aerobic and anaerobic conditions. Applied Biochemistry and Biotechnology. 2012;**166**:933-941. DOI: 10.1007/s12010-011-9481-y

[55] Banerjee S, Misra A, Chaudhury S, Dam B. A *bacillus* strain TCL isolated from Jharia coalmine with remarkable stress responses, chromium reduction capability and bioremediation potential. Journal of Hazardous Materials. 2019;**367**:215-223. DOI: 10.1016/j.jhazmat.2018.12.038

[56] Zeng Q, Hu Y, Yang Y, Hu L, Zhong H, He Z. Cell envelop is the key site for Cr (VI) reduction by *Oceanobacillus oncorhynchi* W4, a newly isolated Cr (VI) reducing bacterium. Journal of Hazardous Materials. 2019;**368**:149-155. DOI: 10.1016/j.jhazmat.2019.01.031

[57] Sevak P, Pushkar B, Mazumdar S. Mechanistic evaluation of chromium bioremediation in *Acinetobacter junii* strain b2w: A proteomic approach. Journal of Environmental Management. 2019;**328**:116978. DOI: 10.1016/j.jenvman.2022.116978

[58] Ma L, Xu J, Chen N, Li M, Feng C. Microbial reduction fate of chromium (Cr) in aqueous solution by mixed bacterial consortium. Ecotoxicology and Environmental Safety. 2019;**170**:763-770. DOI: 10.1016/j.ecoenv.2018.12.041

[59] Tekerlekopoulou AG, Tsiflikiotou M, Akritidou L, Viennas A, Tsiamis G, Pavlou S, et al. Modelling of biological Cr (VI) removal in draw-fill reactors using microorganisms in suspended and attached growth systems. Water Research. 2013;**47**:623-636. DOI: 10.1016/j.watres.2012.10.034

[60] Desai C, Jain K, Patel B, Madamwar D. Efficacy of bacterial consortium-AIE2 for contemporaneous Cr (VI) and azo dye bioremediation in batch and continuous bioreactor systems, monitoring steady-state bacterial dynamics using qPCR assays. Biodegradation. 2009;**20**:813-826. DOI: 10.1007/s10532-009-9269-8

[61] Singh R, Kumar A, Kirrolia A, Kumar R, Yadav N, Bishnoi NR, et al. Removal of sulphate, COD and Cr (VI) in simulated and real wastewater by sulphate reducing bacteria enrichment in small bioreactor and FTIR study. Bioresource Technology. 2010;**102**:677-682. DOI: 10.1016/j.biortech.2010.08.041

[62] Cárdenas-González JF, Acosta-Rodríguez I. Hexavalent chromium removal by a Paecilomyces sp. fungal strain isolated from environment. Bioinorganic Chemistry and Applications. 2010;**2010**:676243. DOI: 10.1155/2010/676243

[63] Xu X, Xia L, Chen W, Huang Q. Detoxification of hexavalent chromate by growing *Paecilomyces lilacinus* XLA. Environmental Pollution. 2017;**225**:47-54. DOI: 10.1016/j.envpol.2017.03.039

[64] Long B, Ye B, Liu Q, Zhang S, Ye J, Zou L, et al. Characterization of *Penicillium oxalicum* SL2 isolated from

indoor air and its application to the removal of hexavalent chromium. PLoS One. 2018;**13**:0191484. DOI: 10.1371/journal.pone.0191484

[65] Igiehon NO, Babalola OO. Fungal bio-sorption potential of chromium in Norkrans liquid medium by shake flask technique. Journal of Basic Microbiology. 2019;**59**:62-73. DOI: 10.1002/jobm.201800011

[66] Kumar V, Dwivedi SK. Hexavalent chromium stress response, reduction capability and bioremediation potential of *Trichoderma* sp. isolated from electroplating wastewater. Ecotoxicology and Environmental Safety. 2019;**185**:1-12. DOI: 10.1016/j.ecoenv.2019.109734

[67] Cui Y, Masud A, Aich N, Atkinson JD. Phenol and Cr (VI) removal using materials derived from harmful algal bloom biomass: Characterization and performance assessment for a biosorbent, a porous carbon, and Fe/C composites. Journal of Hazardous Materials. 2019;**368**:477-486. DOI: 10.1016/j.jhazmat.2019.01.075

[68] Gupta VK, Rastogi A. Biosorption of hexavalent chromium by raw and acid-treated green alga *Oedogonium hatei* from aqueous solutions. Journal of Hazardous Materials. 2009;**163**:396-402. DOI: 10.1016/j.jhazmat.2008.06.104

[69] Megharaj M, Avudainayagam S, Naidu R. Toxicity of hexavalent chromium and its reduction by bacteria isolated from soil contaminated with tannery waste. Current Microbiology. 2003;**47**:0051-0054. DOI: 10.1007/s00284-002-3889-0

[70] Das C, Naseera K, Ram A, Meena RM, Ramaiah N. Bioremediation of tannery wastewater by a salt-tolerant strain of *Chlorella vulgaris*. Journal of Applied Phycology. 2017;**29**:235-243. DOI: 10.1007/s10811-016-0910-8

[71] Costa IG, Terra NM, Cardoso VL, Batista FR, Reis MH. Photoreduction of chromium (VI) in microstructured ceramic hollow fibers impregnated with titanium dioxide and coated with green algae *Chlorella vulgaris*. Journal of Hazardous Materials. 2019;**379**:1-10. DOI: 10.1016/j.jhazmat.2019.120837

[72] Sen S, Dutta S, Guhathakurata S, Chakrabarty J, Nandi S, Dutta A. Removal of Cr (VI) using a cyanobacterial consortium and assessment of biofuel production. International Biodeterioration & Biodegradation. 2017;**119**:211-224. DOI: 10.1016/j.ibiod.2016.10.050

[73] Garnham GW, Green M. Chromate (VI) uptake by and interactions with cyanobacteria. Journal of Industrial Microbiology. 1995;**14**:247-251

[74] Parveen S, Khattar JIS, Singh DP. The cyanobacterium *Synechocystis* sp. PUPCCC 62: A potential candidate for biotransformation of Cr (VI) to Cr (III) in the presence of sulphate. Environmental Science and Pollution Research. 2015;**22**:10661-10668. DOI: 10.1007/s11356-015-4260-x

[75] Long D, Tang X, Cai K, Chen G, Chen L, Duan D, et al. Cr (VI) reduction by a potent novel alkaliphilic halotolerant strain *Pseudochrobactrum saccharolyticum* LY10. Journal of Hazardous Materials. 2013;**256**:24-32. DOI: 10.1016/j.jhazmat.2013.04.020

[76] Ge S, Ge S, Zhou M, Dong X. Bioremediation of hexavalent chromate using permeabilized *Brevibacterium* sp. and *Stenotrophomonas* sp. cells. Journal of Environmental Management. 2015;**157**:54-59. DOI: 10.1016/j.jenvman.2015.04.011

[77] Ran ZHAO, Bi WANG, Cai QT, Li XX, Min LIU, Dong HU, et al. Bioremediation of hexavalent chromium pollution by *Sporosarcina saromensis* M52 isolated from offshore sediments in Xiamen, China. Biomedical and Environmental Sciences. 2016;**29**:127-136. DOI: 10.3967/bes2016.014

[78] Nayak AK, Panda SS, Basu A, Dhal NK. Enhancement of toxic Cr (VI), Fe, and other heavy metals phytoremediation by the synergistic combination of native *Bacillus cereus* strain and *Vetiveria zizanioides* L. International journal of Phytoremediation. 2018;**20**:682-691. DOI: 10.1080/15226514.2017.1413332

[79] Yang W, Hong W, Huang Y, Li S, Li M, Zhong H, et al. Exploration on the Cr (VI) resistance mechanism of a novel thermophilic Cr (VI)-reducing bacteria *Anoxybacillus flavithermus* ABF1 isolated from Tengchong geothermal region, China. Environmental Microbiology Reports. 2022;**14**:795-803. DOI: 10.1111/1758-2229.13070

[80] Gu Y, Chen X, Liu L, Wang S, Yu X, Jia Z, et al. Cr (VI)-bioremediation mechanism of a novel strain *bacillus paramycoides* Cr6 with the powerful ability to remove Cr (VI) from contaminated water. Journal of Hazardous Materials. 2023;**455**:131519. DOI: 10.1016/j.jhazmat.2023.131519

[81] Guria MK, Guha AK, Bhattacharyya M. A green chemical approach for biotransformation of Cr (VI) to Cr (III), utilizing *Fusarium* sp. MMT1 and consequent structural alteration of cell morphology. Journal of environmental. Chemical Engineering. 2014;**2**:424-433. DOI: 10.1016/j.jece.2014.01.016

[82] Sharma S, Malaviya P. Bioremediation of tannery wastewater by chromium resistant novel fungal consortium. Ecological Engineering. 2016;**91**:419-425. DOI: 10.1016/j.ecoleng.2016.03.005

[83] Chang F, Tian C, Liu S, Ni J. Discrepant hexavalent chromium tolerance and detoxification by two strains of *Trichoderma asperellum* with high homology. Chemical Engineering Journal. 2016;**298**:75-81. DOI: 10.1016/j.cej.2016.04.023

[84] Mane PC, Bhosle AB. Bioremoval of some metals by living algae *spirogyra* sp. and *Spirullina* sp. from aqueous solution. International. Journal of Environmental Research. 2012;**6**:571-576

[85] Anjana K, Kaushik A, Kiran B, Nisha R. Bio-sorption of Cr (VI) by immobilized biomass of two indigenous strains of cyanobacteria isolated from metal contaminated soil. Journal of Hazardous Materials. 2017;**148**:383-386. DOI: 10.1016/j.jhazmat.2007.02.051

[86] Mitra S, Sarkar A, Sen S. Removal of chromium from industrial effluents using nanotechnology: A review. Nanotechnology for Environmental Engineering. 2017;**2**:1-14. DOI: 10.1007/s41204-017-0022-y

[87] Yu F, Wang L, Ma H, Pan Y. Zeolitic imidazolate framework-8 modified active carbon fiber as an efficient cathode in electro-Fenton for tetracycline degradation. Separation and Purification Technology. 2020;**237**:116342. DOI: 10.1016/j.seppur.2019.116342

[88] Lyu H, Tang J, Huang Y, Gai L, Zeng EY, Liber K, et al. Removal of hexavalent chromium from aqueous solutions by a novel biochar supported nanoscale iron sulfide composite. Chemical Engineering Journal. 2017;**322**:516-524. DOI: 10.1016/j.cej.2017.04.058

[89] Nithya R, Gomathi T, Sudha PN, Venkatesan J, Anil S, Kim SK. Removal of Cr(VI) from aqueous solution using chitosan-g-poly(butyl acrylate)/silica gel nanocomposite. International Journal of Biological Macromolecules. 2016;**87**:545-554. DOI: 10.1016/j.ijbiomac.2016.02.076

[90] Jin Q, Dai M, Zhan X, Wang S, He Z. Carbon nanotubes and graphene composites used in Cr (VI) detection techniques: A review. Journal of Alloys and Compounds. 2022;**922**:166268

[91] Wang D, Wu C, Zong Z, Ye J, Wu Q, Li R, et al. Carbon nanotubes-based fuel cell for Cr (VI) removal and electricity generation. Langmuir. 2022;**38**:9021-9029. DOI: 10.1021/acs.langmuir.2c01472

[92] Igiri BE, Okoduwa SI, Idoko GO, Akabuogu EP, Adeyi AO, Ejiogu IK. Toxicity and bioremediation of heavy metals contaminated ecosystem from tannery wastewater: A review. Journal of Toxicology. 2018;**2018**(1-16):2568038 DOI: 10.1155/2018/2568038

[93] Suganthi V, Mahalakshmi M, Balasubramanian B. Development of hybrid membrane bioreactor for tannery effluent treatment. Desalination. 2013;**309**:231-236. DOI: 10.1016/j.desal.2012.10.014

[94] Hou Y, Liu H, Zhao X, Qu J, Chen JP. Combination of electroreduction with biosorption for enhancement for removal of hexavalent chromium. Journal of Colloid and Interface Science. 2012;**385**:147-153. DOI: 10.1016/j.jcis.2012.05.056

[95] Moradi M, Moussavi G. Enhanced treatment of tannery wastewater using the electrocoagulation process combined with UVC/VUV photoreactor: Parametric and mechanistic evaluation. Chemical Engineering Journal. 2019;**358**:1038-1046. DOI: 10.1016/j.cej.2018.10.069

[96] Rahman Z, Singh VP. The relative impact of toxic heavy metals (THMs) (arsenic (As), cadmium (Cd), chromium (Cr)(VI), mercury (Hg), and lead (Pb)) on the total environment: An overview. Environmental Monitoring and Assessment. 2019;**191**:1-21. DOI: 10.1007/s10661-019-7528-7

[97] Jiang S, Tang J, Rahimi S, Mijakovic I, Wei Y. Efficient treatment of industrial wastewater with microbiome and synthetic biology. Frontiers in Environmental Science. 2022;**10**:432

[98] Pande V, Pandey SC, Sati D, Bhatt P, Samant M. Microbial interventions in bioremediation of heavy metal contaminants in agroecosystem. Frontiers in Microbiology. 2022;**13**:824084. DOI: 10.3389/fmicb.2022.824084

[99] Bhati R, Sreedharan SM, Rizvi A, Khan MS, Singh R. An insight into efflux-mediated arsenic resistance and biotransformation potential of *Enterobacter cloacae* RSC3 from arsenic polluted area. Indian Journal of Microbiology. 2022;**62**:456-467. DOI: 10.1007/s12088-022-01028-7

[100] Jaiswal S, Singh DK, Shukla P. Gene editing and systems biology tools for pesticide bioremediation: A review. Frontiers in Microbiology. 2019;**10**:87. DOI: 10.3389/fmicb.2019.00087

[101] Ojuederie OB, Babalola OO. Microbial and plant-assisted bioremediation of heavy metal polluted environments: A review. International Journal of Environmental Research and Public Health. 2017;**14**(12):1504. DOI: 10.3390/ijerph14121504

[102] Jung J, Lee SJ. Biochemical and biodiversity insights into heavy metal

ion-responsive transcription regulators for synthetic biological heavy metal sensors. Journal of Microbiology and Biotechnology. 2019;**29**(10):1522-1542. DOI: 10.4014/jmb.1908.08002

[103] Xue Y, Qiu T, Sun Z, Liu F, Yu B. Mercury bioremediation by engineered *pseudomonas putida* KT2440 with adaptationally optimized biosecurity circuit. Environmental Microbiology. 2022;**24**(7):3022-3036. DOI: 10.1111/1462-2920.16038

[104] Naz M, Benavides-Mendoza A, Tariq M, Zhou J, Wang J, Qi S, et al. CRISPR/Cas9 technology as an innovative approach to enhancing the phytoremediation: Concepts and implications. Journal of Environmental Management. 2022;**323**:116296. DOI: 10.1016/j.jenvman.2022.116296

[105] Dangi AK, Sharma B, Hill RT, Shukla P. Bioremediation through microbes: Systems biology and metabolic engineering approach. Critical Reviews in Biotechnology. 2019;**39**(1):79-98. DOI: 10.1080/07388551.2018.1500997

[106] Adekunle A, Rickwood C, Tartakovsky B. Online monitoring of heavy metal–related toxicity using flow-through and floating microbial fuel cell biosensors. Environmental Monitoring and Assessment. 2020;**192**:1-12. DOI: 10.1007/s10661-019-7850-0

[107] Rylott EL, Bruce NC. How synthetic biology can help bioremediation. Current Opinion in Chemical Biology. 2020;**58**:86-95. DOI: 10.1016/j.cbpa.2020.07.004

[108] Basu A, Ali SS, Hossain SS, Asif M. A review of the dynamic mathematical modeling of heavy metal removal with the biosorption process. PRO. 2022;**10**(6):1154. DOI: 10.3390/pr10061154

Chapter 6

Trends in Systematic Techniques for Pollutants Monitoring in the Environmental Water Systems

Napo Ntsasa, Andile Mkhohlakali, Tumelo Mogashane, James Tshilongo and Mokgehle R. Letsoalo

Abstract

The global issue of water quality degradation that exacerbates water scarcity prompted establishment of strategic water management systems. The majority of anthropogenic activities that affect water quality are indispensable and economical. Despite its importance for achieving economic expansion, prosperity, and advancement on a national and international scale, industrialization is a primary cause of adverse environmental effects. There is significant strain on rivers and dams from chemical spills, fertilizers and pesticides, personal hygiene products, wastewater discharge, and land drainage due to lack of proper water treatment infrastructures or wilful activities of illegal disposal. The developing era of scientific research is focused on effective technologies for water-pollution monitoring. This study is interested in exploration of environmental forensics using forensic-analytical techniques such as aerial photography, satellite analytics, and chemometric and spectroscopic techniques to identify sources of water pollutants. This approach is of great importance in connecting the identified pollutants to their emission sources in order to provide efficient long-term strategies for a holistic water management system. The environmental forensics approach in this study can improve comprehension of pollution that will assist national departments and policy makers in charge of pollution control in the geographical locations to ensure that water polluters abide by applicable laws.

Keywords: remote sensing, water quality parameters, chemometric, spectroscopic techniques, water pollution

1. Introduction

The health and sustainability of water ecosystems are seriously threatened by both current and emerging contaminants. When hazardous materials, pollutants, or toxins find their way into water bodies like rivers, lakes, seas, groundwater, or aquifers influences human health, aquatic ecosystems, and environment at risk [1, 2]. Ensuring the safety of human health and ecosystems requires the continual monitoring of pollutants in environmental water systems, such as rivers, dams, and groundwater for sustainable ecosystems, and water resources. Many natural processes and human activities contribute to this ubiquitous problem, which frequently has negative effects on land and aquatic ecosystems [3]. For instance, as human activities expand, so does the number of industries, leading to several possible entry points of pollutants such as chemical spills, fertilizers, pesticides, personal hygiene products, pharmaceuticals, toxic elements, and radioactive materials [4]. The distinct and identifiable point source and non-point sources pollution negatively affect the environment and water ecosystem which makes it difficult to identify their origin [3]. Water pollutants are classified into the category of chemical, biological, and physical pollutants which include chemicals, pesticides, fertilizers, medicines, heavy metals (such as lead and mercury), wastewater treatment plants, sewage discharges, trash, plastics, and debris [5]. Water pollution problems are aggravated by rapid urbanization, population increase, industrialization, and climate change; therefore, coordinated international efforts are required to address these intricate difficulties because of serious environmental issues with significant effects on ecosystems, human health, and socioeconomic well-being [6, 7].

Water-related aims for sustainable development are sought after by international agreements like the Stockholm Convention and the Paris Agreement, as well as by international collaborations like the United Nations Sustainable Development Goals (SDGs) [3, 8]. To prevent water pollution, set standards for water quality, and create systems for monitoring and enforcement, governments and international organizations pass laws and policies [3]. The installation of wastewater treatment systems, the adoption of best management practices in industry and agriculture, and the encouragement of sustainable waste management and land use are examples of pollution prevention and control strategies [9]. Approaches to integrated water resource management seek to maintain water quality, balance conflicting demands for water, and guarantee that current and future generations have fair access to clean water [10]. In order to protect water resources and advance a healthy and sustainable future, addressing this complex issue calls for interdisciplinary approaches, political commitment, and coordinated action at the local, national, and international levels [11]. Most importantly, the innovative scientific approaches to pollution monitoring, water treatment, and sustainable practices present viable ways to lower pollutant inputs, repair damaged water bodies, and strengthen water bodies' resistance to environmental change [12]. The engagement between scientists, decision-makers, and the general public, community could improve the efficacy and transparency of pollution management and monitoring. This chapter explores techniques and technological advances in environmental water monitoring that could be of great importance to identify pollutant origin sources. The trends in systematic techniques such as sampling approaches, remote sensing, spectroscopic techniques, and chemometric methods are discussed to envisage efficient long-term strategies for a holistic water management system.

2. Conventional and modern sampling approaches

Sampling is the process of choosing a sample that is representative of the area being sampled but small enough to be handled and transferred easily in a laboratory environment [6]. Monitoring both organic and inorganic environmental pollutants is a continuous task necessary to determine the quality of many ecological compartments, including the soil, water basins, atmosphere, indoor air, and biota. In research, techniques for sampling are the elements critical to the validity of the research laboratory [13]. Conventional random sampling methods such as stratified, simple random, cluster, and systematic are mostly preferred due to ease of use and accuracy of presentation since sample collection is based on water ecosystem accessibility [12]. With the increasing pollution of water resources which requires rapid and continual monitoring, passive sampling has attracted a considerable approach due to its convenience, sufficient, and reliable approach [14]. The types of passive sampling reported in the literature include diffusive gradient thin-film, polar organic chemical integrative samplers, and chemcatcher to name a few, which have emerged as important tools for monitoring different pollutants in the water ecosystem due to their intriguing/advantages in contrast to active sampling. Polar organic chemical integrative and chemcatcher are two of these samplers that could provide the best balance between large spatial coverage but low temporal resolution and inadequate spatial resolution but higher temporal resolution [14]. These samplers are currently offered for sale and can be used to elute chemicals in the laboratory after they are adsorbed to media in the field to measure the time-average or time-weighted mean concentration of various pollutants in the environmental waters. Vrana et al. [15], reported a comprehensive review on passive sampling for environmental management of inorganic and organic water pollutants [15]. Nitti and co-workers utilized the passive sampling technique for monitoring PCPs in the water matrix, looking at the trend starting from 2016 to 2020 [16].

2.1 Passive sampling techniques

Passive sampling approaches also known as passive dosimetry samplers, is an *in-situ* integrative approach that is the most useful samplers for monitoring the broad range of polar chemicals in water streams [17]. When determining the organic and inorganic contaminants in aqueous, gaseous, and soil environments through passive sampling and/or analyte extraction methods, analyte concentrations are usually measured using weighted averages over the course of sampling and/or extraction times. Although passive samplers have been recommended to be an advanced method of monitoring river water quality, little research has been done on how well they work over a prolonged period of time when compared to high-frequency data.

2.2 Polar organic chemical integrative samplers

Polar organic chemical Integrative samplers are comprised of a solid sequestration phase submerged between a hydrophilic microporous polyethersulfone membrane which has a porosity of 0.1 μm diameter that is protected from direct contact with environmental matrix [18]. These samplers are designed to be more representative of real-time exposure, pinpoint the pollution peaks, and considerably reduce detection limits. The schematic illustration of POCIS is displayed in **Figure 1**. These samplers are suited well for monitoring organic hydrophilic compounds including surfactants, medical drugs, triazine, and caffeine to name a few in aquatic environments [19]. In literature [20], Benard et al. reported a 1-year monitoring of pesticides in five sites using these samplers which provided them with time-weighted average concentration estimates that can be used in risk assessments to determine the biological impact of hydrophilic chemical molecules.

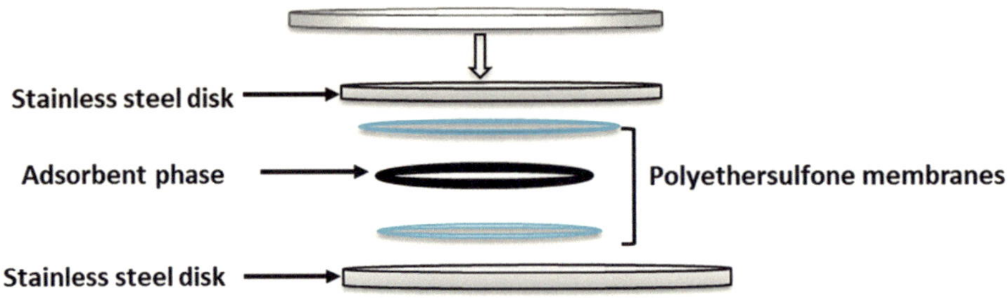

Figure 1.
Schematic representation of a POCIS device.

2.3 Chemcatcher passive sampler

Among the most applicable passive samplers for organic pollutants are chemcatchers [17]. The chemcatcher devices were mainly established for quantification and detection of organic pollutants such as aromatic hydrocarbons in water [21]. These samplers are comprised of receiving phase with a high affinity for organic compounds which is then separated from sampling media (environmental analyte) by diffusion limiting membrane as displayed in **Figure 2**. The receiving phase and diffusion limiting membrane are placed on a strong/rigid polytetrafluoroethylene. Traditionally, an integrated passive (chemcatcher) sampler was composed of a sample body, Empore disk as receiving phase that is saturated with n-octanol which is overlaid with low-density polyethene membrane. There has been extensive research on chemcatcher, and passive samplers due to cost-effectiveness, and reduced time sampling approach [22]. Schafer et al. reported the chemcatcher composed of SDB-XC Empore disk for monitoring of pesticides, i.e. linuron, carbofuran, actochlor, chlorfenvinphos, fenpropidin, α-endosulfan to name a few. The time-weighted average concentration went up to 1 μg/L and a total of 54 detections were obtained [23]. Martinez and co-workers reported the chemctacher-based C18 Empore disk as the receiving phase for monitoring levels of organotin compounds in water. This sampler-based C_{18} Empore produced time-weighted average concentrations of the bioavailable parts of the tin compounds, whose environmental quantities changed significantly over time [24].

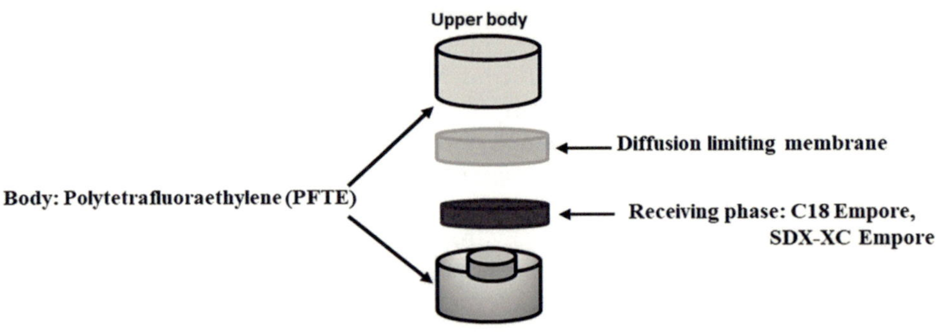

Figure 2.
Schematic illustration of a chemcatcher device.

Ferrow et al., reported the comparison of the chemcatcher sampling device with the high-frequency sampling in two rivers over 16 months, monitoring the acidic herbicides [17]. Aguilar-Martinez et al. reported the monitoring of mercury (Hg) in contaminated rivers using the chemcatcher passive sample [25].

2.4 Diffusive gradient thin-film

Developed in 1993 by William Davison and Hao Zhang of Lancaster University in the United Kingdom, Diffusive Gradient in Thin Films (DGTS) consists of a binding gel layer, an ion-permeable diffusive gel layer, a filter membrane, and a plastic base and cap as demonstrated in **Figure 3**, allowing for the simultaneous measurement of phosphor and iron release in the aquatic system [26, 27]. This device uses binding agents such as zirconia, titanium dioxide (TiO_2), Ferrihydrite, and thiol-functionalized silica solid phase for monitoring of oxyanions [28]. In these devices, the binding agents are significant in adsorbing species of oxyanions such as sulfides (S(III), Arsenic (As (III) [29]. DGTS sampler has been widely studied and extensively utilized not only for heavy metal and organic compounds in water but monitoring the speciation of precious elements. Gao et al., reported environmental monitoring, specifically the speciation of platinum group metal elements (PGEs) and oxyanions [26]. Borros and co-workers have reported the sensitivity improvement of the device for monitoring neutral pesticides in water and lower LOQ of 0.2–1.3 ng/L was obtained [29]. In contrast to another passive sampler device, hydrodynamic variations in the water have less of an impact on DGTS gadgets during the monitoring [30]. Martins et al. have reported the large DGTS devices for monitoring numerous organic pollutants such as pharmaceuticals, pesticides, and hormones [30].

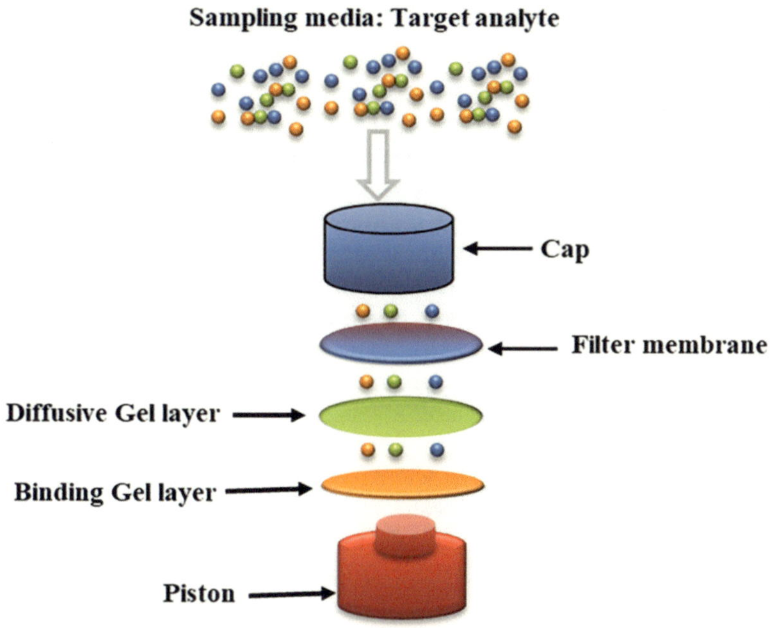

Figure 3.
Schematic illustration diffusive gradient in thin films.

2.5 Active and grab sampling approach

Active sampling is categorized into two approaches, namely active grab sampling and automated grab sampling approach [31]. Active grab sampling is also known as spot sampling whereby the sample of gas, liquid, or soil sample is collected in a pipeline/spot, with the intention of analysis in a laboratory [32]. When using the active grab sampling approach, a large number of samples from a particular site must be collected from a specific site over the course of the sampling process. However, the sampling approach is time-inefficient and expensive [33]. In addition, it is not ideal when levels of pollutants fluctuate over time, even though it can be used to capture a snapshot of the water quality at any particular site at any given time. The issue of pollutant fluctuation over time, instigated active automated sampling which is applied in most efforts dealing with monitoring to enhance the sampling frequency [34].

2.5.1 Advantages and disadvantages of water sampling techniques

The grab sampling devices estimate a compound's TWA concentration, stabilize and pre-concentrate species, reduce matrix interferences, and do not accumulate compounds attached to suspended materials or dissolved organic carbon [18]. They are less influenced by hydrodynamic fluctuations in water and are less expensive than integrative passive sampling. However, they have limitations such as limited calibration data and high hydrodynamic fluctuation of results [4]. The device typically does not accumulate compounds attached to suspended materials or dissolved organic carbon. However, in certain situations, it may collect the most accessible fraction. GDT devices are less affected by water fluctuations [29]. Active (grab) sampling offers advantages like less expensive and easy calibration but also has disadvantages like large sample numbers, time consumption, and high hydrodynamic fluctuation.

3. Remote sensing techniques

Comprehensive metrics of water quality management can be acquired using conventional methods such as laboratory analysis and field sampling. These techniques are incapable of providing large-scale and continuous monitoring because of the high-cost of data acquisition, labor intensity, and low processing efficiency. Moreover, they are easily affected by a wide range of variables including climate hydrology, equipment performance, and technical personnel which instigated the need for water quality management using remote sensing [35]. A thorough comprehension of water-pollution status trends is essential for efficient water resource management with remote sensing enabling large-scale water quality monitoring [36]. Remote sensing has emerged as a critical foundation for water quality evaluation and pollution prevention. The advances in science space have rendered remote sensing an invaluable tool for environmental water management and ensuring water sustainability. In contrast to conventional methods, remote sensing may gather information about the area of interest without physical contact [37]. The primary techniques that vary in the platforms from which the data is gathered are aerial photography and satellite imagery. The remote sensors in both techniques can be mounted on aircraft, unmanned aircraft systems, satellites, drones, or other platforms to capture information about the physical properties such as temperature, reflectance, and emissions. Aerial photography offers higher spatial resolution and flexibility because data is collected at lower altitudes closer to the Earth's surface [38]. In aerial photography, sensors mounted on aircraft such as airplanes, helicopters, or drones offer higher spatial resolution and flexibility because data is collected at lower altitudes closer to the Earth's surface. The fundamental temporal, spectral, and spatial resolution challenges encountered with various satellite sensors can be effectively and adaptably resolved with the use of airborne-based sensors. In contrast to aerial photography, which is costly with limited coverage, spaceborne sensing platforms employ satellites to access large areas and offer low-cost global coverage for water quality monitoring [39]. These remote sensing methods are the most promising for managing water resources, despite their evident disparities. This is because of its capacity to offer a precise overview of WQPs and the concentrations of optically active constituents, which are often found through laborious, time-consuming, and expensive methods [40]. Through the use of spectral signatures from specialized sensors, these techniques monitor water quality indicators spatiotemporally and generate insightful information for decision-makers.

3.1 Types of remote sensors

Many kinds of remote sensors can receive electromagnetic radiation from the monitored area and transform it into a signal that may be recorded and exhibited as an image or as numerical data [40]. Remote sensing techniques employ microwave and optical sensors, which can be further classified into active and passive sensors, to measure a variety of water quality indicators in including temperature, total dissolved solid, transparency, turbidity, concentration of photosynthetic pigments, and colored dissolved organic matter. Microwave sensors have a longer wavelength that can penetrate through haze, dust, clouds, and all forms of precipitation with the exception of severe rainfall other types of sensors [41]. The longer wavelengths allow the detection of microwave radiation in every weather or environmental state at any time which also influences their resistance to atmospheric scattering. The active microwave sensors can record the energy that is reflected or scattered back by sending electromagnetic radiation pulses which is known as microwave radiation. The distance between the target properties, antenna, and the wavelength of the signal received, determines the properties of the reflected energy that the sensor receives at its antenna [41]. Examples of active remote sensing techniques include radio detection and ranging and light detection and ranging, which measure the time interval between an object's emission and return to determine its location, speed, and direction. The energy detected by the passive microwave remote sensors may originate from the surface and be reflected, transmitted, or emitted. The low spatial resolution is the major drawback of passive microwave remote sensors. Optical remote sensors rely on solar illumination for Earth observation and the solar reflection common source of radiation measured by passive sensors [42]. Depending on the spectral resolution of the sensors, the passive optical remote sensors can be classified as multispectral or hyperspectral which have been employed in several studies.

3.2 Data acquisition of optical remote sensors

The opportunity to determine spectral characteristics unique to a monitored parameter is provided by hyperspectral and multispectral satellite sensors. The quantity and spectra of electromagnetic radiation that each band contains distinguish multispectral radiation from hyperspectral radiation. Passive hyperspectral sensors determine solar radiation reflectance over the electromagnetic spectrum, including the visible, near-infrared, mid-infrared, and thermal regions [41]. Hyperspectral imaging, which collects and processes data from numerous narrow, adjacent bands across the electromagnetic spectrum, generates a high-resolution spectrum. Hyperspectral imaging offers enhanced data detail, facilitating more precise investigation to yield accurate identification of targeted parameters. Multispectral imagery is collected using a sensor that measures the reflected energy in specific portions of the electromagnetic spectrum. Due to its low spectral resolution, multispectral imagery may not be able to differentiate between parameters that are closely related. When optical remote sensing techniques are employed for water ecosystem monitoring, the quality parameters are determined by relating them to remotely sensed reflectance using analytical empirical, artificial intelligence or semi-empirical models [41]. These methods are applied in response to variations in the spectral signature backscattered by water and the general approach of the optical remote sensing techniques is presented in **Figure 4**. The empirical methods are less complicated to assess the water quality using statistical correlations between measured water quality and remote sensing spectral values. A suitable waveband combination is associated with the parameters' actual spectral properties. To adjust the spectral radiance to reflectance above the surface irradiance, regression techniques connected to the WQP are

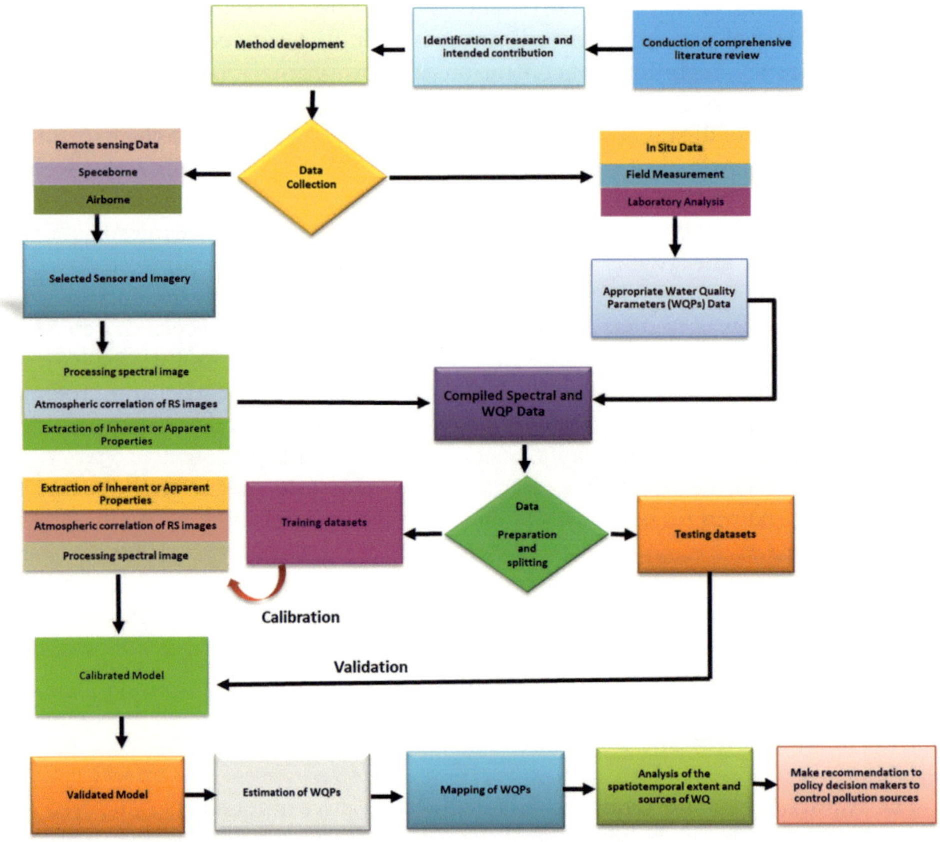

Figure 4.
Remote sensing schematic representation of water quality monitoring.

employed. Regression techniques are used to establish the link, although *in-situ* data is still needed for estimations conducted using empirical models because numerous variables are expected to change between remote sensing missions [43]. Analytical methods utilizing inherent optical properties and apparent optical properties can derive physical correlations among WQPs, underwater light field, and remotely sensed radiance by model reflectance. This method simulates light propagation in the water ecosystem, and link WQPs with reflection using bio-optical and transmission models. The parameters like total dissolved solids can be established because of their correlation with other colored WQPss. In contrast to the aforementioned techniques, artificial intelligence technologies use an implicit algorithm approach to recover WQPs from sediment deposits, water surfaces, and other waper parameters combinations. When compared with conventional statistical methods, artificial intelligence applications are capable of capturing both linear and nonlinear correlations [44]. Numerous artificial intelligence applications have been explored with use of machine learning models such as nonlinear artificial neural network outperforming multilinear regression and support vector machine regression models in water quality monitoring with significant results [44].

4. Spectroscopic technique

Broader range of spectroscopic method has advantages and uses that are exclusive to its application. Researchers can get extensive insights into the species composition, pollution source identification, and quantitative determination of various water pollutant structures by integrating multiple spectroscopic approaches [45]. The sensitivity of spectroscopic methods for identifying pollutants is constantly being improved, both quantitatively and subjectively. Each molecule in acquoes media emits, absorbs, or reflects electromagnetic radiation from light sources, and light intensity characteristics are analyzed to assess the composition of the sample, making it appropriate for detecting pollutants. By using optical properties, the spectrometer can be used to determine the size distribution and particle composition of samples [46]. A light source or laser is typically used in spectroscopy as the emitter, and a spectrometer or detector is used for the spectral analysis. The methods are straightforward, non-invasive, quick to detect, and pollution-free because they do not use any chemical components. Numerous spectroscopic techniques, including ratio-isotope mass spectrometry, chromatography, inductively coupled plasma, and others have been examined for the purpose of evaluating water quality and pollutant quantification [42]. These techniques can facilitate long-term planning for water resource management and conservation by assisting in the identification of pollution hotspots and the assessment of the efficacy of pollution control methods [47].

4.1 Isotope ratio mass spectrometer

The synthesis of intricate knowledge on hydrodynamics, pollution source origin, and water quality can be effortless with the use of isotopic ratio mass spectrometry that leads to more precise forecasts and decision-making [48]. Many different water pollutants have isotopic variations, and each isotopic profile is specific depending on its history and geographical location. An accurate assessment of the relative amount of stable isotopes in a given sample can be obtained by using a specialized analytical technique, such as isotope ratio mass spectrometry (IRMS). The IRMS has a broad variety of applications and current popularity in the forensic sciences makes the highest demands on the reliability of analytical results. To meet a wider range of analytical requirements, IRMS is able to provide precise and sensitive measurements of the isotope signatures of C, N, S, O, and H which owes it to connected peripherals such as gas chromatography (GC) and elemental analyzers (EA). The use of stable isotope ratio analysis as an identification technique has been reported for providing information that distinguishes between anthropogenic and naturally occurring organic contaminants [49]. The gas chromatography–mass spectrometry (GC–MS) has been extensively applied for the quantitative determination of organic contaminant sources. These techniques have a potential limitation for identifying pollution sources because of notable uncertainties associated with distinguishing between contaminant releases that occurred prior to or after the environmental weathering process [50]. The surge in the use of isotopic signatures in high-resolution chromatography lies in the exceptional capabilities for the identification of various sources of contaminants with adequate accuracy and precision [51]. The development of gas chromatography–mass spectrometry (GC-IR-MS) is a specialized tool that can determine the isotopic composition of a native compound to identify various contaminant sources [51]. The use of complementary forensic-analytical techniques forms a major part of this study for the affirmation of pollution source identification. The interpretation of forensic analyses obtained from the combination of aerial photography and analytical techniques is capable of withstanding the intense scientific scrutiny relative to the pollution source identification and quantitative determination of the contaminants. Moreover, persuasion of forensic-analytical will also assure certainty in scrutinizing whether the identified pollutants were determined individually or collectively relative to providing information to forensic investigation of the water ecosystem.

4.2 Hyphenated detection system

Multi-elemental speciation monitoring of environmental water is achievable with the hyphenated detection system high-performance liquid chromatography coupled to inductively plasma mass spectrometry (HPLC/ICP-MS) and ion chromatography coupled to ICP-MS. The online coupling of these techniques provides an efficient method for determining native species for a better understanding of pollution control [6]. Chromatographic techniques allow the separation of species with a synchronous detection and quantification using ICP-MS. The greater speed, precision, and sensitivity of ICP-MS allow rapid sample processing due to its capability to scan for different elements swiftly. This integrated communication is also possible for monitoring non-volatile organic pollutants using gas chromatography—mass spectrometry. Apart from the usual pollutants like organic chemicals and heavy metals, there's now growing concern about new pollutants including microplastics, pharmaceuticals, synthetic compounds, and surfactants [52]. Systematic techniques that exhibited notable capability for monitoring these new organic pollutants include ultra-high pressure liquid chromatography coupled to quadrupole time of flight mass spectrometry, liquid chromatography-electrospray ionization mass spectrometry, HPLC-diode array detector (HPLC-DAD), supercritical-fluid chromatography-DAD, HPLC-MS-MS, and GC-MS-MS [3, 52]. By enabling the identification and measurement of emerging pollutant trace levels in water systems, these techniques aid in the evaluation of possible threats to the environment and public health [53].

Methodological approach	Advantages	Limitations
Biosensors	• High sensitivity • Real-time monitoring • Potential for on-site use	• Limited range of detectable contaminants • Can be prone to biofouling • Requires frequent calibration
Automated sensor networks	• Continuous monitoring • Real-time data • Can be integrated with Internet of Things	• High setup and maintenance costs • Potential for sensor drift • Data management complexities
Molecular techniques (e.g., qPCR, Metagenomics)	• High specificity • Can detect low concentrations • Identifies microbial contaminants	• Requires lab facilities • Higher costs per sample • Time-consuming sample processing
Portable analytical devices (e.g., handheld spectrophotometers)	• On-site analysis • Quick results • User-friendly	• Limited range of analytes • Lower accuracy compared to lab instruments • Regular calibration needed
Nanotechnology-based sensors	• High sensitivity • Can detect a wide range of contaminants • Miniaturized devices	• High development costs • Potential environmental impact of nanomaterials • Regulatory hurdles
Integrated sampling and analysis systems	• Comprehensive data • Can combine physical, chemical, and biological analysis • Reduces sample handling errors	• Complex setup • High operational costs • Requires skilled personnel for operation and maintenance

Table 1.
Various contemporary approaches to water contaminant monitoring.

4.3 Other systematic techniques

The other systematic techniques can offer insightful information on the chemical makeup, pollutant concentration, and general quality of water samples. They also make it possible to detect pollutants with accuracy and sensitivity by using the distinct spectral signatures of various substances, which helps with efficient water quality monitoring and management initiatives. **Table 1** provides some recent advancements in systematic methods of monitoring pollutants in environmental water systems, along with their benefits and drawbacks.

5. Chemometrics techniques

Chemometrics is an advanced method that could support the design of regional environmental management and influence the decisions of the government and research agencies to predict future pollution patterns [54]. Chemometrics comprises a wide range of techniques, both parametric and non-parametric, depending on the dataset's normalcy and characteristics. Moreover, these techniques have credible capability to maximize the efficiency of data variation and derive useful information from the analysis regarding pollution sources. These approaches can be categorized in variable ways and for the purpose of environmental water evaluation and monitoring, factor analysis, principal component analysis, cluster analysis, positive matrix factorization, and artificial neural network machine learning are suitable techniques [54, 55]. Conventional spectroscopic techniques are known to produce experimental data from the environmental analysis which are usually nonlinear and multi-variant due to varied relationships among the numerous contributing variables. Numerous uncertainties may influence reliability of these techniques to produce measurements of data that are accurate for environmental monitoring. The inherent limitations related to the conventional approach prompted the use of chemometric techniques to improve the details of the analysis dataset by statistical and mathematical techniques. For instance, the principal of component analysis is regarded as an accurate statistical tool for apportioning the sources of toxic metals in the environmental water. Moreover, positive matrix factorization is feasible for quantitative determination and source identification of the broader range of environmental pollutants [55]. The use of chemometric techniques in the analysis of environmental data is a novel approach that facilitates the identification of the relationship among environmental factors, pollution sources, and their impact on the environment.

6. Conclusion

The field of systematic methods for monitoring contaminants in environmental water systems is changing, and this change is dynamic, with an emphasis on inclusivity, creativity, and integration. The trends that have been addressed highlight the combined efforts to improve our comprehension of the dynamics of water quality, lessen the threat of pollution, and promote environmentally friendly water management techniques. Real-time monitoring of water quality indices is now possible thanks to the integration of cutting-edge sensor technology, providing stakeholders with timely and useful information for managing pollution and allocating resources. This pattern signals the beginning of a new era in proactive monitoring and adaptive management, wherein prompt response to pollution incidents is combined with ongoing strategy development for managing water quality. Our ability to monitor water bodies at regional and global scales has also increased because of the combination of satellite imagery and remote sensing techniques, which have yielded insightful information about growing pollution hotspots, long-term trends, and spatial patterns. Scientists and policymakers may create focused interventions to address pollution causes and safeguard delicate ecosystems by utilizing the potential of satellite data. The emphasis on newly discovered contaminants is indicative of a progressive strategy for pollution control, which recognizes the dynamic character of pollution hazards and the demand for accurate detection techniques. A proactive approach to reducing threats to human health and ecological integrity can be taken by stakeholders by giving priority to monitoring of new pollutants, like pharmaceuticals and microplastics.

More precise predictions of pollutant movement, source identification, and pollution control measure evaluation are now possible because of developments in data integration and modeling that have made it easier to synthesize complex datasets. In order to promote evidence-based approaches to water quality management, integrated modeling methodologies provide decision-makers with useful tools for scenario analysis, risk assessment, and policy design. The empowerment of citizen science projects has brought water quality monitoring efforts closer to the general people, encouraging public awareness, community engagement, and cooperative problem-solving. Stakeholders can increase the efficacy and transparency of pollution control initiatives by involving the public in the monitoring and management of pollution. This can be achieved by utilizing local resources and knowledge.

Acknowledgements

This Research was funded by Water Research Commission (C2023-2024-01320), Mintek internal Science Vote (ASR-002517 and ASR-024032).

Author details

Napo Ntsasa[1,2], Andile Mkhohlakali[1], Tumelo Mogashane[1], James Tshilongo[1,2] and Mokgehle R. Letsoalo[1*]

1 Analytical Chemistry Division, Mintek, Randburg, South Africa

2 School of Chemistry, University of the Witwatersrand, WITS, South Africa

*Address all correspondence to: refiloel@mintek.co.za

References

[1] Hasan MN, Rahman K, Tajmunnaher BMR. Assessment of ground water quality in the vicinity of Sylhet City, Bangladesh: A multivariate analysis. Sustainable Water Resources Management. 2020;**6**(5):1-16. DOI: 10.1007/s40899-020-00448-x

[2] Dintsi BS, Letsoalo MR, Ambushe AA. Bioaccumulation and human health risk assessment of arsenic and chromium species in water–soil–vegetables system in Lephalale, Limpopo Province, South Africa. Minerals. 2023;**13**(7):1-24

[3] Zainurin SN, Wan Ismail WZ, Mahamud SNI, Ismail I, Jamaludin J, Ariffin KNZ, et al. Advancements in monitoring water quality based on various sensing methods: A systematic review. International Journal of Environmental Research and Public Health. 2022;**19**(21):1-21

[4] Reymond N, Glanzmann V, Huisman S, Plagellat C, Weyermann C, Estoppey N. An improved chemcatcher-based method for the integrative passive sampling of 44 hydrophilic micropollutants in surface water—Part B: Field implementation and comparison with automated active sampling. Science of the Total Environment. 2023;**871**:1-10

[5] Reichenberg F, Mayer P. Two complementary sides of bioavailability: Accessibility and chemical activity of organic contaminants in sediments and soils. Environmental Toxicology and Chemistry. 2006;**25**(5):1239-1245

[6] Letsoalo MR, Godeto TW, Magadzu T, Ambushe AA. Quantitative speciation of arsenic in water and sediment samples from the Mokolo River in Limpopo Province, South Africa. Analytical Letters. 2018;**51**(17):2761-2775. DOI: 10.1080/00032719.2018.1450879

[7] Bhardwaj J, Gupta KK, Gupta R. Towards a cyber-physical era: Soft computing framework based multi-sensor array for water quality monitoring. Drinking Water Engineering and Science. 2018;**11**(1):9-17

[8] Pasika S, Gandla ST. Smart water quality monitoring system with cost-effective using IoT. Heliyon. 2020;**6**(7):e04096. DOI: 10.1016/j.heliyon.2020.e04096

[9] Li H, Eksteen J, Oraby E. Hydrometallurgical recovery of metals from waste printed circuit boards (WPCBs): Current status and perspectives—A review. Resources, Conservation and Recycling. 2018;**139**(August):122-139. DOI: 10.1016/j.resconrec.2018.08.007

[10] Khatri P, Kumar Gupta K, Kumar GR. Raspberry pi-based smart sensing platform for drinking-water quality monitoring system: A python framework approach. Drinking Water Engineering and Science. 2019;**12**(1):31-37

[11] Shao Y, Hollert H, Tarcai Z, Deutschmann B, Seiler TB. Integrating bioassays, chemical analysis and in silico techniques to identify genotoxicants in surface water. Science of the Total Environment. 2019;**650**(September):3084-3092

[12] Letsoalo MR, Mamo MA, Ambushe AA. Simultaneous quantitative speciation of selected toxic elements in water using high performance liquid chromatography coupled to inductively coupled plasma-mass spectrometry (HPLC-ICP-MS). Physics and Chemistry of the Earth. 2021;**124**(March):103011. DOI: 10.1016/j.pce.2021.103011

[13] Khan N. Critical review of sampling techniques in the research process in the world. SSRN Electronic Journal. 2020;**9**:1-27. DOI: 10.2139/ssrn.3572336

[14] Hale SE, Oen AMP, Cornelissen G, Jonker MTO, Waarum IK, Eek E. The role of passive sampling in monitoring the environmental impacts of produced water discharges from the Norwegian oil and gas industry. Marine Pollution Bulletin. 2016;**111**(1-2):33-40. DOI: 10.1016/j.marpolbul.2016.07.051

[15] Vrana B, Allan IJ, Greenwood R, Mills GA, Dominiak E, Svensson K, et al. Passive sampling techniques for monitoring pollutants in water. TrAC—Trends in Analytical Chemistry. 2005;**24**(10):845-868

[16] Nitti F, Kapitan OB, Ola PD, Siswanta D. Passive Sampling Techniques for Monitoring of Pharmaceuticals and Personal Care Products in Water Matrix: Trends from 2016 to 2020. Singapore: Springer Nature; 2022. pp. 17-44. DOI: 10.1007/978-981-19-1847-6_2

[17] Farrow LG, Morton PA, Cassidy R, Floyd S, McRoberts WC, Doody DG, et al. Evaluation of Chemcatcher® passive samplers for pesticide monitoring using high-frequency catchment scale data. Journal of Environmental Management. 2022;**324**(June):116292. DOI: 10.1016/j.jenvman.2022.116292

[18] Suchana S, Passeport E. Implications of polar organic chemical integrative sampler for high membrane sorption and suitability of polyethersulfone as a single-phase sampler. Science of the Total Environment. 2022;**850**(December 2021):157898. DOI: 10.1016/j.scitotenv.2022.157898

[19] Links DA. Polar organic chemical integrative sampler (POCIS): Application for monitoring organic micropollutants in wastewater effluent and surface water †. Journal of Environmental Monitoring. 2012;**14**:626-635

[20] Bernard M, Boutry S, Guibal R, Morin S, Lissalde S, Guibaud G, et al. Multivariate tiered approach to highlight the link between large-scale integrated pesticide concentrations from polar organic chemical integrative samplers and watershed land uses. Journal of Agricultural and Food Chemistry. 2023;**71**(7):3152-3163

[21] de la Cal A, Kuster M, de Alda ML, Eljarrat E, Barceló D. Evaluation of the aquatic passive sampler chemcatcher for the monitoring of highly hydrophobic compounds in water. Talanta. 2008;**76**(2):327-332

[22] Edmiston PL, Carter E, Toth K, Hershberger R, Hill N, Versluis P, et al. Field evaluation of the sentinel ™ integrative passive sampler for the measurement of perfluoroalkyl and polyfluoroalkyl substances in water using a modified organosilica adsorbent. Monitoring and Remediation. 2022;**43**:38-54

[23] Schäfer RB, Paschke A, Vrana B, Mueller R, Liess M. Performance of the Chemcatcher® passive sampler when used to monitor 10 polar and semi-polar pesticides in 16 central European streams, and comparison with two other sampling methods. Water Research. 2008;**42**(10-11):2707-2717

[24] Aguilar-Martınez R, Palacios-Corvillo MA, Greenwood R, Mills GA, Vrana B, Gomez MMG. Calibration and use of the chemcatcher® passive sampler for monitoring organotin compounds in water. Analytica Chimica. 2008;**618**:157-167

[25] Aguilar-Martínez R, Gómez-Gómez MM, Greenwood R, Mills GA,

Vrana B, Palacios-Corvillo MA. Application of chemcatcher passive sampler for monitoring levels of mercury in contaminated river water. Talanta. 2009;**77**(4):1483-1489

[26] Gao Y, Zhou C, Galceran J, Rey-Castro C, Puy J, Baeyens W, et al. Developments in the diffusive gradients in thin-films technique for the speciation of oxyanions and platinum group elements in aquatic systems. TrAC—Trends in Analytical Chemistry. 2022;**147**(May 2021):116513. DOI: 10.1016/j.trac.2021.116513

[27] Sans-Duñó J, Cecilia J, Galceran J, Puy J, Baeyens W, Gao Y. Back accumulation of diffusive gradients in thin-films devices with a stack of resin discs to assess availability of metal cations to biota in natural waters. Environmental Science & Technology. 2023;**57**(20):7840-7848

[28] Pouran H, Colodrero P. Assessment of ATR-FTIR spectroscopy with multivariate analysis to investigate the binding mechanisms of Ag and TiO_2 nanoparticles to Chelex®-100 or Metsorb™ for the DGT technique. Analytical Methods. 2020;**12**:959-969

[29] Martins de Barros R, Rougerie J, Ballion T, Buzier R, Simon S, Guibal R, et al. Sensitivity improvement of o-DGT for organic micropollutants monitoring in waters: Application to neutral pesticides. Talanta Open. 2022;**6**:1-9

[30] Martins R, Barros D, Rougerie J, Guibal R, Lissalde S, Guibaud G. Interest of a new large diffusive gradients in thin films (L-DGT) for organic compounds monitoring: On-field comparison with conventional passive samplers. Environmental Pollution. 2023;**323**:1-7

[31] Tadić Đ, Manasfi R, Bertrand M, Sauvêtre A, Chiron S. Use of passive and grab sampling and high-resolution mass spectrometry for non-targeted analysis of emerging contaminants and their semi-quantification in water. Molecules. 2022;**27**:1-17

[32] Ji Y, Zheng F, Du J, Huang Y, Bi W, Duan HF, et al. An effective and efficient method for identification of contamination sources in water distribution systems based on manual grab-sampling. Water Resources Research. 2022;**58**(11):32784

[33] Koskiaho J, Tattari S. The e ff ect of sampling frequency and strategy on water quality modelling driven by high-frequency monitoring data in a boreal catchment. Journal of Hydrology. 2019;**579**(September):124186

[34] Cassidy R, Jordan P. Limitations of instantaneous water quality sampling in surface-water catchments: Comparison with near-continuous phosphorus time-series data. Journal of Hydrology. 2011;**405**(1-2):182-193

[35] Yan Y, Wang Y, Yu C, Zhang Z. Multispectral remote sensing for estimating water quality parameters: A comparative study of inversion methods using unmanned aerial vehicles (UAVs). Sustainability. 2023;**15**(13):1-18

[36] Tesfaye A. Remote sensing-based water quality parameters retrieval methods: A review. East African Journal of Environment and Natural Resources. 2024;**7**(1):80-97

[37] Yang H, Kong J, Hu H, Du Y, Gao M, Chen F. A review of remote sensing for water quality retrieval: Progress and challenges. Remote Sensing. 2022;**14**(8):1-21

[38] Huang C, Chen Y, Zhang S, Wu J. Detecting, extracting, and monitoring surface water from space using optical sensors: A review. Reviews of Geophysics. 2018;**56**(2):333-360

[39] Whiteside TG, Esparon AJ, Bartolo RE. A semi-automated approach for quantitative mapping of woody cover from historical time series aerial photography and satellite imagery. Ecological Informatics. 2020;**55**(October 2019):101012. DOI: 10.1016/j.ecoinf.2019.101012

[40] Bangira T, Matongera TN, Mabhaudhi T, Mutanga O. Remote sensing-based water quality monitoring in African reservoirs, potential and limitations of sensors and algorithms: A systematic review. Physics and Chemistry of the Earth. 2024;**134**(March 2023):103536. DOI: 10.1016/j.pce.2023.103536

[41] Adjovu GE, Stephen H, James D, Ahmad S. Overview of the application of remote sensing in effective monitoring of water quality parameters. Remote Sensing. 2023;**15**(7):1-35

[42] Ma T, Zhang D, Li X, Huang Y, Zhang L, Zhu Z, et al. Hyperspectral remote sensing technology for water quality monitoring: Knowledge graph analysis and frontier trend. Frontiers in Environmental Science. 2023;**11**(August):1-19

[43] Benediktsson JA, Wu Z. Distributed computing for remotely sensed data processing. Proceedings of the IEEE. 2021;**109**(8):1278-1281

[44] Niu WJ, Feng ZK, Feng BF, Min YW, Cheng CT, Zhou JZ. Comparison of multiple linear regression, artificial neural network, extreme learning machine, and support vector machine in deriving operation rule of hydropower reservoir. Water (Switzerland). 2019;**11**(1):1-17

[45] Zhang R, Zhang X, Shi J, Feng M, Wang X, Crump D, et al. Polychlorinated diphenyl sulfides: An emerging class of persistent, bioaccumulative, and toxic substances in the environment. Environmental Toxicology and Chemistry. 2021;**40**(10):2657-2666

[46] Radzevičius A, Dapkienė M, Sabienė N, Dzięcioł J. A rapid uv/v is spectrophotometric method for the water quality monitoring at on-farm root vegetable pack houses. Applied Sciences. 2020;**10**(24):1-15

[47] Manoj M, Kumar VD, Arif M, Bulai ER, Bulai P, Geman O. State of the art techniques for water quality monitoring systems for fish ponds using IoT and underwater sensors: A review. Sensors. 2022;**22**(6):1-17

[48] Rico A, Van den Brink PJ, Leitner P, Graf W, Focks A. Relative influence of chemical and non-chemical stressors on invertebrate communities: A case study in the Danube River. Science of the Total Environment. 2016;**571**(August):1370-1382

[49] Duncan PB, Morrison RD, Vavricka E. Forensic identification of anthropogenic and naturally occurring sources of perchlorate. Environmental Forensics. 2005;**6**(2):205-215

[50] Lu J. Identification of forensic information from existing conventional site-investigation data [Internet]. In: Introduction to Environmental Forensics. 3rd ed. United State: Elsevier Ltd; 2015. pp. 149-164. DOI: 10.1016/B978-0-12-404696-2.00006-0

[51] Douglas GS, Emsbo-Mattingly SD, Stout SA, Uhler AD, McCarthy KJ. Hydrocarbon fingerprinting methods. In: Introduction to Environmental Forensics. 3rd ed. United State: Elsevier Ltd.; 2015. pp. 201-309

[52] Letsoalo MR, Sithole T, Mufamadi S, Mazhandu Z, Sillanpaa M, Kaushik A,

et al. Efficient detection and treatment of pharmaceutical contaminants to produce clean water for better health and environmental. Journal of Cleaner Production. 2023;**387**(December 2022):135798. DOI: 10.1016/j.jclepro.2022.135798

[53] Shahabuddin M, Uddin MN, Chowdhury JI, Ahmed SF, Uddin MN, Mofijur M, et al. A review of the recent development, challenges, and opportunities of electronic waste (e-waste). International Journal of Environmental Science and Technology. 2023;**20**(4):4513-4520. DOI: 10.1007/s13762-022-04274-w

[54] Inobeme A, Nayak V, Mathew TJ, Okonkwo S, Ekwoba L, Ajai AI, et al. Chemometric approach in environmental pollution analysis: A critical review. Journal of Environmental Management. 2022;**309**(January):114653. DOI: 10.1016/j.jenvman.2022.114653

[55] Platikanov S, Baquero D, González S, Martín-Alonso J, Paraira M, Cortina JL, et al. Chemometric analysis for river water quality assessment at the intake of drinking water treatment plants. Science of the Total Environment. 2019;**667**:552-562. DOI: 10.1016/j.scitotenv.2019.02.423

Chapter 7

Membrane Filtration for Wastewater Treatment – Fouling Mitigation

Chetan Prakash Sharma, Ze Zhu and Avner Ronen

Abstract

Membrane technology is at the forefront of addressing critical challenges in water scarcity and environmental conservation through advanced wastewater treatment. This paper reviews membrane fouling mitigation strategies in wastewater treatment, highlighting recent advances and future directions. The review underscores the significance of membrane-based processes due to their high efficiency, adaptability, and sustainability. It discusses various membrane classifications, operational modes, and configurations, particularly focusing on cutting-edge developments in membrane-based technologies in wastewater treatment. Membrane fouling, characterized by the undesirable accumulation of particles, microorganisms, and organic compounds on membrane surfaces, impedes membrane performance, leading to permeability loss and increased energy demand. To counteract this, fouling mitigation strategies such as utilizing anti-fouling membrane materials have been at the forefront of research. These strategies aim to maintain membrane efficacy and longevity, thereby ensuring the sustainability of membrane-based wastewater treatment systems. Future research anticipates a paradigm shift toward smart membrane systems, utilizing external potentials and advanced material properties to address fouling and optimize performance, representing the next frontier in sustainable wastewater management.

Keywords: membrane, wastewater treatment, fouling, fouling mitigation, membrane surface modifications, nanoparticle coatings

1. Introduction

Water scarcity is a pressing global issue, profoundly impacting both human health and environmental sustainability. With the ever-increasing human population and its correlated demands on water resources, traditional water sources are being depleted at an alarming rate. This depletion is further exacerbated by factors such as economic development, population explosion, climate change, and expansion of irrigated agriculture [1]. Concurrently, the generation of wastewater from domestic, industrial, and agricultural activities has escalated, creating a critical need for effective wastewater treatment methods [2]. Membrane technology, as a versatile and efficient separation process, has emerged as a key solution in addressing these challenges.

Membrane technology is not only integral in the treatment of wastewater but also plays a crucial role in the purification and formulation of products across a wide array of sectors including water, energy, food, and environmental industries [3].

1.1 Fundamentals of membrane technology

The fundamentals of membrane technology in wastewater treatment revolve around membrane classification, materials and fabrication techniques, operation principles and systems [4]. A membrane is a molecular sieve built into the structure of a film with/without small pores or fine mesh to facilitate the separation of small particles and molecules [5]. Membranes are typically classified based on their pore size, driving force, morphology, and material composition, which can be either organic or inorganic. The pore size is a critical determinant in membrane selection, influencing its suitability for separating different components in a mixture [6]. The operational advantage of membrane technology in wastewater treatment is primarily its high efficiency in treating feed mixtures that are difficult to handle with conventional separation methods (e.g., adsorption or settling). This efficiency is due to the selective nature of the membrane, allowing for the differentiation of components based on size or affinity [7]. Its principle lies in the use of a semi-permeable barrier, which separates undesired components from a feed solution under various potential gradients, for example, pressure, temperature, electrical, or concentration differences. This technology stands out for its ability to produce stable products without the addition of chemicals and its relatively low energy consumption, making it a sustainable option for wastewater treatment [8]. Additionally, its modularity and scalability allow for flexible application across various treatment capacities, while its compact and simple design eases implementation and maintenance [9].

1.2 Classification of membranes

In the field of wastewater treatment, membrane-based technologies have become central solutions owing to their effectiveness and versatility in addressing diverse treatment requirements. These technologies include a range of membrane processes, specifically microfiltration (MF), ultrafiltration (UF), nanofiltration (NF), and reverse osmosis (RO), each distinguished by unique functionalities and applications [10]. **Figure 1** presents the range of membrane processes including the average diameter of the pores and the relative size of different solutes removed by each membrane class. Microfiltration and ultrafiltration are used in the separation of emulsions, silts, and oils from wastewater [11]. Their efficacy in these applications is enhanced when combined with other technologies, broadening their implementation scope. For instance, modified ultrafiltration membranes, developed from materials like polyvinylidene fluoride (PVDF) and titania nanotubes (TNT), have demonstrated significant color removal efficiencies in industrial wastewater treatments, such as those from palm oil mills [8]. Nanofiltration plays a vital role in rejecting divalent ions while allowing the passage of monovalent ions like Na^+ and Cl^- [12]. This makes NF particularly effective in recovering heavy metals from water, as most heavy metals are bivalent or higher [10]. NF effectively softens water by selectively removing hardness-causing ions, such as calcium and magnesium, while retaining beneficial minerals and consuming less energy than traditional methods. The technique's ease of operation, low energy consumption, high efficiency, and reliability make it a promising option for treating various types of wastewater. Reverse osmosis uses dense

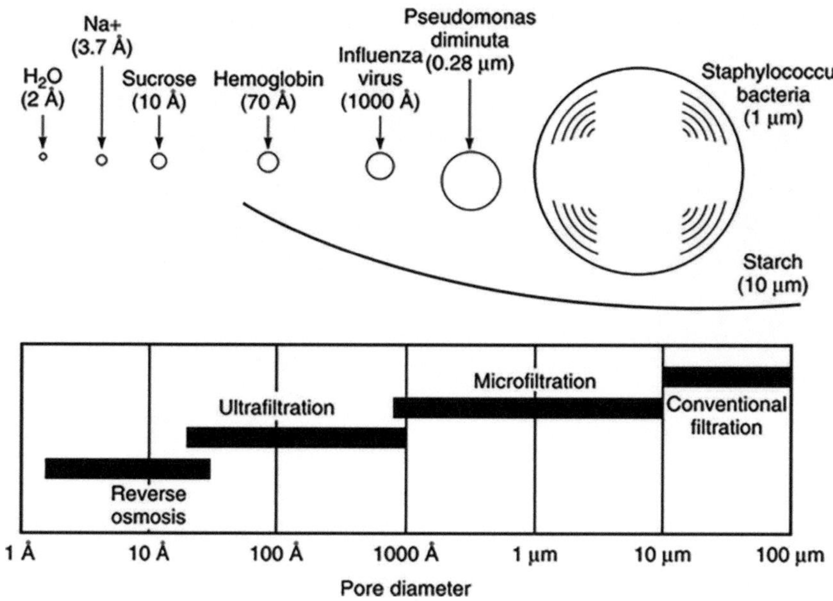

Figure 1.
Traditional filtration, microfiltration, ultrafiltration, and reverse osmosis represent a spectrum of related membrane-based processes, primarily distinguished by the average diameter of the pores in their respective membrane filters. This schematic illustrates the relative size of different solutes removed by each membrane class. Modified from baker [3].

(non-porous) membranes for water separation and is widely implemented for seawater desalination and for the management of micropollutants such as perfluoroalkyl substances (PFASs) in aqueous solutions, which are detected in various industrial products and have been associated with health and environmental risks [3]. In addition to these specific applications, the integration of different membrane technologies is often preferable to overcome the limitations of each process, conserve energy, and increase overall performance efficiency [9]. The coaupling of MF or UF with NF, and RO technologies, for instance, was shown to enhances treatment of oil-produced water, meeting environmental legislative requirements [4].

1.3 Operational modes and membrane modules

Membrane technology primarily operates by two distinct modes: dead-end and cross-flow filtration. Dead-end filtration (**Figure 2**) involves the direct flow of the feedwater perpendicular to the membrane surface. As elucidated in the studies by Wang et al. [13], this mode can be employed in gravity-driven membrane (GDM) systems for drinking water treatment. The primary characteristic of dead-end filtration is the accumulation of a contaminant layer or 'cake' on the membrane's surface, which can lead to increased resistance and reduced flow over time. This mode is generally more straightforward and less energy-intensive but requires regular maintenance to manage the cake buildup and prevent membrane fouling, resulting in a 'batch mode' operation pattern. Cross-flow filtration (**Figure 2**) introduces the feedwater parallel to the membrane surface, creating a shear that minimizes the accumulation of contaminants. The cross-flow mechanism allows for a continuous operation with a more stable flux, as the tangential flow constantly removes the particles from the

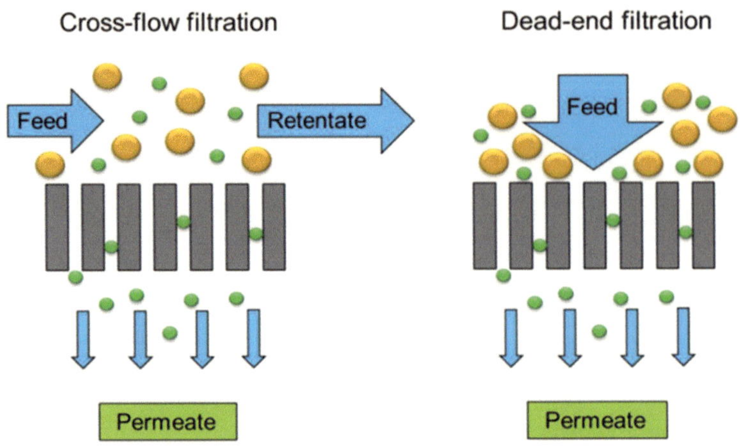

Figure 2.
Configurations of cross-flow filtration (left) and dead-end filtration (right). Adapted from Wang et al. [13].

membrane surface. However, this mode typically demands higher energy inputs due to the need for continuous flow across the membrane. This mode, as highlighted by Díaz et al. [14], is particularly effective in applications like anaerobic membrane bioreactors (AnMBRs), where the prevention of membrane fouling is crucial for maintaining flux. The choice between dead-end and cross-flow filtration is often dictated by the specific requirements of the process, including the nature of the feedwater, desired purity levels, and operational costs [15].

The selection of membrane modules (**Table 1**) is pivotal, governed by their design and operational characteristics. Spiral Wound (SW), Hollow Fiber (HF), Tubular, and Flat-sheet modules offer unique advantages [12]. SW modules are industry standards in RO/NF, HF is predominant in wastewater treatment, Tubular modules excel in high-solids wastewater treatment, and Flat-sheet designs are favored for small to medium-scale applications with turbid wastewater [8]. Furthermore, the choice of membrane module is informed by quality requirements, fouling tendency, and operational lifespan [15]. Quality requirements hinge on the specific water treatment needs, while the fouling tendency, a critical consideration, varies with feedwater type and includes strategies like hydrodynamic operation and chemical cleaning for prevention [3]. Operational lifespan, an economic factor, varies across membrane types, influencing the overall efficiency and cost-effectiveness of water treatment systems [8].

Property	Plate-and-Frame	Tubular	Spiral Wound	Hollow Fiber	Ref.
Packing Density ft^2/ft^3 (m^2/m^3)	45–150 (148–492)	6–120 (20–374)	150–380 (492–1247)	150–1500 (492–4924)	[8]
Potential for fouling	Moderate	Low	High	Very High	[12]
Ease of Cleaning	Good	Excellent	Poor	Poor	[3]
Relative Manufacturing cost	High	High	Moderate	Low	[15]

Table 1.
Basic properties of various membrane modules.

2. Membrane-based technologies for wastewater treatment

Advancements in wastewater treatment and reuse technologies are essential for achieving optimal water quality, a critical factor in various applications. Accordingly, a vast spectrum of treatment processes are tailored for water reuse applications, each designed to enhance effluent quality to meet diverse requirements for water reuse scenarios [3]. Traditional wastewater treatment methodologies typically rely on biological processes to eliminate biochemical oxygen demand (BOD), total suspended solids (TSS), and nutrients such as nitrogen in terms of ammonia and nitrate, and phosphorus. However, in membrane-based technologies used for wastewater treatment, implementing mainly UF and MF membranes, the role of membranes becomes increasingly pivotal, not only in removing residual suspended solids for tertiary treatment but also in augmenting the efficiency of disinfection methods as UF membranes are selective to bacteria, parasites, and some viruses [10]. Their adaptability, coupled with environmental sustainability and cost-effectiveness, positions membrane-based processes at the forefront of advanced wastewater treatment solutions, aligning with the growing demand for high-quality, reusable water (by tertiary treatment). The next chapter reviews technologies coupling membranes with the biological, electrical and thermal processes to improve wastewater treatment.

2.1 Membrane bioreactors

Membrane Bioreactors (MBRs) have revolutionized wastewater treatment by integrating biodegradation and advanced filtration technologies. At their core, MBRs combine activated sludge treatment MF or UF, using porous membranes to achieve superior separation of liquid and solid phases [16]. This results in several significant enhancements compared to conventional activated sludge processes [6], including reducing the overall footprint of the treatment system due to compact membrane modules, enhancing the quality of effluent, and providing thorough retention of bacterial flocs and suspended solids, thereby ensuring effective disinfection [8] and improving the system's ability to operate under high mixed liquor-suspended solids concentrations and extended solid retention times (SRT), thus enabling improved wastewater quality and the treatment of specific organic pollutants, particularly those nitrogen-based, while reducing the bioreactor size [4].

The development of MBR systems has seen significant technological advancements including the introduction of submerged membrane configuration, which involved placing the membrane directly within the bioreactor and employing aeration to minimize fouling, significantly lowering operational expenses and streamlining the treatment process, thereby catalyzing the widespread acceptance of MBR technology (**Figure 3**) [8]. Submerged MBRs come in various configurations, including vertical flat plates and hollow fibers, each with unique advantages and considerations regarding fouling and cleaning [17]. The operational aspects of MBRs, such as solid and hydraulic retention times, are critical and influence the nature of the activated sludge [16]. These systems can operate at lower HRT, resulting from decoupling them from SRT [18]. The intensive biological degradation capability of MBRs, coupled with their ability to retain biomass, facilitates efficient pollutant removal and, in some cases, complete nitrification. Furthermore, MBRs can effectively remove nitrogen and phosphorus through denitrification and chemical or biological processes, respectively [4]. As a result, MBRs have found applications across various industries including for treating high-strength wastewater, that pose challenges to conventional treatment

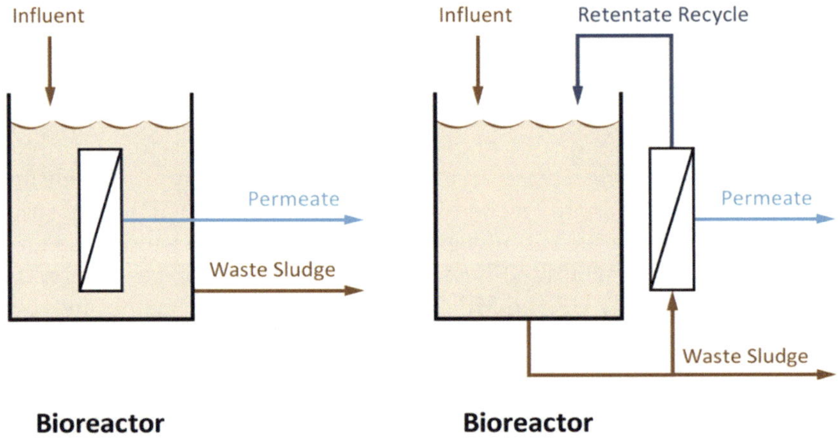

Figure 3.
Schematic configuration of side-stream (non-submerged) and submerged MBRs.

processes [18]. Additionally, MBR can be paired with advanced treatment methods, e.g., activated carbon, post-ozonation, or reverse osmosis, for further purification and reuse, including groundwater recharge and surface water augmentation [6].

While aerobic MBRs are typically considered for wastewater treatment, anaerobic membrane bioreactors (AnMBR) are a good alternative due to their low energy demand. AnMBRs combine anaerobic biological treatment with membrane filtration [19] with varying configurations (**Figure 4**). AnMBR processes leverage the inherent advantages of anaerobic treatment, including low energy requirements and biogas production, with the added benefit of membrane-based solid-liquid separation [21]. This results in a system with a smaller footprint, higher organic removal rates, and improved effluent quality compared to conventional treatment methods [20]. AnMBRs are particularly adept at handling municipal wastewater, industrial effluents, and high-strength organic wastes, making them versatile for various applications [22]. Recent advancements in AnMBR include integration with forward osmosis (FO) systems for pretreatment, which shows promise in reducing biofouling and enhancing membrane longevity [16]. AnMBRs also employ various membrane materials and configurations, such as hollow fiber, flat sheet, and multi-tubular membranes, each with distinct performance characteristics and operational parameters [8]. The integration of AnMBR with other treatment systems, such as microalgae cultivation and microbial fuel cells, offers promising avenues for nutrient recovery and energy efficiency improvements [19]. One of the key challenges in AnMBR technology is membrane fouling, which can hamper operational efficiency and productivity [20]. Fouling is typically caused by the accumulation of organic and inorganic matter and biological colloidal particles on the membrane surface. As AnMBRs have limited/no air scouring, they are more prone to fouling accumulation. Various strategies have been developed to mitigate fouling, such as increasing HRT and SRT and employing physical and chemical cleaning methods [21].

Apart from the biological treatment of wastewater by MBRs, we present several membrane technologies based on driving forces related to thermal and electrical gradients which are not very common but may have applications in wastewater treatment.

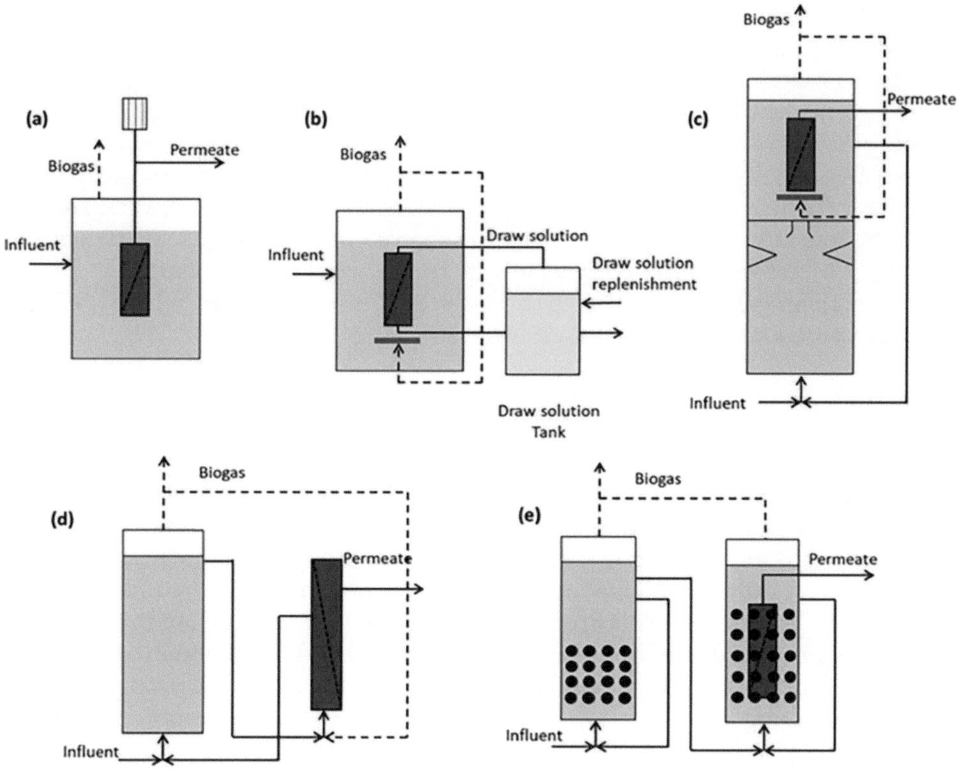

Figure 4.
AnMBR configurations for wastewater treatment. (a) Rotating membranes; (b) anaerobic osmotic membrane bioreactor; (c) membrane coupled at the top of the AnMBR-UASB; (d) gas-lift AnMBR; (e) two-stage anaerobic fluidized membrane bed bioreactor. Adapted from Vinardell et al. [20].

2.2 Membrane distillation

Membrane Distillation (MD) is a thermal membrane process that involves the separation of substances based on their partial vapor pressure [4]. Water vapor is transported across a hydrophobic microporous membrane for wastewater treatment, driven by a vapor pressure gradient [8]. MD is especially effective for feed solutions with higher salinity and utilizes a low-grade thermal energy based gradient (around 50–70°C) to create the necessary vapor pressure difference across the membrane [23]. The technology has several advantages, including the ability to use solar heating and waste heat from industrial processes to drive the separation process [15]. MD operates under very low hydrostatic pressure compared to RO or even MF, leading to reduced membrane fouling as only water vapor is transported across the membrane, and therefore, bacteria are not driven to the membrane by permeated drag force [6]. Furthermore, due to the elevated temperatures, some bacterial strains do not survive, and only thermophilic bacteria will eventually be able to form biofilms [24]. MD can efficiently separate nonvolatile materials from volatile ones, with feed product separation reaching 100% [3]. However, MD faces challenges as membrane wetting, resulting from fouling (scaling and biofouling), leading to rejection decline and possible damage to the membrane [25]. Avoiding such processes requires extensive feed water pretreatment, which increases operational costs [8]. Moreover, simultaneous heat and mass transfer in MD can cause temperature polarization, negatively impacting the driving force and reducing permeate flux [4]. MD has been applied in various contexts, such as treating water from the Arabian Gulf and table olive wastewater, showing promise in achieving high permeate flux and energy utilization [23].

Incorporating MD into wastewater treatment from aquaponics and hydroponics systems offers significant potential for enhancing water management and sustainability [26]. For example, for coupling aquaponic systems, MD treated water from the hydroponic units can be purified and reused in recirculating aquaculture systems (RAS) [27]. Moreover, MD's relatively low thermal energy requirement (in comparison with distillation) aligns well with these systems' energy profiles, where waste heat from other processes can be harnessed [4].

2.3 Electrodialysis

Electrodialysis (ED) is a membrane separation process that relies on the movement of ions across selective membranes under the influence of an electrical field [28]. It consists of an array of alternating cation and anion exchange membranes (**Figure 5**). When a feed solution containing various ions passes through the system, cations move toward the cathode and anions toward the anode, traversing their respective exchange membranes [29]. This process effectively separates ions from the feed solution, concentrating them in different compartments. ED's application in wastewater treatment is multifaceted [8]. For example, it can selectively remove phosphates from wastewater by using a combination of selective and nonselective membranes [11], demonstrating the versatility of ED in targeting specific ions for the removal and recovery of compounds with economic value. Another example of selective removal is the removal of arsenic from brackish water [28]. The selectivity in removing monovalent anions over di- and multivalent anions is due to steric effects and different charge interactions [11]. This selectivity allows for the optimization of the process for specific applications, such as nutrient removal from wastewater. For instance, a more than 70% removal rate for monovalent ions can be achieved, compared to a removal

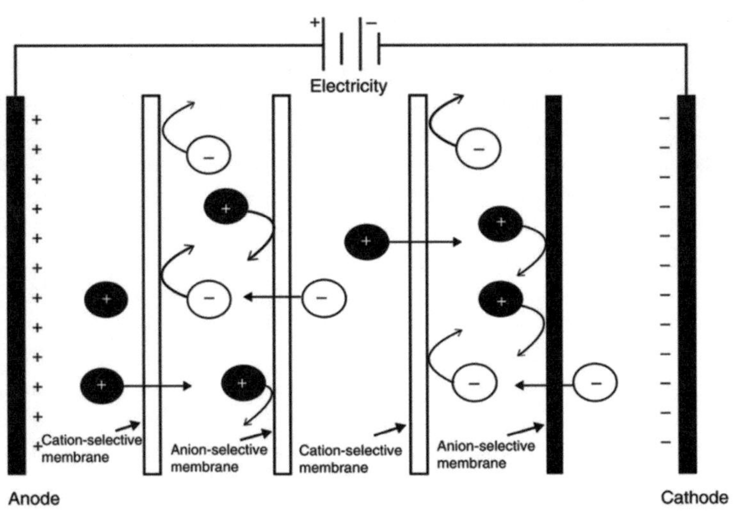

Figure 5.
A basic electrodialysis system for wastewater treatment. (Source: [28]).

rate below 50% for divalent ions [29]. This highlights ED's effectiveness in selective ion removal and its potential for customization in various wastewater treatment scenarios [10]. Additionally, ED can be used to treat concentrated brine from RO in desalination plants or treat saline water, thus reducing brine discharge and using an additional water source [3]. The energy requirements for ED in industrial wastewater applications vary, typically ranging from 6 to 11 kW h/m^3 of feed stream, depending on the water reclamation rate and water composition [29].

2.4 Integration of membranes in aquaculture wastewater treatment

Aquaculture effluents present unique challenges due to high levels of nutrients and suspended solids, necessitating efficient treatment for environmental protection and water reuse [27]. Membrane technology, including MF, UF, and MBRs, plays a crucial role in aquaculture wastewater treatment [10]. These systems efficiently remove fine solids, bacteria, and nutrients, enabling water reuse in aquaculture systems. Studies show that ceramic membranes with Ti–Mn/TiO_2/Al_2O_3 coatings effectively remove suspended solids, achieving up to 100% turbidity removal and reducing Total Ammonia Nitrogen (TAN) to below 0.1 mg/L [30]. MBRs and capillary microfiltration effectively manage organic loads, with a Chemical Oxygen Demand (COD) removal efficiency of 52.1% [13]. The use of superhydrophobic membranes in MD is crucial for treating high-salinity aquaculture wastewater [4] and demonstrates superior wetting resistance, which is essential for long-term operation [3]. However, challenges such as membrane fouling and operational optimization remain significant. For instance, while the integration of MBRs in RAS has notably enhanced water quality and reduced the need for water exchange, it has not substantially improved

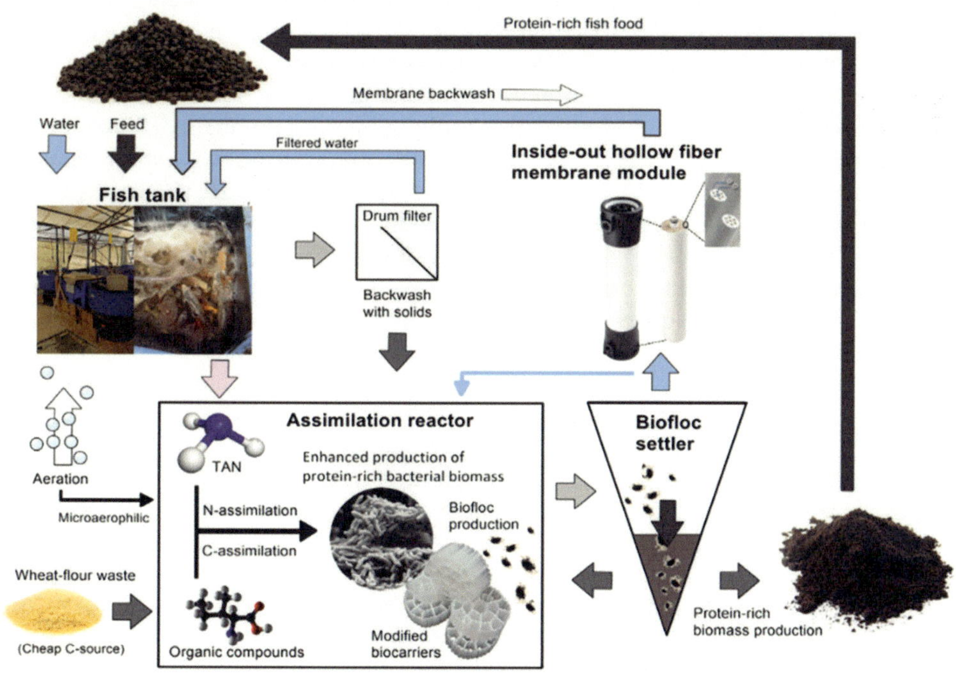

Figure 6.
A novel treatment of intensive fish wastewater by ultrafiltration membrane technology.

fish survival or growth, underscoring the necessity for additional research to refine the efficiency and cost-effectiveness of these systems [30, 31].

Another innovative water treatment technology coupling membrane with RAS integrates nutrient assimilation into microbial biomass production (**Figure 6**), offering environmental benefits over conventional RAS, which relies on nitrification. Using a microaerophilic membrane assimilation reactor (MMAR-RAS), this new approach reduces greenhouse gas emissions and nutrient discharge, lowers energy demands, and utilizes nutrients to produce protein-rich biomass [32]. This biomass is then converted into high-protein fish feed with health-promoting properties [33]. However, the MMAR-RAS faces challenges due to the accumulation of fine suspended solids. The integration of ultrafiltration membrane technology is proposed to address this issue, enhancing the efficiency and sustainability of the system.

3. Membrane fouling

Although membranes have many advantages, as described above, all membrane processes suffer from fouling, which is predominantly caused by the physical and chemical interactions between the membrane surfaces and foulants. Mass transport and operational conditions (including temperature, pH, ionic strength, permeate flux, and cross-flow velocity) may contribute to foulant attachment, accumulation, or adsorption on the surfaces of membranes [34]. Generally, fouling could result from the presence of foulants such as organic matter, colloids, microorganisms, ions, or a mixture of them in the treated wastewater. As the membranes are typically porous, inorganic fouling (scaling) is not considered a significant fouling contributor.

The interactions between foulants and membrane surface can be physical or chemical depending on the properties and other operational parameters, resulting in pore blocking, cake formation, organic adsorption, or biological membrane fouling as shown in **Figure** 7. Depending on the interaction between the membrane surface and the foulants present in wastewater, membrane fouling can be classified as reversible or irreversible. In the case of reversible fouling, membrane flux can be restored through backwashing or by using a particular chemical cleaning (e.g., chlorine or surfactants) including by cleaning in place (CIP). Nevertheless, the chemisorption or pore plugging due to the presence of foulants on the membrane surface may result in permanent, irreversible membrane fouling. In this case, the reduction in membrane performance (permeate flux loss) cannot be recovered chemically or hydrodynamically [36]. The following subsection provides an overview of different kind of membrane fouling and their impact on membrane performance.

3.1 Colloidal and particulate fouling

Colloidal or particluar membrane fouling occurs when the particles present in the feedwater deposit onto the surface of the membrane, forming a fouling cake layer [35]. Cake formation often depends on the nature of the particles present in the wastewater. The presence of the cake layer significantly affects the efficacy of the membrane performance and leads to a decrease in membrane flux or an increase in transmembrane pressure (TMP). The formation of a cake layer may also result in a gel layer and elevated ion concentration near the membrane, leading to a phenomenon described as concentration polarization (CP), which may further lead to inorganic scaling if ions concentration exceeds their solubility limit [36]. Colloids are particles

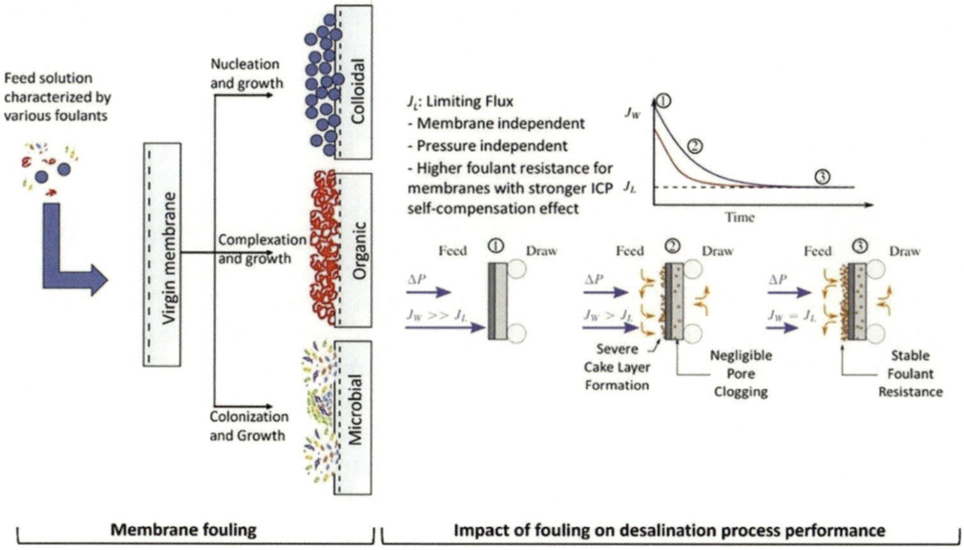

Figure 7.
The process of membrane fouling and its impact on the membrane process performance. (Source: [35]).

ranging from nanometers to micrometers in size and are one of the common foulants in aquatic environments. They include silt, clay, small organic compounds, and silica [37]. A specific case of particulate fouling occurs by colloids resulting from their small diameter. Collidial transport is governed by convection, Brownian diffusion, shear-induced diffusion, inertial lift, gravitational settling, and lateral migration [38]. Generally, membrane fouling by colloidal particles is attributed to three main factors: pore blockage, cake formation, and concentration polarization [39]. Whenever colloids contact the porous membrane surface, they first deposit onto the surface and block its pores; as additional particles continue to further deposition onto this initial layer, a cake layer forms. This cake layer controls the overall transport and efficiency of membrane performance, resulting in an increment in the TMP and a decrement in permeate flux. Colloidal fouling may also lead to CP, resulting in back diffusion and possible scaling [10, 36, 40].

3.2 Organic fouling

Organic fouling occurs from the presence of organic matter in the treated wastewater, this inlcudes proteins, humic acids, other organic acids and additional organic components such as natural organic matter (NOM). These compounds can accumulate on the membrane's surface or inside membrane pores, leading to a decrease in flux and possible formation of a "gel layer" when the applied pressure compacts the NOM on the membrane's surface. The presence of bacteria and biofilms may also lead to further extraction of organic matter and recution in filtration abilities. Overall, organic fouling may result in a gel/cake layer formation, physical obstruction of pores and CP. Organic fouling is also affected by operating variables such as temperature, pressure, and flow rate in addition to membrane characteristics such as surface charge, pore size, and hydrophobicity [24, 41]. As organic coumpound in the feed water vary between sources and the types of wastewater, impacting foulant types and concentrations, it is vital to have an in-depth understanding on the properties of the feed to allow designing efficient strategies (e.g., coagulation and separation, dissolved air floatations) that specifically target the organic compounds in order to prevent membrane fouling.

3.3 Biological fouling

Biofouling is considered the 'achiless heel' of membrane filtration and is a significant issue in membrane applications. Biofouling is characterized by the buildup of microorganisms at the interface between phases, for example, the phase between liquid (wastewater) and solids (a polymeric membrane). Bacterial cells deposit, develop, and metabolize on the membranes, leading to surface conditioning, adherence, microcolonies, and maturation of mature biofilms, as shown in **Figure 8** [42]. Biofilms consist of multiple layers of microbial cells (dead and active) and extracellular materials secreted by them. Cell biomass and extracellular polymeric substances (EPS) comprise 50–80% of biofilms' organic matter and proteins [43]. Biofouling can affect membrane performance in all membrane types, leading to a decrease in membrane permeability, increased pressure requirements, and elevated salt passage due to CP. Proteins and carbohydrates are two main macromolecular components contributing to biofilm production. In addition, shorter polymeric structures are secreted by bacteria or break down from the EPS to from transparent exopolymer particles (TEPs). TEPs are small particles and do not tend to agglomerate, so they are challenging to remove using conventional pretreatment procedures. Therefore, small bacterial precursors produce fouling on RO membranes even after removing large particles [44]. The use of biocides was shown only slightly efficiecnt for biofouling mitigation as the EPS provides as a buffer and protective layer for bacteria once it is formed. Additional approaches such as pretreatment of the wastewater aimed at lowering bacteria and nutrient concentrations. Low levels of biofilm formation rate (BFR) and assimilable organic carbon (AOC) have been linked to lower risks of biofouling [45]. Phosphate limiting is an additional technique for managing biofouling of spiral wound RO membranes. Overall, understanding biofilm characteristics and behavior is necessary to design more effective methods for managing membrane biofouling [46]. It should be noted that membrane fouling resulting from organic and biolofical factors is a significant problem for all MBRs as bacterial and organic matter concentrations in the reactors are very high.

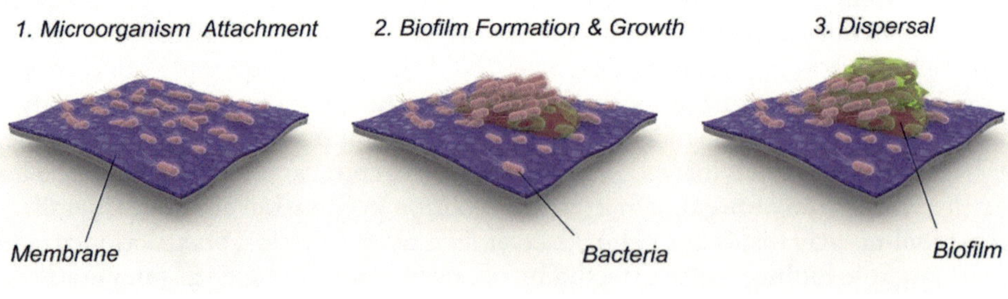

Figure 8.
Biofilm formation steps, (1) attachment to the surface, often preceded by a surface conditioning step and/or by deposition of organic material, (2) growth and creation of microcolonies, and (3) expansion of the biofilm to other areas, often followed by a detachment of part of the biofilm and dispersal. (Source: [42]).

4. Fouling mitigation strategies

As discussed in the previous section, membrane fouling is mainly caused by ions, particles, and bacteria in wastewater systems. However, wastewater typically contains a mixture of these foulants. Consequently, numerous initiatives have been undertaken to tackle the issue of membrane fouling. For example, effective per-treatment of wastewater, timely backwashing of membranes, chemical cleaning, optimization of operating conditions, membrane surface modification, development of advanced membrane materials, and in-line fouling monitoring.

Three promising fouling mitigation approaches, which are membrane surface modification, optimization of operation conditions, and advanced fouling mitigation strategies, are discussed in detail as follows.

4.1 Membrane surface modifications

Membrane surface modification is considered a common strategy to reduce membrane fouling. During membrane modification, controlling surface roughness, membrane surface charge, hydrophilicity, and hydrophobicity play a significant role in reducing possible fouling tendency. Generally, membranes with hydrophilic and smooth surfaces demonstrate less fouling [47]. Among the surface modification possibilities, a few approaches are described in **Table 2**. The following section discusses three approaches to reduce fouling by surface modification.

4.1.1 Application of anti-fouling coatings

Membrane coating using polymers promoting anti-fouling has been explored. The polymers are designed to reduce bacterial and organic matter adhesion by increasing surface hydrophilicity or proving anti-fouling abilities via higher surface negative charge, and surface smoothness.

For example, Zhao et al. used polydopamine (PDA) to modify the membrane's surfaces and improve its hydrophilicity. However, this method often results in decreased water permeance due to adding a layer with lower permeability on the pristine membrane. Therefore, a pressure-assisted coating method was developed to tune the inner structure of membranes, enhancing water permeance and salt rejection [58]. In another research, Jee et al. modified a spiral-wound RO membrane with a polysiloxane coating with 3-glycidoxypropyltrimethoxysilane (GPPTMS) to increase its hydrophilicity, resulting in improvment in fouling resistance. The study found that fouling decreased with GPPTMS concentrations over 1.0 wt% [59]. Xu et al. showed that loose nanofiltration membranes (LNMs) are promising for dye/salt separation, but membrane fouling hinders real-world applications (**Figure 9**). Therefore, they prepared a new anti-fouling LNM, by co-deposition of diacrylamide (AMTHBA) and polydopamine (PDA) onto polysulfone (PSf) membrane, thus improving the membrane's hydrophilicity. The optimum PSf/PDA-AM 1.5 LNM achieved high water permeance, excellent dye rejection, and low salt retention. It also exhibited outstanding anti-fouling properties against organic foulants and bacteria [60]. Qiu et al. reported fabrication of a modified PES membrane where [a] cationic antibacterial agent, quaternary ammonium compound (QAC), was integrated into a MgAl layered double hydroxide (QAC/LDH) composite, enhancing its hydrophilicity, water flux, and resistance to organic fouling. The PES-QLDH membrane showed superior antibiofouling performance, inhibiting bacteria adhesion and growth. Its synergistic

Membrane Surface Modifications	Modification method	Membrane details	Modifier	Foulant	Parameters changed	Comments on characteristics	Ref.
Application of anti-fouling coatings	Dip coating	Commercial PA TFC membrane by Woongjin Chemical Co., Ltd. (Korea)	MMA-HPOEM	BSA, *E. coli*	Increased hydrophilicity and the results demonstrated that the coating layer was durable to chemical cleaning.	Coating resistant to chemical cleaning; lower initial flux; enhanced anti-fouling characteristics	[48]
	Layer-by-layer assembly	Commercial RO membrane (ES-20; Nitto Denko, Japan)	Polyelectrolyte	BSA	Enhanced hydrophilicity and surface smoothness	Enhanced hydrophilicity and surface smoothness	[49]
	Spray coating	Commercial PA RO membrane	TiO_2 NPs	Humic acid	Increased membrane hydrophilicity and negative charge	Increased membrane hydrophilicity and negative charge	[50]
	In-situ modification	Commercial PA TFC membrane	P(NIPAM-co-Am)	BSA	Increased membrane surface hydrophilicity	Higher permeability and resistance to BSA fouling; improved cleaning efficiency	[51]
Embedment of polymeric brushes	Interfacial polymerization with TMC	Polysulfone ultrafiltration substrates	Hydrophilic zwitterionic striped structure	Humic acid (HA), BSA and sodium alginate (SA)	Improved hydrophilicity	High selectivity and outstanding anti-fouling ability	[52]
	Grafting onto polydopamine activated membrane	PES Ultrafiltration 20 kDa	Thiolated zwitterionic polyurethane	BSA	Improved hydrophilicity	Exceptional fouling resistance	[53]
	Layer-by-layer assembly	Commercial NF membrane (NF 90)	Amino-Quinone networks-Phytic acid pseudo Zwitterionic	BSA	Improved hydrophilicity	Excellent fouling resistance and high membrane cleaning efficiency	[54]

Membrane Surface Modifications	Modification method	Membrane details	Modifier	Foulant	Parameters changed	Comments on characteristics	Ref.
Surface enhancement using nanoparticles	Vacuum filtration	Polyacrylonitrile (PAN) ultrafiltration substrate	TiO_2	Na_2SO_4	Crumpled and thinner PA layer with more uniform structure	An increased permeability by 20.8% than that of TFC membrane and a Na_2SO_4 rejection of 98.2%	[55]
	Brush painting	Poly(ether sulfone) microfiltration	SWCNT	Na_2SO_4	Ultrathin PA layer with higher cross-linking degree, and increased negative surface zeta potential	High water permeability of 40 L·m − 2 ·h − 1 ·bar−1, high Na_2SO_4 rejection of 96.5%	[56]
	Spray coating	Ultrafiltration PSF membrane	MXene	Salt rejection	Thinner polyamide layer with increased hydrophilicity and roughness	Nearly double fold of water flux, higher selective separation factor, outstanding salt rejection (>97%)	[57]

Table 2.
Studies related to membrane surface modification using different modification methodologies.

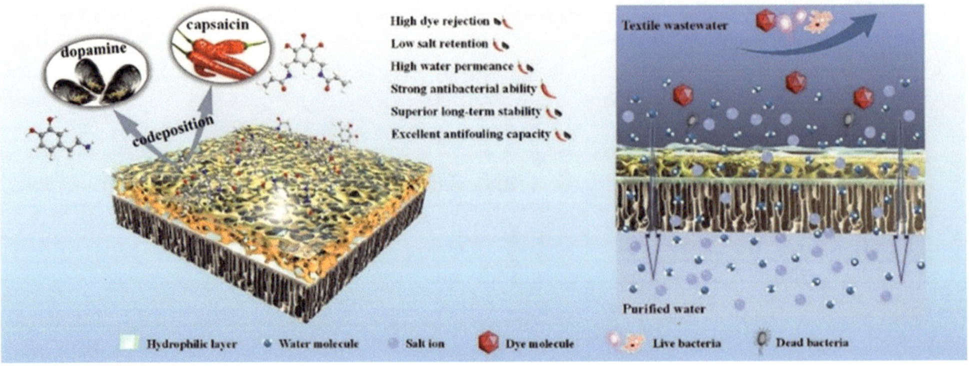

Figure 9.
Representation of membrane modification and its applicability for anti-fouling capability. (Source: [60]).

effect resulted in high biocidal activity, making it a promising strategy for wastewater treatment [61].

4.1.2 Embedment of polymeric brushes

Industrial membranes are fabricated from robust, hydrophobic or slightly hydrophobic polymers such as polyethersulfone (PES), polyvinylidene fluoride (PVDF), and polysulfone (PS), which are more prone to fouling. Consequently, hydrophilic polymer brushes can impact fouling potential resulting from improved hydrophilicity and steric interferences [62]. Novel brushes are based on zwitterionic polymers with two opposing charged moieties, yet their hydrophilicity, chemical stability, and "salting-in" action make them useful for anti-fouling applications. They are often used in grafting to create a hydration layer that diminishes fouling and raises the membrane's hydrophilicity. Water molecules were shown to form bonds with zwitterionic brush chains as they include both positively and negatively charged moieties. Furthermore, these materials promote unhindered mobility and prevent the accumulation of foulants [62]. For example, Song et al. reported the utilization of novel thiolated zwitterionic polyurethane (TZP) and its grafting onto a polydopamine-activated membrane surface for the preparation of engineering anti-fouling nanofiltration (NF) membranes. The chemical structure was confirmed by ^{1}H nuclear magnetic resonance and the grafting improved membrane hydrophilicity. The pore size and ion selectivity of the NF membranes were optimized to achieve divalent anion retention, and excellent fouling resistance (**Figure 10a**). **Figure 10c, d**, presents the grafted polymer brushes onto the engineered NF membrane and the repulsion of organic fouling agent (Bovine Serum Albumin) away from the membrane surface. After grafting, the formation of a hydration layer also improved water flux by about 10–20% (**Figure 10b**) [53].

In addition, Deng et al. fabricated an anti-fouling thin-film composite nanofiltration membrane using zwitterionic polyethyleneimine (PEI) moieties. The membrane showed stronger hydration capacity and thicker hydration layer, improved water permeability by 11.6%, and improved ion selectivity. The zwitterionic PEI grafted on the TFC NF membrane mitigated electrostatic attraction and improved anti-fouling properties [63]. Zhang et al. proposed a difunctional nanofiltration membrane with anti-fouling and antibacterial properties (**Figure 11**) by grafting a random terpolymer with quaternary ammonium and zwitterionic moieties fabricated by free radical

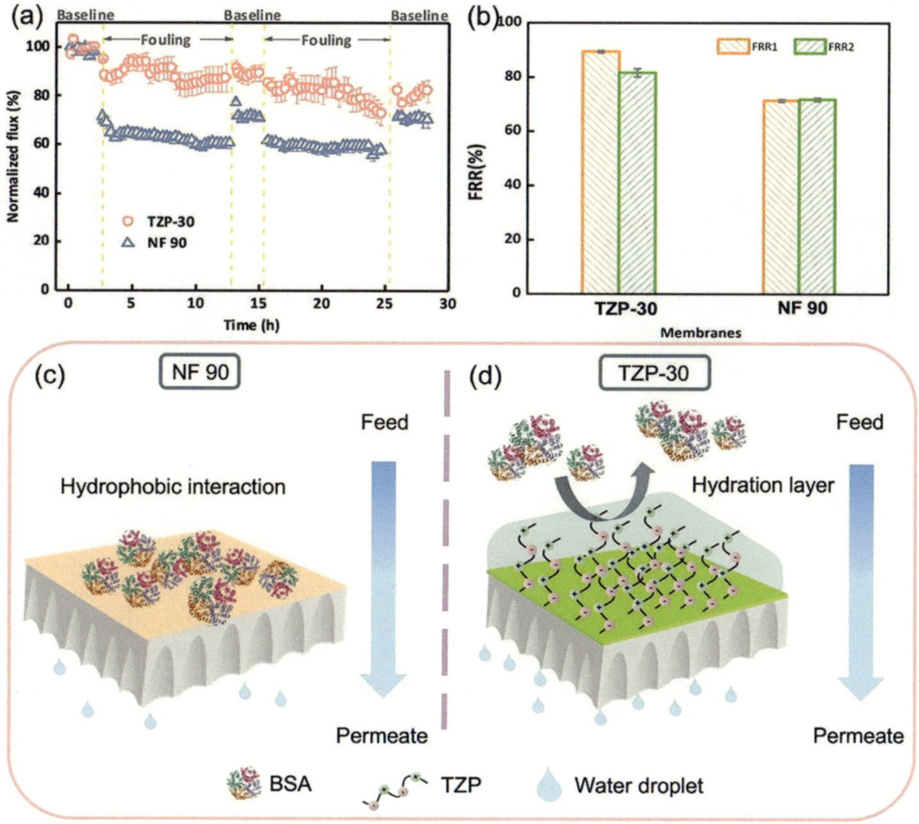

Figure 10.
(a) Normalized water flux of the TZP-30 and NF 90 membranes as a function of cumulative filtration time during two cycles of dynamic fouling tests. Synthetic wastewater, including 0.94 mM NH_4Cl, 0.45 mM KH_2PO_4, 0.5 mM $CaCl_2$, 0.5 mM $NaHCO_3$, 2.0 mM NaCl, 0.6 mM $MgSO_4$, and 100 mg L^{-1} BSA, was used as feed solution, and the initial flux of membrane was set at 50 L m^{-2} h^{-1}; (b) flux recovery ratios of the TZP-30 and NF 90 membranes after each fouling cycle; (c–d) schematic illustration of the interaction between foulants and membrane surface: (c) NF 90 membrane and (d) TZP modified membrane. (Source: [53]).

polymerization. The modified membrane inhibits protein and bacterial adhesion and shows high bacteria-inhibiting efficiency [64].

4.1.3 Surface enhancement using nanoparticles

Apart from polymer-based modifications, research efforts have been directed toward creating nano-enhanced membrane surfaces by leveraging the hydrophilic properties of nanomaterials. These include carnon based Np such as Graphene Oxide (GO), Carbon Nanotubes (CNTs), and Metal-Organic Frameworks (MOFs), Metal Nanoparticles (Metal NPs, including metal nanoparticles and metal oxides), and Mxenes. Nanomaterials are recognized for their high aspect ratio and outstanding electrical and thermal conductivities. Researchers worldwide have explored the incorporation of these nanomaterials into membrane filtration processes. Several recent review articles delve into the applications of these nanomaterials in enhancing membrane performance [65–70].

For example, recently, Firouzjaei et al. [71] prepared a thin-film nanocomposite (TFN) membrane by incorporating graphene oxide and silver-based metal-organic

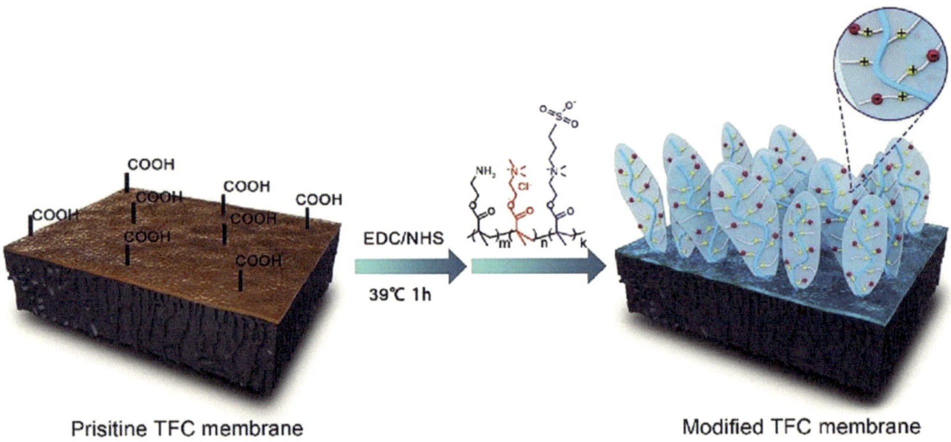

Figure 11.
Schematic illustration of grafting bifunctional terpolymer to the TFC nanofiltration membrane surface. (Source: [64]).

framework (Ag-MOF) into the TFC selective layer (**Figure 12**). This study showed significant improvement in membranes' anti-biofouling and anti-fouling properties. Also, the modified TFN membrane exhibits higher hydrophilicity, water permeability, and elimination of of *Escherichia coli*. Furthermore, in a study by Suresh et al., a straightforward chemical grafting and deposition technique was showcased to enhance the physical and separation properties of a thin film composite RO membrane. The modification involved the complexation of tannic acid (TA) with silver nanoparticles (Ag NPs), resulting in improved chlorine resistance and reduced biofouling. The TFC-TA/Ag membrane exhibited a notable 60% enhancement in flux and NaCl rejection compared to the unmodified TFC membrane [72].

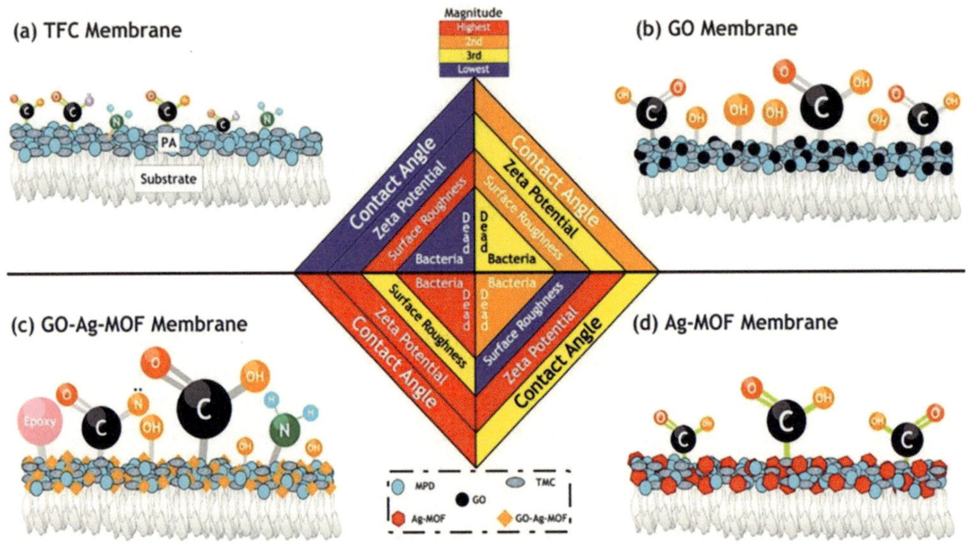

Figure 12.
Parameters contributing to anti-biofouling and anti-fouling properties of the membranes: (a) TFC, (b) GO TFN, (c) GO–Ag-MOF TFN, and (d) Ag-MOF TFN. (Source: [71]).

4.2 Operational approaches

In addition to membrane surface modification strategies, operational parameters play a crucial role in influencing membrane performance. Factors such as temperature, applied hydraulic pressure, pH, controlled mixing, feed aeration, and the implementation of backwash cycles can collectively contribute to enhancing membrane lifespan [40]. Hydrodynamic conditions, such as cross-flow velocity and flux, significantly impact membrane fouling. High cross-flow velocity induces high shear rates, promoting particle back diffusion and reducing concentration polarization, thus minimizing fouling development. Resulting from the impact of permeate drag force on particles and bacterial attachment to the membranes, fouling occurrance was corrolated with elevated peameare flux. In addition, elevated flux also requires a higher TMP and thus, increases the hydraulic resistance of the foulant layer, which could lead to the formation of a gel layer and CP [40]. In additon, aie/gas scouring is an effective way to reduce and control membrane fouling, mainly in MF and UF systems. The mechanism of fouling minimization depends on the bubble size, where macrobubbles and air slugs are implemented in MBRs to mechanically vibrate and shake the membrane (hollo fibers), promoting local shear and limiting pore blocking and cake formation. Microbubbles are spherical and tend to spread evenly over the bulk solution while contacting the foulants to form a bubble-foulants flock ahead of the foulant-membrane contact [73]. Wu et al. examined the impact of gas slug flow on MF in an external tubular membrane reactor. The air slug flow characteristics ranged from 30 to 120 L h^{-1} with an interval of 30 L h^{-1}. The shear stress was found to be 6.37×10^{-3} Pa with aeration rate of 90 L h^{-1} and the TMP was effectively controlled by increasing the aeration rate, with the optimal rate around 90 L h^{-1} [74].

4.3 Advanced fouling mitigation techniques

In recent years, novel approaches have been developed to mitigate membrane fouling, including using ultrasonic waves, heated membrane surfaces, electrically conductive membranes and more. Two examples are presented below.

Ultrasonic technology: Ultrasonic waves and energy have been used in membrane filtration systems for fouling mitigation and *in-situ* membrane cleaning. Ultrasonic waves can form acoustic cavitation, a phenomenon where bubbles grow and collapse due to pressure fluctuations, generating heat, shear forces, turbulence, and microstreaming. Low-frequency, low-power ultrasonic fields were shown to significantly enhance permeate flux in whey and milk MF [75–77]. Camara et al. reported that the combination of PVDF membranes with ultrasound (US) can significantly improve the permeate flux. Results show that in a feed solution containing 0.075 wt% of solid (750 mg/L TSS), application of 20 kHz and 0.29 W cm^{-2} of US can inprove the permeate flux by 19.7%. In another work, Ehsani et al. presented the utilization of in-situ ultrasounds for fouling control and membrane cleaning during the tretment of skimmed milk solution using UF membranes. The improved filtration efficiencies was obtained using a combination of high US power and low US frequencies. At optimum operational conditions of 100 W, 28 kHz, and 0.5 wt% (5000 mg/L) of particle mass with 1 L/min of feed flow rate, above 67% improvement was achieved in the permeate flux [78].

Electrically Conductive Membranes (ECMs): Electrically conducting membranes are a class of specialized membranes that exhibit electrical conductivity, offering unique properties and applications across various industries. These membranes are designed to harness the benefits of electrical fields for specific purposes, including sensing, separation, and fouling mitigation. ECMs can be fabricated by coating or infusing conductive polymers like polyaniline, polypyrrole, or polythiophene. These polymers provide electrical conductivity while maintaining the flexibility and processability of the membrane. Incorporating NPs such as CNTs, graphene, or carbon black into the membrane structures enhances electrical conductivity while offering high surface area and excellent electrical properties. ECMs are highly efficient in organic fouling, biofouling, and inorganic fouling mitigation based on the externally applied potential that creates an electrostatic repulsion force between the negatively charged foulants and the membrane, reducing foulants' attachment to the membrane. In addition, biofouling mitigation was shown to be efficient by reactive oxygen species (ROS) produced at the membrane's surface and local pH shift [25, 79–83]. A recent study showed the ability to mitigate biofouling in AnMBR, where the ECM serves as a cathode and a platinum plate as the counter electrode. Bacterial inactivation was estimated under varying cathodic potentials, and biofouling formation was assessed during 48 hours of filtration. The system was efficient in fouling mitigation, mainly attributed to bacterial inactivation and biofilm mitigation from extreme local pH conditions [84]. In addition, Wang et al. [85] reported the utilization of a novel in-situ combination approach for fabricating a highly conductive carbon nanotube (CNT)/polyvinylidene fluoride (PVDF) membrane (as shown in **Figure 13**). The CNT/PVDF membrane had high electrical conductivity and permeability and exhibited superior anti-fouling performance in the presence of external electric fields at a 0 V/2 V-pulsed mode.

5. Summary and future outlook

Membrane technology has emerged as a crucial innovation in the field of wastewater treatment, addressing complex challenges related to water scarcity and environmental conservation. With advanced separation capabilities, this technology employs various membranes such as microfiltration, ultrafiltration, nanofiltration, and reverse osmosis to effectively eliminate a broad spectrum of contaminants. The inherent features of membrane technology, including low energy requirements,

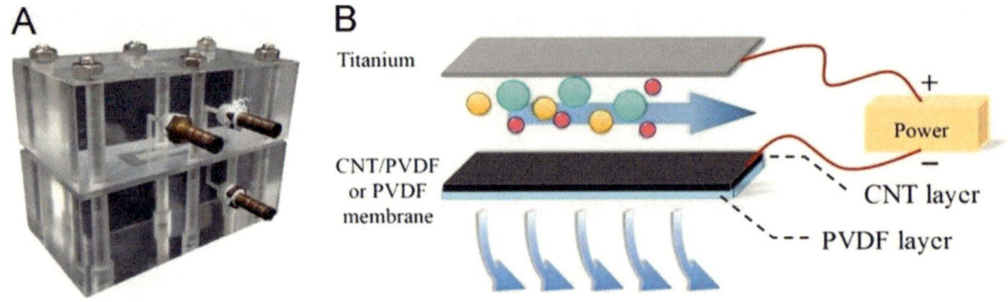

Figure 13.
(A) A photograph of the filtration cell. (B) Schematic of the electrically enhanced cross-flow filtration cell for evaluating the anti-fouling performance of the PVDF and CNT/PVDF membranes. A titanium foil was used as an anode. (Source: [85]).

reduced reliance on chemical processes, and adaptability to diverse pollutant profiles, underscore its significance in contemporary water treatment methodologies. This technological advancement not only enhances water reuse strategies but also plays a pivotal role in promoting sustainable water resource management and environmental stewardship. However, membrane fouling in wastewater treatment poses significant limitations, characterized by the accumulation of unwanted materials on membrane surfaces, leading to decreased performance and increased operational costs. Recent scientific literature highlights diverse strategies for fouling mitigation including physical and chemical cleaning, operational optimization, and material engineering. Innovations in membrane surface modification, such as incorporating hydrophilic coatings and nano-enhanced materials, show promise in reducing fouling in addition to operational adjustments like hydraulic pressure management and implementing backwash cycles that contribute to fouling prevention and membrane longevity. These multifaceted approaches are essential for maintaining the efficiency and sustainability of membrane technologies in wastewater treatment.

The future of membrane technologies in wastewater management is defined by the integration of advanced anti-fouling strategies, innovative material science, and process engineering to significantly enhance membrane longevity and operational efficiency. Recent advancements, particularly in developing electrically conductive membranes (ECMs), demonstrate potential in addressing the persistent challenge of membrane fouling. The integration of ECMs with Joule heating capabilities, as highlighted by Liao et al. [86] serves to mitigate fouling while offering real-time fouling monitoring (**Figure 14**). Building on this foundation, there is an emerging inclination to explore novel materials, such as nanocomposite-based membranes exhibiting anti-fouling properties and ultra-high flux efficiency for complex separations. Furthermore, the design of hybrid processes that couple membrane bioreactors with advanced oxidation processes or bio-electrochemical systems is gaining traction. These hybrid systems aim to leverage the strengths of each component technology to enhance treatment efficiency and system stability.

In summary, the future outlook for membrane technologies in sustainable wastewater management includes:

1. Enhancement of membrane anti-fouling properties through stable hydrophilic coatings and novel material integration.

2. Optimization of ECMs for improved fouling resistance, operational stability, and energy efficiency.

3. Development of hybrid membrane systems that synergize various treatment technologies to achieve superior performance.

4. Refined fabrication and characterization techniques to understand and improve membrane structure-function relationships.

5. Integration of real-time monitoring systems to optimize performance.

Collective efforts in these areas are poised to revolutionize wastewater treatment, pushing the boundaries of efficiency, sustainability, and resource recovery.

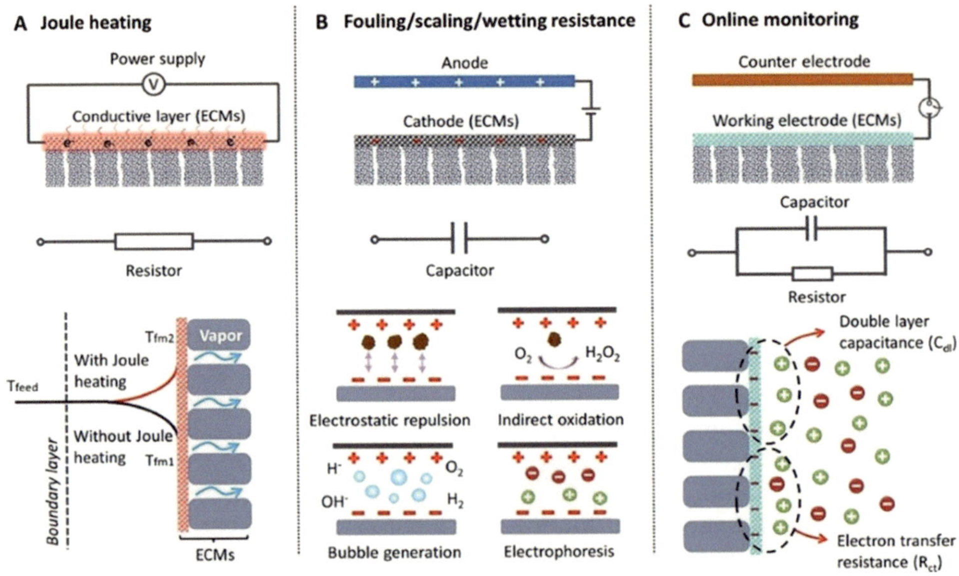

Figure 14.
Suggested future research directions. (Source: [86]).

Acknowledgements

This work was supported by NSF-BSF funding, 2022607 (to A.R). The authors also acknowledge the postdoctoral fellowships from the Jacob Blaustein Center for Scientific Cooperation of Ben Gurion University of the Negev.

Conflict of interest

There is no conflict of interest to declare.

Author details

Chetan Prakash Sharma†, Ze Zhu† and Avner Ronen*
Zuckerberg Institute for Water Research, Jacob Blaustein Institutes for Desert Research, Ben Gurion University of the Negev, Sde Boker Campus, Israel

*Address all correspondence to: avnerr@bgu.ac.il

† These authors contributed equally to this chapter.

References

[1] Richardson K, Steffen W, Lucht W, Bendtsen J, Cornell SE, Donges JF, et al. Earth beyond six of nine planetary boundaries. Science Advances. 2023;**9**:1-17. DOI: 10.1126/sciadv.adh2458

[2] Calvin K, Dasgupta D, Krinner G, Mukherji A, Thorne PW, Trisos C, et al. Climate Change 2023: Synthesis Report. Geneva, Switzerland: IPCC; 2023. DOI: 10.59327/IPCC/AR6-9789291691647

[3] Baker RW. Membrane Technology and Applications. Newark, California: Wiley; 2004. DOI: 10.1002/0470020393

[4] Ahmed SF, Mehejabin F, Momtahin A, Tasannum N, Faria NT, Mofijur M, et al. Strategies to improve membrane performance in wastewater treatment. Chemosphere. 2022;**306**:135527. DOI: 10.1016/j.chemosphere.2022.135527

[5] Li C, Sun W, Lu Z, Ao X, Li S. Ceramic nanocomposite membranes and membrane fouling: A review. Water Research. 2020;**175**:115674. DOI: 10.1016/j.watres.2020.115674

[6] Deowan SA, Bouhadjar SI, Hoinkis J. Membrane Bioreactors for Water Treatment. Sawston, Cambridge: Elsevier Ltd; 2015. DOI: 10.1016/B978-1-78242-121-4.00005-8

[7] Younas M, Rezakazemi M. Introduction to membrane technology. In: Membrane Contactor Technology. Weinheim, Germany: Wiley; 2022. pp. 1-16. DOI: 10.1002/9783527831036.ch1

[8] Ezugbe EO, Rathilal S. Membrane technologies in wastewater treatment: A review. Membranes (Basel). 2020;**10**:89. DOI: 10.3390/membranes10050089

[9] Peters T. Membrane technology for water treatment. Chemical Engineering and Technology. 2010;**33**:1233-1240. DOI: 10.1002/ceat.201000139

[10] Shehata N, Egirani D, Olabi AG, Inayat A, Abdelkareem MA, Chae KJ, et al. Membrane-based water and wastewater treatment technologies: Issues, current trends, challenges, and role in achieving sustainable development goals, and circular economy. Chemosphere. 2023;**320**:137993. DOI: 10.1016/j.chemosphere.2023.137993

[11] Wang LK, Chen JP, Hung Y-T, Shammas NK, editors. Membrane and Desalination Technologies. Totowa, NJ: Humana Press; 2011. DOI: 10.1007/978-1-59745-278-6

[12] Singh R, Hankins NP. Introduction to Membrane Processes for Water Treatment. Amsterdam, Netherlands: Elsevier B.V.; 2016. DOI: 10.1016/B978-0-444-63312-5.00002-4

[13] Wang Q, Tang X, Liang H, Cheng W, Li G, Zhang Q, et al. Effects of filtration mode on the performance of gravity-driven membrane (GDM) filtration: Cross-flow filtration and dead-end filtration. Water (Switzerland). 2022;**14**:190. DOI: 10.3390/w14020190

[14] Díaz H, Azócar L, Torres A, Lopes SIC, Jeison D. Use of flocculants for increasing permeate flux in anaerobic membrane bioreactors. Water Science and Technology. 2014;**69**:2237-2242. DOI: 10.2166/wst.2014.153

[15] Sadr SMK, Saroj DP. Membrane Technologies for Municipal Wastewater Treatment. Sawston, Cambridge:

Elsevier Ltd; 2015. DOI: 10.1016/B978-1-78242-121-4.00014-9

[16] Le-Clech P. Membrane bioreactors and their uses in wastewater treatments. Applied Microbiology and Biotechnology. 2010;**88**:1253-1260. DOI: 10.1007/s00253-010-2885-8

[17] Dezotti M, Lippel G, Bassin JP. Advanced Biological Processes for Wastewater Treatment: Emerging, Consolidated Technologies and Introduction to Molecular Techniques. Cham, Switzerland; 2017. DOI: 10.1007/978-3-319-58835-3

[18] Tan X, Acquah I, Liu H, Li W, Tan S. A critical review on saline wastewater treatment by membrane bioreactor (MBR) from a microbial perspective. Chemosphere. 2019;**220**:1150-1162. DOI: 10.1016/j.chemosphere.2019.01.027

[19] Aslam A, Khan SJ, Shahzad HMA. Anaerobic membrane bioreactors (AnMBRs) for municipal wastewater treatment- potential benefits, constraints, and future perspectives: An updated review. Science of the Total Environment. 2022;**802**:149612. DOI: 10.1016/j.scitotenv.2021.149612

[20] Vinardell S, Astals S, Peces M, Cardete MA, Fernández I, Mata-Alvarez J, et al. Advances in anaerobic membrane bioreactor technology for municipal wastewater treatment: A 2020 updated review. Renewable and Sustainable Energy Reviews. 2020;**130**:109936. DOI: 10.1016/j.rser.2020.109936

[21] Shahid MK, Kashif A, Rout PR, Aslam M, Fuwad A, Choi Y, et al. A brief review of anaerobic membrane bioreactors emphasizing recent advancements, fouling issues and future perspectives. Journal of Environmental Management. 2020;**270**:110909. DOI: 10.1016/j.jenvman.2020.110909

[22] Maaz M, Yasin M, Aslam M, Kumar G, Atabani AE, Idrees M, et al. Anaerobic membrane bioreactors for wastewater treatment: Novel configurations, fouling control and energy considerations. Bioresource Technology. 2019;**283**:358-372. DOI: 10.1016/j.biortech.2019.03.061

[23] Khan SJ, Hankins NP, Shen LC. Submerged and Attached Growth Membrane Bioreactors and Forward Osmosis Membrane Bioreactors for Wastewater Treatment. Amsterdam, Netherlands: Elsevier B.V.; 2016. DOI: 10.1016/B978-0-444-63312-5.00011-5

[24] Laqbaqbi M, Sanmartino JA, Khayet M, García-Payo C, Chaouch M. Fouling in membrane distillation, osmotic distillation and osmotic membrane distillation. Applied Sciences. 2017;7:334. DOI: 10.3390/app7040334

[25] Anvari A, Kekre KM, Azimi Yancheshme A, Yao Y, Ronen A. Membrane distillation of high salinity water by induction heated thermally conducting membranes. Journal of Membrane Science. 2019;**589**:117253. DOI: 10.1016/j.memsci.2019.117253

[26] Zhu Z, Yogev U, Goddek S, Yang F, Keesman KJ, Gross A. Carbon dynamics and energy recovery in a novel near-zero waste aquaponics system with onsite anaerobic treatment. Science of the Total Environment. 2022;**833**:155245. DOI: 10.1016/j.scitotenv.2022.155245

[27] Zhu Z. Nutrient Dynamics and Bioresource Recovery in Novel Zero-Waste Multi-Loop Aquaponic Systems. Netherlands: Wageningen University; 2023. DOI: 10.18174/590685

[28] Van der Bruggen B. Advances in Electrodialysis for Water

Treatment. Sawston, Cambridge: Elsevier Ltd; 2015. DOI: 10.1016/B978-1-78242-121-4.00006-X

[29] Kabay N, Arar Ö, Bunani S. Water Treatment by Electromembrane Processes. Amsterdam, Netherlands; 2016. DOI: 10.1016/B978-0-444-63312-5.00008-5

[30] Chen S, Yu J, Wang H, Yu H, Quan X. A pilot-scale coupling catalytic ozonation-membrane filtration system for recirculating aquaculture wastewater treatment. Desalination. 2015;**363**:37-43. DOI: 10.1016/j.desal.2014.09.006

[31] Zhou L, Tan SH, Ahmad AL, Low SC. High-flux strategy for electrospun nanofibers in membrane distillation to treat aquaculture wastewater: A review. Journal of Chemical Technology and Biotechnology. 2021;**96**:3259-3272. DOI: 10.1002/jctb.6828

[32] Yogev U, Gross A. Reducing environmental impact of recirculating aquaculture systems by introducing a novel microaerophilic assimilation reactor: Modeling and proof of concept. Journal of Cleaner Production. 2019;**226**:1042-1050. DOI: 10.1016/j.jclepro.2019.04.003

[33] Nayak S, Yogev U, Kpordzaxor Y, Zhu Z, Gur N, Gross A, et al. From fish excretions to high-protein dietary ingredient: Feeding intensively cultured barramundi (Lates calcarifer) a diet containing microbial biomass (biofloc) from effluent of an aquaculture system. Aquaculture. 2023;**562**:738780. DOI: 10.1016/j.aquaculture.2022.738780

[34] Li Q, Elimelech M. Organic fouling and chemical cleaning of nanofiltration membranes: Measurements and mechanisms. Environmental Science & Technology. 2004;**38**:4683-4693. DOI: 10.1021/es0354162

[35] Mahlangu OT, Nthunya LN, Motsa MM, Morifi E, Richards H, Mamba BB. Fouling of high pressure-driven NF and RO membranes in desalination processes: Mechanisms and implications on salt rejection. Chemical Engineering Research and Design. 2023;**199**:268-295. DOI: 10.1016/j.cherd.2023.09.037

[36] Guo W, Ngo HH, Li J. A mini-review on membrane fouling. Bioresource Technology. 2012;**122**:27-34. DOI: 10.1016/j.biortech.2012.04.089

[37] Yiantsios SG, Karabelas AJ. The effect of colloid stability on membrane fouling. Desalination. 1998;**118**:143-152. DOI: 10.1016/S0011-9164(98)00110-6

[38] Huang H, Spinette R, O'Melia CR. Direct-flow microfiltration of aquasols. I. Impacts of particle stabilities and size. Journal of Membrane Science. 2008;**314**:90-100. DOI: 10.1016/j.memsci.2008.01.040

[39] Lim AL, Bai R. Membrane fouling and cleaning in microfiltration of activated sludge wastewater. Journal of Membrane Science. 2003;**216**:279-290. DOI: 10.1016/S0376-7388(03)00083-8

[40] Goh PS, Lau WJ, Othman MHD, Ismail AF. Membrane fouling in desalination and its mitigation strategies. Desalination. 2018;**425**:130-155. DOI: 10.1016/j.desal.2017.10.018

[41] Sisay EJ, Al-Tayawi AN, László Z, Kertész S. Recent advances in organic fouling control and mitigation strategies in membrane separation processes. A Review. Sustainability. 2023;**15**:13389. DOI: 10.3390/su151813389

[42] Firouzjaei MD, Seyedpour SF, Aktij SA, Giagnorio M, Bazrafshan N, Mollahosseini A, et al. Recent advances in functionalized polymer membranes

for biofouling control and mitigation in forward osmosis. Journal of Membrane Science. 2020;**596**:117604. DOI: 10.1016/j.memsci.2019.117604

[43] Deng L, Guo W, Ngo HH, Zhang H, Wang J, Li J, et al. Biofouling and control approaches in membrane bioreactors. Bioresource Technology. 2016;**221**:656-665. DOI: 10.1016/j.biortech.2016.09.105

[44] Ke X, Hongqiang R, Lili D, Jinju G, Tingting Z. A review of membrane fouling in municipal secondary effluent reclamation. Environmental Science and Pollution Research. 2013;**20**:771-777. DOI: 10.1007/s11356-012-1147-y

[45] Bui X-T, Chiemchaisri C, Fujioka T, Varjani S. Biofouling Detection on Reverse Osmosis Membranes BT - Water and Wastewater Treatment Technologies. Singapore: Springer; 2019. DOI: 10.1007/978-981-13-3259-3

[46] Lopez-Vazquez CM, Brdjanovic D, Volcke E, van Loosdrecht MCM, Wu D, Chen G. Biological Wastewater Treatment - Examples and Exercises. London, UK: IWA; 2023. DOI: 10.2166/9781789062304

[47] Asadollahi M, Bastani D, Musavi SA. Enhancement of surface properties and performance of reverse osmosis membranes after surface modification: A review. Desalination. 2017;**420**:330-383. DOI: 10.1016/j.desal.2017.05.027

[48] Choi H, Park J, Tak T, Kwon YN. Surface modification of seawater reverse osmosis (SWRO) membrane using methyl methacrylate-hydroxy poly(oxyethylene) methacrylate (MMA-HPOEM) comb-polymer and its performance. Desalination. 2012;**291**:1-7. DOI: 10.1016/j.desal.2012.01.018

[49] Ishigami T, Amano K, Fujii A, Ohmukai Y, Kamio E, Maruyama T, et al. Fouling reduction of reverse osmosis membrane by surface modification via layer-by-layer assembly. Separation and Purification Technology. 2012;**99**:1-7. DOI: 10.1016/j.seppur.2012.08.002

[50] Kim JS, Kim DH, Hwang DK. Efficiency enhancement of bifacial $Cu_2ZnSnSe_4$ thin-film solar cells on indium tin oxide glass substrates by suppressing In-Sn diffusion with Mo interlayer. Journal of Power Sources. 2018;**400**:9-15. DOI: 10.1016/j.jpowsour.2018.08.001

[51] Yu S, Liu X, Liu J, Wu D, Liu M, Gao C. Surface modification of thin-film composite polyamide reverse osmosis membranes with thermo-responsive polymer (TRP) for improved fouling resistance and cleaning efficiency. Separation and Purification Technology. 2011;**76**:283-291. DOI: 10.1016/j.seppur.2010.10.017

[52] Lin B, Tan H, Liu W, Gao C, Pan Q. Preparation of a novel zwitterionic striped surface thin-film composite nanofiltration membrane with excellent salt separation performance and anti-fouling property. RSC Advances. 2020;**10**:16168-16178. DOI: 10.1039/d0ra00480d

[53] Song F, Tu C, Meng F, Wang L, Xia L, Zhao S. Engineering a novel zwitterionic polyurethane nanofiltration membrane with enhanced fouling resistance for water reclamation. Desalination. 2023;**551**:116440. DOI: 10.1016/j.desal.2023.116440

[54] Xia L, Song F, Meng F, Wang L, Zhao S. Engineering anti-fouling nanofiltration membranes using amino-quinone networks-phytic acid pseudo zwitterionic clusters for water treatment. ACS EST Water. 2023;**3**:1923-1934. DOI: 10.1021/acsestwater.3c00015

[55] Chi M, Zheng P, Wei M, Zhu A, Zhong L, Zhang Q, et al. Polyamide composite nanofiltration membrane modified by nanoporous TiO_2 interlayer for enhanced water permeability. Journal of Industrial and Engineering Chemistry. 2022;**115**:230-240. DOI: 10.1016/j.jiec.2022.08.004

[56] Gao S, Zhu Y, Gong Y, Wang Z, Fang W, Jin J. Ultrathin polyamide Nanofiltration membrane fabricated on brush-painted single-walled carbon nanotube network support for ion sieving. ACS Nano. 2019;**13**:5278-5290. DOI: 10.1021/acsnano.8b09761

[57] Wang A, Xu H, Fu J, Lin T, Ma J, Ding M, et al. Enhanced high-salinity brines treatment using polyamide nanofiltration membrane with tunable interlayered MXene channel. Science of the Total Environment. 2023;**856**:158434. DOI: 10.1016/j.scitotenv.2022.158434

[58] Zhao DL, Zhao Q, Lin H, Chen SB, Chung TS. Pressure-assisted polydopamine modification of thin-film composite reverse osmosis membranes for enhanced desalination and anti-fouling performance. Desalination. 2022;**530**:115671. DOI: 10.1016/j.desal.2022.115671

[59] Jee KY, Shin DH, Lee YT. Surface modification of polyamide RO membrane for improved fouling resistance. Desalination. 2016;**394**:131-137. DOI: 10.1016/j.desal.2016.05.013

[60] Xu L, Pang Z, Yu H, Guo M, Yan X, Jiang X, et al. Anti-fouling loose nanofiltration membranes prepared via the fast codeposition of capsaicin-mimic/polydopamine for efficient dye/salt separation. Desalination. 2023;**565**:116809. DOI: 10.1016/j.desal.2023.116809

[61] Qiu Z, Chen J, Dai R, Wang Z. Modification of ultrafiltration membrane with antibacterial agent intercalated layered nanosheets: Toward superior antibiofouling performance for water treatment. Water Research. 2022;**219**:118539. DOI: 10.1016/j.watres.2022.118539

[62] Huyen DTT, Ray SS, Kwon YN. Zwitterionic material for construction of an antifouling polyamide thin film composite membrane. Desalination. 2024;**572**:117118. DOI: 10.1016/j.desal.2023.117118

[63] Deng L, Li S, Qin Y, Zhang L, Chen H, Chang Z, et al. Fabrication of anti-fouling thin-film composite nanofiltration membrane via surface grafting of polyethyleneimine followed by zwitterionic modification. Journal of Membrane Science. 2021;**619**:118564. DOI: 10.1016/j.memsci.2020.118564

[64] Zhang X, He F, Yang Z, Xie M, Ren H, Wu D, et al. Engineering polyamide nanofiltration membrane with bifunctional terpolymer brushes for anti-fouling and antimicrobial properties. Desalination. 2023;**558**:116642. DOI: 10.1016/j.desal.2023.116642

[65] Daer S, Kharraz J, Giwa A, Hasan SW. Recent applications of nanomaterials in water desalination: A critical review and future opportunities. Desalination. 2015;**367**:37-48. DOI: 10.1016/j.desal.2015.03.030

[66] Xie T, Li F, Chen K, Zhao S, Chen Y, Sun H, et al. Fabrication of novel thin-film nanocomposite polyamide membrane by the interlayer approach: A review. Desalination. 2023;**554**:116509. DOI: 10.1016/j.desal.2023.116509

[67] Zhao S, Liao Z, Fane A, Li J, Tang C, Zheng C, et al. Engineering anti-fouling reverse osmosis membranes: A review. Desalination. 2021;**499**:114857. DOI: 10.1016/j.desal.2020.114857

[68] Ong CS, Goh PS, Lau WJ, Misdan N, Ismail AF. Nanomaterials for biofouling and scaling mitigation of thin film composite membrane: A review. Desalination. 2016;**393**:2-15. DOI: 10.1016/j.desal.2016.01.007

[69] Noor U, Farid MF, Sharif A, Saleem A, Nabi Z, Mughal MF, et al. Functionalized carbon 1D/2D nanomaterials for effective water desalination: Synthesis, applications and cost issues. An overview. Desalination. 2024;**571**:117086. DOI: 10.1016/j.desal.2023.117086

[70] Yin J, Deng B. Polymer-matrix nanocomposite membranes for water treatment. Journal of Membrane Science. 2015;**479**:256-275. DOI: 10.1016/j.memsci.2014.11.019

[71] Firouzjaei MD, Shamsabadi AA, Aktij SA, Seyedpour SF, Sharifian M, Rahimpour A, et al. Exploiting synergetic effects of graphene oxide and a silver-based metal-organic framework to enhance anti-fouling and anti-biofouling properties of thin-film nanocomposite membranes. ACS Applied Materials & Interfaces. 2018;**10**:42967-42978. DOI: 10.1021/acsami.8b12714

[72] Suresh D, Goh PS, Ismail AF, Mansur SB, Wong KC, Asraf MH, et al. Complexation of tannic acid/silver nanoparticles on polyamide thin film composite reverse osmosis membrane for enhanced chlorine resistance and anti-biofouling properties. Desalination. 2022;**543**:116107. DOI: 10.1016/j.desal.2022.116107

[73] Levitsky I, Tavor D, Gitis V. Micro and nanobubbles in water and wastewater treatment: A state-of-the-art review. Journal of Water Process Engineering. 2022;**47**:102688. DOI: 10.1016/j.jwpe.2022.102688

[74] Wu Y, Wang J, Zhang H, Ngo HH, Guo W, Zhang N. The impact of gas slug flow on microfiltration performance in an airlift external loop tubular membrane reactor. RSC Advances. 2016;**6**:109067-109075. DOI: 10.1039/c6ra19903h

[75] Qasim M, Darwish NN, Mhiyo S, Darwish NA, Hilal N. The use of ultrasound to mitigate membrane fouling in desalination and water treatment. Desalination. 2018;**443**:143-164. DOI: 10.1016/j.desal.2018.04.007

[76] Naji O, Al-juboori RA, Khan A, Yadav S, Altaee A, Alpatova A, et al. Ultrasound-assisted membrane technologies for fouling control and performance improvement: A review. Journal of Water Process Engineering. 2021;**43**:102268. DOI: 10.1016/j.jwpe.2021.102268

[77] Camara HWD, Doan H, Lohi A. *In-situ* ultrasound-assisted control of polymeric membrane fouling. Ultrasonics. 2020;**108**:106206. DOI: 10.1016/j.ultras.2020.106206

[78] Ehsani M, Doan H, Lohi A, Zhu N, Abdelrasoul A. Experimental and statistical study of an *in-situ* ultrasound (US)-assisted ultrafiltration process: Application to fouling control in skimmed milk filtration. Journal of Water Process Engineering. 2022;**49**:103171. DOI: 10.1016/j.jwpe.2022.103171

[79] Duan W, Chen G, Chen C, Sanghvi R, Iddya A, Walker S, et al. Electrochemical removal of hexavalent chromium using electrically conducting carbon nanotube/polymer composite ultrafiltration membranes. Journal of Membrane Science. 2017;**531**:160-171. DOI: 10.1016/j.memsci.2017.02.050

[80] Duan W, Ronen A, Walker S, Jassby D. Polyaniline-coated carbon

nanotube ultrafiltration membranes: Enhanced anodic stability for in situ cleaning and electro-oxidation processes. ACS Applied Materials & Interfaces. 2016;**8**:22574-22584. DOI: 10.1021/acsami.6b07196

[81] Ronen A, Walker SL, Jassby D. Electroconductive and electroresponsive membranes for water treatment. Reviews in Chemical Engineering. 2016;**32**:533-550. DOI: 10.1515/revce-2015-0060

[82] Ronen A, Duan W, Wheeldon I, Walker S, Jassby D. Microbial attachment inhibition through Low-voltage electrochemical reactions on electrically conducting membranes. Environmental Science & Technology. 2015;**49**:12741-12750. DOI: 10.1021/acs.est.5b01281

[83] Thamaraiselvan C, Ronen A, Lerman S, Balaish M, Ein-Eli Y, Dosoretz CG. Low voltage electric potential as a driving force to hinder biofouling in self-supporting carbon nanotube membranes. Water Research. 2018;**129**:143-153. DOI: 10.1016/j.watres.2017.11.004

[84] Qi S, Grossman AD, Ronen A, Bernstein R. Low-biofouling anaerobic electro-conductive membrane bioreactor: The role of pH changes in bacterial inactivation and biofouling mitigation. Journal of Membrane Science. 2022;**662**:120960. DOI: 10.1016/j.memsci.2022.120960

[85] Wang S, Liang S, Liang P, Zhang X, Sun J, Wu S, et al. In-situ combined dual-layer CNT/PVDF membrane for electrically-enhanced fouling resistance. Journal of Membrane Science. 2015;**491**:37-44. DOI: 10.1016/j.memsci.2015.05.014

[86] Liao X, Lim YJ, Khayet M, Liao Y, Yao L, Zhao Y, et al. Applications of electrically conductive membranes in water treatment via membrane distillation: Joule heating, membrane fouling/scaling/wetting mitigation and monitoring. Water Research. 2023;**244**:120511. DOI: 10.1016/j.watres.2023.120511

Chapter 8

Metal Organic Frameworks Used as Antibiotic Removal Agents in Water

Miguel Martell Mendoza, Carlos Alberto Méndez Cuesta, Miguel Angel Zavala Sánchez, Erick Cuauhtemoc Pérez Montiel, Alfonso Mata Bermúdez and Cuauhtemoc Pérez González

Abstract

Currently, there are approximately 250 antibiotics registered for use in human and veterinary medicine worldwide, which, as a result of inadequate management and poor disposal of waste, among other bad practices in their management, are more common in wastewater; this situation has begun to attract more attention and be an interesting topic for the proposal of solutions. In the search for options to solve this problem, structured materials are being actively studied; one material that has received significant attention is the metal organic framework (MOF) type. The use of MOFs with various topologies and characteristics are demonstrating great effectiveness in the elimination of different types of contaminants from water, such as medications, such as anti-inflammatories, antibiotics, explosives, and heavy metals. Different antibiotics are used as models in these studies, from which it follows that the nature of the drugs that can be eliminated from effluents is very varied, which allows us to see that it can be an interesting option. Another objective pursued by these studies is to seek to establish the number of cycles in which these materials can be used to carry out the capture of drugs, as well as the type of kinetics that follow the adsorption of these materials in the MOFs.

Keywords: MOF, antibiotic, water removal, wastewater, recyclability, bioremediation

1. Introduction

Antibiotics are molecules with several functional groups like aromatic rings, heterocyclic, and aliphatic chains and might be conjugated with sugar fragments (type A) [1]. These compounds have been widely used in medicine, animal husbandry, and agriculture due to their low cost and potential to inhibit or eliminate the growth of pathogenic microorganisms [2]. Moreover, several reports talk about other properties of these molecules, such as antifungal, antiparasitic, antioxidant, and antitumoral agents, which results in great interest in their development and applications [3]. However, antibiotics also represent a problem for the environment because of their bad deposition and treatment after their administration. Many reports have found that the accumulation of these pharmaceutical products in water carries a lot of problems in public health because of microbial resistance to the constant exposition of people to these compounds, and this might have toxic effects on the respiratory, digestive, and reproductive systems, among others [2, 4]. Mainly, microbial resistance has increased not only to the indiscriminate use of antibiotics to treat other infections that do not have a correlation with bacteria diseases but also to the high concentration of antibiotics in ecosystems leading to multi-resistance bacteria (MRB). An example of MRB is *S. aureus*, which was inhibited by methicillin, but nowadays, with the resistance gene mecA found in new strains of this bacteria, the effect of this antibiotic is very low. The last is a worrying fact because the increase in MRB is notorious for public health [4].

Therefore, this has resulted in the development of new strategies to solve this problem; one of these solutions is the obtention and synthesis of new materials that can adsorb and remove these compounds from water, one of the most important ecosystems. Metal organic frameworks (MOFs) are hybrid materials that have homogeneous porosity, great stability, high surface areas, and multidimensional structures that can capture and release substances via several stimulations. The synthesis of these materials can be driven with a wide range of metal and organic ligands that can be functionalized with carboxylates, phenolates, imidazolates, and sulfonates, among others, and their stability in water can be enhanced with the conjugation with other molecules like polietilenglicole, sugar or silices [5]. The attractive features of these materials are the highly combinatory possibilities for their synthesis and architecture; for this reason, the synthesis of selective materials that can capture and degrade or reuse the molecule adsorbed is possible.

MOFs have been used for various purposes; due to their characteristics, they have found application for gas storage, as catalysts in different types of reactions, as detectors of various types of compounds, and as drug carriers; this last application has aroused the interest for the use of these materials in the removal of drugs in bodies of water, for a few years to date, research on this topic has aroused great interest, demonstrated in the gradual increase in publications of applications of these reticular materials in water treatment.

This work addresses several examples of the applicability of MOFs as water removal agents and demonstrates that MOFs can be suitable and selective platforms for the removal of many different antibiotics found in water.

2. MOFs used have water removal agents

2.1 Materials Institute of Lavoisier (MIL)

MILs are a class of MOFs based on linkers with benzoic di or tri-carboxylates combined with several metals like Fe, Al, Cr, Ti, etc. These MOFs are very attractive because of their water stability, high porosity, and superficial areas of up to 1000 m^2. These features allow these families to be used in several applications like gas capture and capture/release of drugs. The last is one of the features that made it a great candidate to remove antibiotics from water. However, several reports have pointed out that the use of MILs as composite materials enhances their properties, allows the materials to be more selective, and increases the number of life cycles. The next table enlists several reports of MIL-based materials with great removal potential from different antibiotics (**Table 1**).

MOF	Antibiotic	Experiment	Results	Ref
In_2S_3@MIL-125(Ti)	Tetracycline (TC)	Absorption of the drug on aqueous media and its photodegradation in a tetracycline solution of 100 mg/L and photodegradation in visible light	The efficiency of Aditive mass of MIL 125 (Ti diuring synthesis process, was denoted has MLS-5 in the removal of low concentrations of tetracycline is 99 and 70% at high concentrations. The amount of photodegradation with MLS-1 was 63.3%	[6]
MIL-101(HCl) MIL-101(HF)	Oxytetracycline (OTC)	Absorption of drug in aqueous media	The efficiency of absorption in MIL-101 (HCl) was 115.34 mg/g, which was higher than other materials like ZIF-8	[7]
MIL-100(Fe) MIL-101(Fe) MIL-53(Fe)	Tetracycline (TC)	Absorption of drug on aqueous media and its photodegradation	Initial concentration of 50 mg/L. The recovery was 96.6% in MIL-101(Fe), and 87% was photodegraded under visible light.	[8]
CuCo/MIL-101	Tetracycline (TC)	Absorption of drug on aqueous media	Recovery of 82.9% in a solution of 100 mg/L	[9]
MIL-125(Ti)	Tetracycline hydrochloride (TCH)	0.03 g of MIL-125(Ti) was added to a 50 mL TCH solution (35–280 mg/L), then was shaken at 250 rpm. At certain intervals, the concentration of TCH was determined by UV–visible spectrophotometer at 355 nm.	321.5 mg/g at 30°C, and high adsorption capacity in the pH range of 2–11).	[10]
Bi_2WO_6/NH_2-MIL-88B(Fe) heterostructure	Tetracycline (TC)	Absorption and degradation in aqueous media	Degradation efficiency was 89.4% in 1.5 h with solar illumination. The pseudo-second-order model filled well with the data obtained, and the isotherm followed the Freundlich model (R_2 = 0.99803). Also showed a longer lifetime (140.72 ns) compared to MOF (132.05 ns) and Bi_2WO_6 (136.39 ns). The photodegradation was carried over simulated hospital wastewater, and 60.5% TC was obtained under solar light in 6.5 h	[11]

MOF	Antibiotic	Experiment	Results	Ref
MIL-125(Ti) MIL-53(Fe)/Carbon nanotube (CNT) @Alg composite microbeads.	Tetracycline (TC)	Absorption of tetracycline in aqueous media	Adsorption of TC augmented when concentration of CNT up to 15 wt%. The material showed a maximum adsorption capacity of 294.12 mg/g at 25°C and pH 6.	[12]
AgI/D-MIL-53(Fe)	Tetracycline (TC)	Absorption and degradation of tetracycline	Compared to the degradation efficiency of TC, the MIL-53(Fe) showed a 50.8%, compared with AgI/MIL-53(Fe) increased to 85.1% and which further increased to 96.2% by AgI/D-MIL-53(Fe)	[13]
MIL-101(Cr).	Sulfadiazine, Sulfamethazine, Sulfachloropyridazine, and Sulfamethoxazole.	Absorption of drugs from bottled water, tap water, river water, and wastewater.	Using 40 mg of MOF, the results showed were: *Sulfadiazine:* • Bottled: not detected. • Tap: 1.1 μg/L. • River: 2.2 μg/L. • Residuals: 6.7 μg/L. *Sulfamethazine:* • Bottled: not detected. • Tap: 0.3 μg/L. • River: 0.7 μg/L. • Residuals: 1.5 μg/L. *Sulfachloropyridazine:* • Bottled: not detected. • Tap: 0.8 μg/L. • River: 2.0 μg/L. • Residuals: 1.7 μg/L. *Sulfamethoxazole:* • Bottled: 0.7 μg/L. • Tap: 7.8 μg/L. • River: 27.5 μg/L. • Residuals: 28.2 μg/L.	[14]
MIL-53	Trimethoprim (TMP) Sulfamethoxazole (SMX).	Absorption of sulfonamides in aqueous media	From a solution of 100 mg/L of sulfonamides, the absorption capacity was >87%.	[15, 16]

MOF	Antibiotic	Experiment	Results	Ref
NH_2-MIL-68@COF	Sulfadimethoxine (SDM) Sulfamethoxazole (STZ) Sulfamonomethoxine (SMX) Sulfamethazine (SMZ) Sulfamerazine (SMR) Sulfadiazine (SDZ)	Absorption of drugs from water, milk, and meat	Recovery of compounds was in a range from 68.9 to 103.8% in samples of 1 mg/mL	[17]
MIL-101(Cr)	Sulfadimethoxine (SDM), Sulfamonomethoxin (SMM) Sulfachloropyridazine (SCP).	Absorption of SDM, SMM, and SCP in aqueous media	Adsorption capacities >48.46 mg/g, for all compounds.	[18]
MIL-101(Fe)	Sulfadiazine (SDZ), Sulfapyridine (SPD), Sulfameter (SME), Sulfamethoxazole (SMX), Sulfisoxazole (SIX), Sulfabenzamide (SB) Sulfadimethoxine (SDM)	Absorption of SDZ, SPD, SME, SMX, SIX, SB, SDM in aqueous media	Absorption capacity >75.4% for all compounds.	[19]
MIL-101(Cr) Modified with Urea.	Dimetridazole (DMZ) metronidazole (MNZ)	The solutions were prepared from a stock solution of 20 mg/L water/methanol (90/10 v/v). The pH of each solution was 6.30; the mixture was shaken using an incubator shaker at a constant speed (250 rpm) 25 C, and the concentration was determined using UV-spectrometry, as mg/g MOF	Q_o = 185 mg/g MOF Q_o = 188 mg/g MOF Reused four times.	[20]
Fe_3O_4@SiO_2-MIL-53	Amoxicillin (AMX)	They performed sonophotodegradation of AMX including pH, sonophotocatalyst mass, radiation, and AMX levels as parameters.	The 5 pH was chosen because of a high efficiency of 94% of degradation with an AMX quantity of 50 mg/L. Increasing the addition of sonophotocatalyst amount enhanced the removal efficiency from 71–100%. The UV power of 36 W was the best condition to increase the degradation up to 100% in 25 min. The constant rate was 0.204 min^{-1}. The main oxidant agents were OH. and H^+	[21]

MOF	Antibiotic	Experiment	Results	Ref
Mil-125 (Ti)/PANAM G1.0	Cefixime	They performed purification of water with cefixime antibiotic (100 ppm).	The performance was 80.04%, and the principal interactions were amine and hydroxyl groups in the Mil-Den structure.	[22]
MIL-53(Al)	Amoxicillin (AMX)	They determined adsorption in standard solutions (0.5–30 mg/L) obtained from diluting the stock solution at pH 7.5. The ionic strength effect was studied separately by adding 1 mL of 0.01 mol L^{-1}g/l NaCl, 1 mL of 0.1 mol L^{-1} NaCl, 1 mL of 1 mol L^{-1} NaCl, and 1 mL of deionized water. The pH effect was achieved by adding 2 mg of MIL-53(Al) into flasks containing 20 mL of 50 mg/L AMX solution within a pH range of 3.5–11.5.	The adsorption kinetic was a pseudo-second-order model. Adsorption capacity in a solution of 300 mg was 283.9 ± 1.4 mg/g Hydrogen bonding and π–π interaction/stacking were the main interactions. There no influence in adsorption by ionic strength.	[23]
Fe_3O_4@SiO_2@MIL-53-NH_2 (Al) nanocomposite (FeSi@MN NC)	Ampicillin (AMP)	Photocatalyst for photodegradation of AMP solution with 100 mg/L, absorbance was at 320 nm, and they used a mixture of 0.1–0.8 g/L	Fe_3O_4@SiO_2@MIL-53-NH_2 degradation efficiency was about 70% after 120 min.	[24]
MIL-101 (Cr)	Ciprofloxacin(CIP)	The effects of the amount of adsorbent, initial CIP concentration, temperature, and pH of antibiotic solutions were determined; the amount of adsorbent was 1–15 mg. The time interval is 1–120 min. The pH range was 1–12. CIP of 5–40 mg/L Concentration. The adsorption experiments were carried out in batches. The shaking speed is constant, 120 rpm. After the adsorption, the CIP solutions were analyzed with UV–Vis.	The maximum CIP adsorption capacity at 298 K was 63.28 mg/g. The increasing temperature caused a decrease	[25]

MOF	Antibiotic	Experiment	Results	Ref
Metal organic framework sorbents (MIL-100(Fe), MOF-235(Fe)), Fe_3O_4 nanoparticles, and metal organic framework loaded on iron oxide nanoparticles (Fe_3O_4@MIL-100(Fe) and Fe_3O_4@MOF-235(Fe))	Ciprofloxacin (CIP)	The sorption equilibrium studies were conducted with 50 mL of a CIP solution (250 mg/L) and 0.5 g/L of the sorbent at 298 K. The sorbent and sorbate were mixed on a shaking bath at 250 rpm until equilibrium was attained in 5 h. After, the mixture was centrifuged at 3000 rpm for 3 min, and the amount of CIP Remaining was measured at absorbance at 275 nm.	The removal of CIP was 278.39 mg/L	[26]
MIL-101(Cr)–SO_3H	Norfloxacin (NFX) Ofloxacin (OFX) Enoxacin (ENX)	0.010 g of adsorbent was added into 100 mL aqueous solution of adsorbate. The initial pH values of the NFX, OFX, and ENX solutions were adjusted using HCl or NaOH. The suspension was filtered after being shaken at 30°C for a specific time, and the concentration of adsorbate in the supernatant was measured using a UV–Vis spectrophotometer	The adsorption efficiency of the absorbent was studied using adsorption kinetic tests in a time range of 0–1440 min. The final removal efficiency can reach more than 99%.	[27]
Biochar functionalized MIL–53(Fe) MIL–BDPR	Ofloxacin (OFX) Ciprofloxacin (CPX)	Adsorption process was performed through batch adsorption experiments with an adsorbent dose of 50 mg with 100 mL of single antibiotic solutions with different concentration 50–150 mg/L, contact time 0–120 min, pH 3–9, and temperature 25–45°C. Samples were analyzed using a UV–vis spectrometer.	The adsorption process is pH, initial antibiotic concentration, and temperature dependent. Regeneration efficiency can still reach up to 92.74 and 93.64% for OFX and CPX.	[28]

MOF	Antibiotic	Experiment	Results	Ref
NH_2-MIL-125@TpPa-SO_3H. $TpPa$-SO_3H, NH_2-MIL-125, and NH_2-MIL-125@TpPa-SO_3H	Enoxacin, Norfloxacin Ciprofloxacin Sparfloxacin.	Different concentrations (20–200 mg/L) were dissolved in ultrapure water containing 10% methanol. Then, HCl and NaOH were used to adjust the pH of the solution. In the experiments, 5 mg of NH_2-MIL-125@TpPa-SO_3H was added to 40 mL. The mixtures were shaken at 150 rpm at temperatures (30, 40, and 50°C) for 7 h. HPLC filtered and analyzed the supernatant.	Enoxacin 444.7 mg/g Norfloxacin 455.7 mg/g Ciprofloxacin 457.5 mg/g Sparfloxacin 429.1 mg/g. Reused eight cycles.	[29]

Table 1.
MOFs type MIL used as water removal agent.

MOF	Antibiotic	Experiment	Results	Ref
ZIF-8	Oxitetracycline (OTC) Tetracycline (TC) Chlortetracycline (CTC)	Detection in water and milk by a chromatographic method using ZIF-8 as solid phase. Using a flow rate of 0.3 mL min^{-1} for 3 minutes and a mobile phase of 10% MeOH-20% ACN-70% 0.02 mol L^{-1} oxalic acid solution	Recovery from 50 μgL^{-1} OTC: 80–90% TC: 80–90% CTC: 90–100%	[5]
UiO-66/ZIF-8/PDA@CA	Tetracycline (TC)	Solutions of TC at a pH from 3 to 12, stirred for 60 min at 150 rpm, with 0.01 o 0.08 mg of UiO-66/ZIF-8/PDA@CA was added to TC solution at range from 50 to 300 mg/L. Temperature range from 25 to 55°C, and the unabsorbed TC concentration was evaluated using spectrophotometry at 354 nm.	58.54 mg/g 119.03 mg/g 212.5 mg/g 286.57 mg/g Reusable eight times	[30]
CUST-563 Co@C-600	Oxytetracycline (OTC)	In a 25 mL flask, MOF was added by stirring 20 mg/L of OTC in MeOH for 20 min under dark conditions, and the sample was taken every minute.	The degradation rate of OTC 89%. This system could be reused 10 times, and the activity was above of 80%.	[31]
2-(4-carboxyphenyl)-1H-benzo[d] imidazole-5-carboxylic acid Zn-(CBC)-	Oxytetracycline (OTC)	Zn-CBC MACPs were added to the 10 mL water with OTC (0, 0.6, 1.2, 2.4, 3, 5, 10 mg/L) and stirred at 1000 rpm at 25°C. The concentration of OTC was determined on UV–vis.	5.8 mg/g at pH = 8.0	[32]
NiFe-MOF@AHC (NiFeopAHC)	Tetracycline (TC)	Absorption of Tetracycline in aqueous media	The maximum adsorption removal (Qmax) of NiFeopAHC for TC (568.1 mg/g) was 2.7 times higher than that of AHC.	[15]
$MgFe_2O_4$@UiO-66(Zr) (MFeO@UiO)	Tetracycline (TC)	Absorption of tetracycline in aqueous media	The composite material had a removal efficiency of ca. ~94% for TC within 45-min and 120-min under visible light irradiation, which is higher compared with pristine ferrite and UiO-66(Zr).	[33]

MOF	Antibiotic	Experiment	Results	Ref
3NiO/g-C_3N_4	Tetracycline (TC)	Photodegradation under visible light	100% of TC can be degraded in 60 min by 3NiO/g-C3N4 activating Peroxyd MonoSulfate (PMS) activation under visible light, being 6.64-time that of single g-C3N4	[34]
ZnS@In_2S_3 rhombic dodecahedron (ZnS@In_2S_3 RD 4)	Tetracycline (TC)	Photodegradation of Tetracycline	100 mL, 20 mg/L TC-HCl solution can be completely degraded within 20 min, and the apparent reaction rate constant reaches 0.284 min, which far exceeds the majority of the previously reported sulfide photocatalyst	[35]
Eu/Zr- MOF	Tetracycline (TC)	Absorption of tetracycline in aqueous media	Eu/Zr-MOF exhibits high stability and excellent TC adsorption capacity as high as 289 mg g^{-1} in aqueous solution.	[36]
Fein/C-700	Chlortetracycline (CTC)	Absorption of chlortetracycline in aqueous media	Adsorption capacities in a mixed antibiotic system 255.1, 143.3, and 169.6 mg/g for CTC, ciprofloxacin, and sulfamethoxazole, in water 372.9, 383.3, 390.6, and 366.7 mg/g in a different kind of water, and 366.7, 354.2, 360.4, 314.6, and 301.0 mg/g in five runs.	[37]
$Fe_3O(BDC)_3$ (Fe-MOF)	Sulfadiazine (SDZ) and sulfamethazine (SMZ)	Absorption of sulfadiazine and sulfamethazine in aqueous media	The adsorption of SDZ and SMZ sulfonamides on material nanocomposite was a thermodynamically spontaneous process. With high maximum adsorption capacities (223.0 mg/g for SDZ and 135.1 mg/g for SMZ),	[38]
Cu-based MOFs derivative (CuxO@C composite)	Sulfamerazina (SMR)	Absorption of sulfamerazine in aqueous media	The absorption efficiency of SMR was 100% in the CuxO@C/air system at pH of 4.0, air flow rate of 100 mL·min $^{-1}$ in CuxO@C dosage of 1 g ·L^{-1} in a time of 30 min.	[39]

MOF	Antibiotic	Experiment	Results	Ref
Magnetic Ti^{3+}-TiO_2/Ar-Fe_2O_3 type II heterojunction	Norfloxacin (NOR)	Degradation under visible light	Ti^{3+}-TiO_2/Ar-Fe_2O_3 degrades 97.80% of NOR in 5 min in a solution of 30 mg/L	[40]
p-Ag_3PO_4/n-ZnO/C heterojunction with IRMOF-3	Norfloxacin (NOR)	Photodegradation of norfloxacin	The degradation reached 89.48% under the conditions tested, which is three times faster than pure Ag_3PO_4, and also the material has better catalytic capacity than Ag_3PO_4.	[41]
α-Fe_2O_3/ZIF-67	Ciprofloxacin (CIP)	Degradation of ciprofloxacin	The CIP was completely degraded in 30 min when placing 1 mg of catalyst in a CIP dissolution of 10 mg/L	[42]
Chitosan coated Fe_3O_4@Cd-MOF microspheres (Fe_3O_4@Cd-MOF@CS)	Amoxicillin (AMX)	Absorption of amoxicillin in aqueous media	Under the best conditions, 75% of AMX was removed from a dissolution of AMX with a concentration of 50 ppm using only 50 mg of material. The maximum adsorption capacity was ~103.09 mg/g.	[43]
Zirconium–porphyrin-based metal–organic framework (MOF), PCN-224	Sulfamerazine (SMR)	Absorption of sulfamerazine in aqueous media	Considering the maximum concentrations of antibiotics in water, PCN-224 exhibited the highest adsorption capacity (Q0) for NDX and OFL compared to other adsorbents and MOFs, 671 mg/g for NDX (50 ppm) and1030 mg/g for OFL (125 ppm)	[44]
Fe_3O_4@SiO_2	Sulfadiazine (SDZ) Sulfamerazine (SMR) Sulfadimidine (SDD) Sulfisoxazole (SIZ) Sulfathiazole (STZ).	Absorption of drugs in tap water, river water, and rainwater	Absorption capacity >83.4% in samples of 5 mL of tap water, river water, and rainwater	[45]
HKUST-1	Sulfachloropyridazine (SCP)	Absorption of drug in aqueous media	Adsorption capacity of 384 mg/g at 298 K in A solution of sulfachloropyridazine with a concentration of 500 ppm, adding 100 mg of MOF.	[46]

MOF	Antibiotic	Experiment	Results	Ref
Mn(cam)(bpy).	Sulfamethazine (SDD), Sulfamerazine (SMR) Sulfanitran (SN) Sulfadiazine (SDZ), Sulfadimethoxine (SDM), Sulfamonomethoxine (SMT), Sulfachloropyridazine (SCD), Sulfamethoxazole (SMZ), Sulfamethizole (SMI) Phthalylsulfathiazole (PST).	Absorption of sulfonamides from tap water samples	Absorption capacity >71.53% in a solution of a mixture of sulfonamides with a concentration of 20 mg/L	[47]
PCN-224	Sulfadiazine (SD) Sulfathiazole (ST) Sulfapyridine (SP) Sulfamerazine (SM 1) Sulfamethazine (SM 2) Sulfamethoxazole (SMX).	Absorption of sulfonamides from tap and drinking water.	Absorption capacity >87.3% was obtained in a solution of sulfonamides in samples de tap and drinking water	[48]
Fe_3O_4 @COFs	Sulfadiazine (SDZ) Sulfamerazine (SMR), Sulfamethazine (SMZ), Sulfamonomethoxine (SMX) Sulfamethoxazole (STZ) Sulfadimethoxine (SDM).	Absorption of drugs from Jingsi Lake at the campus of Wuhan Institute of Technology	Absorption capacity in a range from 63.5 to 107.3%	[49]
SCAU-1 y SNW-1	Sulfadiazine (SDZ) Sulfadimidine (SMZ) Sulfamethoxydiazine (SME) Sulfadoxine (SDO) Sulfachloropyridazine (SCP).	Absorption of SDZ, SMZ, SME, SDO, and SCP in aqueous media (drinking water, river water, lake water, and wastewater).	Absorption capacity >5.8 μg/mL.	[50]
[H]@ UiO-66-Br	Sulfadiazine (SD) Sulfamerazine (SM) Sulfamethazine (SMT) Sulfamethoxypyridazine (SMPD) Sulfamethoxazole (SMX), Sulfamonomethoxine (SMM) Sulfamethoxypyrazine (SMPZ).	Absorption of SD, Sm, SMT, and SMPD in aqueous media.	Absorption capacity >94%.	[51]

DOI: http://dx.doi.org/10.5772/intechopen.1005000

MOF	Antibiotic	Experiment	Results	Ref
ZnO-Co_3O_4	Sulfadiazine	Absorption of sulfadiazine.	Absorption capacity >96%.	[52]
Fe-MOF	Sulfadiazine (SDZ) Sulfamethazine (SMZ).	Absorption of SDZ and SMZ.	Absorption capacity >135.1 mg/g.	[38]
UiO-66 UiO-66-BC	Sulfamethoxazole (SMX).	Absorption of SMX in aqueous media.	Absorption capacity >106.93 mg/g at 298 K.	[53]
CoZn-ZIF CoZn-ZIF-melamine	Sulfamethoxazole (SMX).	Absorption of SMX in aqueous media.	Degradation capacity >97.09%.	[54]
VNU-1	Sulfamethoxazole	Absorption of sulfamethoxazole in aqueous media.	Adsorption of 74.8% sulfamethoxazole after stirring for 1 h.	[55]
HP-NU-902-X	Sulfadiazine (SDZ) Sulfapyridine (SPY) Sulfamethoxydiazine (SMD) Sulfachloropyridazine (SCP) Sulfabenzamide (SBZ) Sulfamethazine (SMT).	Absorption of SDZ, SPY, SMD, SCP, SBZ, and SMT in aqueous media.	Absorption capacity >70.34%.	[56]
AlFum Bio-MOF.	Sulfadiazine.	Absorption of sulfadiazine in aqueous media in an ultrasonic bath.	Average recoveries >84%.	[57]
BUT-12 BUT-13	Nitrofurazone (NZF) Nitrofurantoin (NFT)	The adsorption isotherms were obtained by mixing 15 mg MOFs with 50 mL NZF or NFT solution of different concentrations (100 to 1600 mg/L) at 298 K with stirring for 4 h.	NZF and NFT were removed in 70–75%, respectively, and this system is reused.	[58]
Al-TCPP	Ornidazole (ONZ) Metronidazole (MNZ)	5 mg Al-TCPP was added to 10 ml solution containing ONZ or MNZ. The adsorption experiments were carried out at 30 C at 150 rpm on an orbital shaker, and then the adsorbed mixture was filtered using a 0.22 mm membrane. The concentration was determined at 320 nm and 317 nm by UV–Vis spectroscopy	Q_o = 320.5 mg/g MOF Q_o = 609.8 mg/g MOF	[59]
AgN/MOF-5 (1:3)	Metronidazole (MNZ)	Solution 100 mL of MNZ at different concentrations (10, 30, 50, 80, and 100 mg/L) and 50 mg of AgN/MOF-5	298 K Q_0 = 101.9 mg/g 308 K Q_0 = 154.5 mg/g	[60]

MOF	Antibiotic	Experiment	Results	Ref
		were shaken at 250 rpm for 1 h, at pH of 6.60 tight with HCl 0.1 M. Then, the solution was filtered. Concentration of MNZ was determined at pH different (4, 5, and 6) and temperatures of 298, 308, and 318 K.	318 K Q_0 = 191.1 mg/g The system was reusable three times.	
MOF-1	Beta-lactam	An aggregate called amoxicillin@MOF1 was obtained after immersing crystals in saturated Dimethylformamide (DMF) solutions of amoxicillin for 72 h. With interactions of amoxicillin, except nitrogen and oxygen. Purification by selective degradation of four-member ring. The uptake efficiency of MOF 1 for amoxicillin, ceftriaxone and clindamycin was evaluated by HPLC-UV analyses in water, with a great selectivity between other ligands.	They showed the appearance of 1 new peak increased as the relative area to amoxicillin original peak after 4 hours. Ceftriaxone also shows a degradation product by the hydrolysis at the sulfur atom of the carbamimidothioate group after 12 h. The conversion was more than 80% after 24 h.	[61]
3 MOF Zn II based with $(Zn_6(IDC)_4(OH)_2(Hprz)_2)n$ (imidazole-4,5-dicarboxylate)	Amoxicillin (AMX)	They measured the adsorption capacity of three antibiotics, amoxicillin, ampicillin, and cloxacillin, dissolved in desionized water, 100 mL of the solution with a concentration of 60 ppm in a reactor.	MOF-VII exhibits a removal of 92% of AMX within 240 min, 89% CLX and 88% AMP. This great adsorption capability is attributed to the smaller size of MOF-VII particles and morphology obtained by this method of synthesis (Sonochemical). Also surface area increased the adsorption capability between MOF-I and MOF-VII, which exhibit a great absorption and more surface area. A comparison between crystals obtained by solvothermal synthesis of MOF showed a decrease in concentration of just 38.5%, which was inefficient for purification.	[62]

MOF	Antibiotic	Experiment	Results	Ref
Zn-MOF (Zinc nitrate and 1,4 -benzendicarboxylic acid) (MC-0.3)	Ceftazidime	Adsorption experiment in a 150 mL solution of ceftazidime in Erlenmeyer, they compared the effect of pH, temperature, and amount of adsorbent on the adsorption process.	The adsorption has pseudo-first-order and a capacity of 371.18 mg/g. At pH 4 the capacity increases up to 404.28 mg/g due to the presence of carboxyl groups and better adsorption. NaCl increases the adsorption.	[63]
Silver organic framework loaded with sulfanilamide to remove AMX (Ag-MOF-NH_2)	Amoxicillin (AMX)	They adsorbed AMX by Ag-MOF-NH_2 in the prepared analytical water. They also applied the MOF in real wastewater samples collected from Industrial Wastewater Treatment Plant in Port Said, Egypt and spiked with 1 mg/L of a solution standard of AMX.	The adsorption capacity was 2.4 mmol/g. The interactions were π–π stacking or aromatic-aromatic interconnections among the benzene rings of AMX and Ag-MOF-NH_2, with a pH of 7.5, which was the best. Ethanol was the best option to restore the adsorbent by elimination of the adsorbate. In real wastewater they achieved a removal rate of 85.4–98.19%.	[64]
β-lactamase@ZIF-8	Amoxicillin (AMX)	They studied the catalytic activity of β-lactamase@ZIF-8 against AMP in different concentrations 4, 20, 40, 100 μL of 25 μg/mL.	At 10 min AMP was degraded by about 45% and reached 94% in 60 min. In contrast, only β-lactamase AMP was degraded 30% in 10 min.	[65]
Fe_3O_4@Cd-MOF Fe_3O_4@Cd a MOF@chitosan (CS) prepared via coating of Fe_3O_4@Cd-MOF with CS biopolymer in the microsphere shape.	Amoxicillin (AMX)	They studied the adsorption of AMX for an aqueous solution. 50 mg of microsphere Fe_3O_4@Cd-MOF@CS. The microspheres were immersed in 100 mL of AMX 50 ppm solution.	The adsorption of AMX was increased after 50 min, reaching 75.40 mg/g, and the equilibrium of adsorption was reached after 240 min. The percentage of the AMX removal increased from 22.07% for pH 3 to 73.92% for pH 8. Main interactions were hydroxyl, amine and aromatic rings, which can interact with π-π stacking.	[43]
ZnMOF@MGO Zn metal organic framework doped magnetic graphene oxide composite	Cefixime	They determined the extraction of cefixime with ZnMOF@MGO. 15 mg of ZnMOF@MGO.	The absorption of cefixime was up to 95.1%. The best results were in phosphate buffer (15 mM and pH 6), 15 mg of ZnMOF@MGO.	[66]

MOF	Antibiotic	Experiment	Results	Ref
Zirconium–porphyrin-based metal–organic framework (MOF), PCN-224	Nalidixic acid (NDX) Ofloxacin	NDX with concentrations of 20, 25, 30, 40, and 50 pp. was dissolved in 1 L of water. Then, the pH was adjusted to 7. 25 mg PCN-224 was put into each solution. The samples were stirred at different times (0, 10, 20, 30, 60, 120, 240, 360, 540, and 720 min) at 450 rpm at 25°C. After each adsorption time, it was extracted and filtered. The concentration of antibiotic was analyzed in 200–500 nm.	NDX = 671 mg/g OFL = 1030 mg/g	[43]
Cadmium-based MOFs (Fe_3O_4@Cd-MOF) modified with chitosan	Amoxicillin (AMX)	The adsorption was studied in a conical glass vessel. The pH of the adsorption solutions was adjusted by using NaOH or HCl solution. Fe_3O_4@Cd-MOF@CS 50 mg was immersed in 100 mL of AMX with 50 ppm. The vessel was sealed and shaken at 150 rpm at ambient temperature. He remained AMX in the solution, which was measured using UV–Vis spectroscopy.	The maximum adsorption capacity was found ~103.09 mg/g. Recycling capability was five cycles.	[43]
Carbon nano sheets (CNS)@ZIF67	Ciprofloxacin (CIP)	A ciprofloxacin concentration of 0.1–100 ppm at pH = 7. The composites are used here as an adsorbent (40 mg) for an aqueous solution of antibiotics at room temperature (303 K). The mixture was agitated for 5 to 120 minute at 120–130 rpm in a room temperature. The analysis of aliquot solution is done by UV–Vis spectrometer. The influence of pH on antibiotic adsorption was studied by pH values in the acidic and basic regions.	The pristine MOF and CNS do not show any impressive result.	[67]

MOF	Antibiotic	Experiment	Results	Ref
UiO-66 and UiO-66-NH_2	Norfloxacin Ciprofloxacin oxytetracycline hydrochloride	The adsorption capability was tested under dark in a water bath shaker at 180 rpm under room temperature. The adsorbents were added to 20 mL of 20 ppm norfloxacin solution for 2 h. After the adsorption time was over, they were separated using centrifugation and filtration; the optical absorbance was measured by UV–vis spectrometer.	The amine-functionalized MOF shows the best adsorption capacity, 93%, followed by UiO-66 at 88%, respectively. It has been found that UiO-66-NH_2 adsorbs 92% of OxyTCH and 89% of CIP.	[68]
Aluminum-based metal organic framework (MOF)	Tetracycline (TC) Ciprofloxacin (CIP)	Stock solutions of TC and CIP were dissolved in EtOH and ultrapure water, respectively. The batch adsorption studies were conducted by immersing an adsorbent (10 mg) in a 50 mL glass vial containing 30 mL of TC or CIP solution with an initial concentration of 10 mg/L, and the solution was shaken at a speed of 150 rpm. After adsorption, the adsorbent materials were separated by filtration. The residual concentrations of TC and CIP were measured. Using HPLC.	Hydrogel beads, Alg@MOF-rGO, showed maximum adsorption capacities of 43.76 and 40.76 mg/g for TC and CIP. The maximum adsorption capacity was at neutral pH.	[69]
HKUST-1-derived carbon (HDC-1100)	Ciprofloxacin (CIP)	In the batch adsorption experiments, 10 mg of HDC-1100 was added to a triangular beaker with 20 mL of a CIP solution with CIP concentrations (10–500 mg/L). The samples were shaken at 150 rpm for 48 h. The concentration of the CIP HPLC.	Within 48 h, HDC-1100 reach adsorption for 30 mg/L CIP. The maximum adsorption capacity was as high as 592 mg/g	[70]

MOF	Antibiotic	Experiment	Results	Ref
$P_2W_{18}O_{62}$/Fe-MIL-101/NiF_2O_4	Tetracycline (TC) Ciprofloxacin (CIP)	Batch adsorption was conducted by adding 40 mg of adsorbent for each 50 mg/L and agitating vigorously. After filtration, the supernatant was assessed using UV–Vis at λ = 275–328 and 273–360 nm.	TC had an adsorption efficiency of approximately 90% within 20 min, whereas CIP had a higher adsorption efficiency of around 100% within 10 min. The structure t unaffected by the presence of solvents, drugs, or changes in pH.	[71]
MOF/hydrogel. Three-dimensional zeolitic imidazolate framework-8 (ZIF-8)/polysaccharide (sodium alginate-kappa-carrageenan, SC) hydrogel beads	Ciprofloxacin (CIP)	A CIP solution of 200 mg/L and 5 mg of the adsorbents into a 30 mL, and shaken at 160 rpm.	The adsorption capacity value in the one-step method was 951.61 ± 33.74 mg/g, reused four times.	[72]

Table 2.
MOFs beyond carboxylates used in antibiotic removal and degradation.

2.2 Different variants of metal organic framework (MOFs)

Many other MOFs have been prepared and synthetized for several applications. However, MOFs based on ligands with catalytic structures like porphyrins, oxometalates, quantum dots, and imidazolates cannot only remove antibiotics from water but also degrade the drug and eliminate it more easily. Many materials based on composites with catalytic entities have been used for bioremediation of pollution in water. The next examples are MOFs based on other ligands beyond carboxylates that had allowed the removal of antibiotics and degraded them with good results (**Table 2**).

3. Discussion

Nowadays, pollution is becoming one of the great challenges facing humanity. This deterioration of the environment comes from many human activities, one of which has begun to arouse interest is pollution produced by medicines. In recent years, efforts have begun to carry out studies that demonstrate the presence of different medications in the environment; one type of drug that is being detected is antibiotics.

This situation contributes, on the one hand, to water contamination, which generates toxicity problems for the living beings of the ecosystem that is contaminated, but it is also helping the phenomenon of resistance; for these two reasons, the investigation of possible solutions to this problem, a line of work on the subject is the use of MOFs to eliminate the presence of antibiotics in water. MOFs offer several characteristics that make them suitable for this purpose: they have a large surface area, adjustable pore sizes when prepared, the nature of secondary building units (SBUs) is very varied, in the last two decades, the work around the preparation of these reticular materials has been very active, which is reflected in the little more than 84 thousand structures registered in the Cambridge Crystallographic Data Centre.

In the period from 2013 to 2023, a large number of works have been published with the use of different MOFs to eliminate some type of antibiotic; however, the most used type of MOF is the MIL type, and the vast majority of works use a model of water prepared with an antibiotic, they have also determined the mechanism by which the antibiotic is eliminated from the water, in general, it is by adsorption based on electrostatic interactions between the MOF and the antibiotic, the efficiency achieved in these cases varies, but it ranges from 80% to the total elimination of the antibiotic in the water.

They have carried out degradation tests of the antibiotic using photochemical methods that have given good results with the total disappearance of the antibiotic.

Although there is still a lot to advance on the subject, these years have shown very interesting advances on the subject, which opens the possibility of reaching a solution to this problem.

4. Conclusions

Metal organic frameworks have a wide variety of applications due to their features, which allow materials to be modified and synthesized for different purposes. This has been demonstrated by several reports that put MOFs as emergent materials with some great absorption capacities. Therefore, using MOFs to remove and/or degrade molecules like antibiotics might be a real solution to the problem of the presence of antibiotics in water. The research and synthesis of new MOFs for solving environmental problems have to be an investigation line that must be attendant, and also composites materials with some matrix like zeolites, silices, and polyoxometalates can be conjugated with MOFs for the obtain of materials with enhanced absorption and selectivity.

In this review, a wide variety of MOFs are reported, which were used to eliminate antibiotics from samples of solutions with the antibiotics in different concentrations and achieve the elimination of quantities greater than 80% of the test antibiotics, and reusable, a situation that opens the possibility of more studies and converts these materials into a solution to the problem of antibiotic contamination in water.

Conflict of interest

The authors declare no conflict of interest.

Author details

Miguel Martell Mendoza[1], Carlos Alberto Méndez Cuesta[1], Miguel Angel Zavala Sánchez[1], Erick Cuauhtemoc Pérez Montiel[1], Alfonso Mata Bermúdez[2] and Cuauhtemoc Pérez González[1*]

1 Department of Biological Systems, Metropolitan Autonomous University-Xochimilco, CDMX, México

2 Department of Health Care, Metropolitan Autonomous University-Xochimilco, CDMX, México

*Address all correspondence to: cperezg@correo.xoc.uam.mx

References

[1] Supuran CT. Special issue: Sulfonamides. Journal of Synthetic Chemistry and Natatural Product Chemistry. 2017;**22**:1-5. DOI: 10.3390/MOLECULES22101642

[2] Zhu M, Pang X, Wan J, Xu X, Wei X, Hua R, et al. Potential toxic effects of sulfonamides antibiotics: Molecular modeling, multiple-spectroscopy techniques and density functional theory calculations. Ecotoxicology and Environmental Safety. 2022;**243**:113979. DOI: 10.1016/J.ECOENV.2022.113979

[3] Azevedo-Barbosa H, Dias DF, Franco LL, Hawkes JA, Carvalho DT. From antibacterial to antitumour agents: A brief review on the chemical and medicinal aspects of sulfonamides. Mini Reviews in Medicinal Chemistry. 2020;**20**:2052-2066. DOI: 10.2174/1389557520666200905125738

[4] Sharma I, Kaur J, Poonia G, Mehta SK, Kataria R. Nanoscale designing of metal organic framework moieties as efficient tools for environmental decontamination. Nanoscale Advances. 2023;**5**:3782. DOI: 10.1039/D3NA00169E

[5] Yang XQ, Yang CX, Yan XP. Zeolite imidazolate framework-8 as sorbent for on-line solid-phase extraction coupled with high-performance liquid chromatography for the determination of tetracyclines in water and milk samples. Journal of Chromatography. A. 2013;**1304**:28-33. DOI: 10.1016/J.CHROMA.2013.06.064

[6] Wang H, Yuan X, Wu Y, Zeng G, Dong H, Chen X, et al. In situ synthesis of In2S3@MIL-125(Ti) core–shell microparticle for the removal of tetracycline from wastewater by integrated adsorption and visible-light-driven photocatalysis. Applied Catalysis B: Environmental. 2016;**186**:19-29. DOI: 10.1016/J.APCATB.2015.12.041

[7] Hu T, Lv H, Shan S, Jia Q, Su H, Tian N, et al. Porous structured MIL-101 synthesized with different mineralizers for adsorptive removal of oxytetracycline from aqueous solution. RSC Advances. 2016;**6**:73741-73747. DOI: 10.1039/C6RA11684A

[8] Wang D, Jia F, Wang H, Chen F, Fang Y, Dong W, et al. Simultaneously efficient adsorption and photocatalytic degradation of tetracycline by Fe-based MOFs. Journal of Colloid and Interface Science. 2018;**519**:273-284. DOI: 10.1016/J.JCIS.2018.02.067

[9] Jin J, Yang Z, Xiong W, Zhou Y, Xu R, Zhang Y, et al. Cu and Co nanoparticles co-doped MIL-101 as a novel adsorbent for efficient removal of tetracycline from aqueous solutions. Science of the Total Environment. 2019;**650**:408-418. DOI: 10.1016/J.SCITOTENV.2018.08.434

[10] Nie L, Yang Y, Fang C, Chen H, Xin S. Co-doped metal–organic framework MIL-125(Ti) derivate for efficient adsorption of tetracycline hydrochloride from water. Applied Surface Science. 2023;**640**:158390. DOI: 10.1016/J.APSUSC.2023.158390

[11] Kaur M, Mehta SK, Devi P, Kansal SK. Bi2WO6/NH2-MIL-88B(Fe) heterostructure: An efficient sunlight driven photocatalyst for the degradation of antibiotic tetracycline in aqueous medium. Advanced Powder Technology. 2021;**32**:4788-4804. DOI: 10.1016/J.APT.2021.10.025

[12] Omer AM, Abd El-Monaem EM, El-Subruiti GM, Abd El-Latif MM,

Eltaweil AS. Fabrication of easy separable and reusable MIL-125(Ti)/MIL-53(Fe) binary MOF/CNT/alginate composite microbeads for tetracycline removal from water bodies. Scientific Reports. 2021;**11**:1-14. DOI: 10.1038/S41598-021-03428-Z

[13] Jiang D, Zhou Y, Wang P, Huang C, Zhu J, Liu Z. Acid-modulated porous AgI/D-MIL-53(Fe) composites for the removal of antibiotic contaminant. Journal of Materials Science. 2021;**56**: 15782-15797. DOI: 10.1007/S10853-021-06321-2/FIGURES/12

[14] Dai X, Jia X, Zhao P, Wang T, Wang J, Huang P, et al. A combined experimental/computational study on metal-organic framework MIL-101(Cr) as a SPE sorbent for the determination of sulphonamides in environmental water samples coupling with UPLC-MS/MS. Talanta. 2016;**154**:581-588. DOI: 10.1016/J.TALANTA.2016.03.042

[15] Jais FM, Ibrahim S, Chee CY, Ismail Z. Solvothermal growth of the bimetal organic framework (NiFe-MOF) on sugarcane bagasse hydrochar for the removal of dye and antibiotic. Journal of Environmental Chemical Engineering. 2021;**9**:106367. DOI: 10.1016/J.JECE.2021.106367

[16] Van Tran T, Nguyen DTC, Nguyen HTT, Nanda S, Vo DVN, Do ST, et al. Application of Fe-based metal-organic framework and its pyrolysis products for sulfonamide treatment. Environmental Science and Pollution Research International. 2019;**26**: 28106-28126. DOI: 10.1007/S11356-019-06011-2

[17] Chen Z, Yu C, Xi J, Tang S, Bao T, Zhang J. A hybrid material prepared by controlled growth of a covalent organic framework on amino-modified MIL-68 for pipette tip solid-phase extraction of sulfonamides prior to their determination by HPLC. Microchimica Acta. 2019;**186**:1-11. DOI: 10.1007/s00604-019-3513-7

[18] Peng J, Wu E, Wang N, Quan X, Sun M, Hu Q. Removal of sulfonamide antibiotics from water by adsorption and persulfate oxidation process. Journal of Molecular Liquids. 2019;**274**:632-638. DOI: 10.1016/J.MOLLIQ.2018.11.034

[19] Qin P, Han L, Zhang X, Li M, Li D, Lu M, et al. MIL-101(Fe)-derived magnetic porous carbon as sorbent for stir bar sorptive-dispersive microextraction of sulfonamides. Mikrochimica Acta. 2021;**188**:1-11. DOI: 10.1007/S00604-021-04993-W

[20] Seo PW, Khan NA, Jhung SH. Removal of nitroimidazole antibiotics from water by adsorption over metal–organic frameworks modified with urea or melamine. Chemical Engineering Journal. 2017;**315**:92-100. DOI: 10.1016/J.CEJ.2017.01.021

[21] Al-Musawi TJ, Kozlitina IA, Moradi M, Rahimpoor R, Mengelizadeh N, Hjazi A, et al. Enhanced sonophotocatalytic degradation of amoxicillin antibiotics using Fe3O4@SiO2/PAEDTC surrounded by MIL-101(Fe) in aquatic environment under the COVID-19 pandemic. Journal of Photochemistry and Photobiology A: Chemistry. 2024; **446**:115140. DOI: 10.1016/J.JPHOTOCHEM.2023.115140

[22] Khosravi MJ, Hosseini SM, Vatanpour V. Polyamidoamine dendrimers-mil-125(Ti) MOF embedded polyethersulfone membrane for enhanced removal of heavy metal, antibiotic and dye from water. Journal of Environmental Chemical Engineering. 2022;**10**:108644. DOI: 10.1016/J.JECE.2022.108644

[23] Imanipoor J, Mohammadi M, Dinari M, Ehsani MR. Adsorption and desorption of amoxicillin antibiotic from water matrices using an effective and recyclable MIL-53(Al) metal-organic framework adsorbent. Journal of Chemical & Engineering Data. 2021;**66**: 389-403. DOI: 10.1021/ACS. JCED.0C00736/SUPPL_FILE/ JE0C00736_SI_001.PDF

[24] Sohrabnezhad S, Pourahmad A, Karimi MF. Magnetite-metal organic framework core@shell for degradation of ampicillin antibiotic in aqueous solution. Journal of Solid State Chemistry. 2020;**288**:121420. DOI: 10.1016/J.JSSC.2020.121420

[25] Bayazit ŞS, Danalıoğlu ST, Abdel Salam M, Kerkez KÖ. Preparation of magnetic MIL-101 (Cr) for efficient removal of ciprofloxacin. Environmental Science and Pollution Research International. 2017;**24**:25452-25461. DOI: 10.1007/S11356-017-0121-0

[26] Moradi SE, Haji Shabani AM, Dadfarnia S, Emami S. Effective removal of ciprofloxacin from aqueous solutions using magnetic metal–organic framework sorbents: Mechanisms, isotherms and kinetics. Journal of the Iranian Chemical Society. 2016;**13**: 1617-1627. DOI: 10.1007/S13738-016-0878-Y/TABLES/4

[27] Guo X, Kang C, Huang H, Chang Y, Zhong C. Exploration of functional MOFs for efficient removal of fluoroquinolone antibiotics from water. Microporous and Mesoporous Materials. 2019;**286**:84-91. DOI: 10.1016/J. MICROMESO.2019.05.025

[28] Chakhtouna H, Benzeid H, Zari N, el kacem QA, Bouhfid R. Microwave-assisted synthesis of MIL–53(Fe)/biochar composite from date palm for ciprofloxacin and ofloxacin antibiotics removal. Separation and Purification Technology. 2023;**308**:122850. DOI: 10.1016/J.SEPPUR.2022.122850

[29] Lin S, Zhao Z, Lv J, Guan L, Du H, Liang S, et al. Designing a novel metal-organic framework@covalent organic framework composite for the selective removal of fluoroquinolones: Adsorption behaviors and theoretical investigation. Applied Surface Science. 2023;**609**: 155433. DOI: 10.1016/J. APSUSC.2022.155433

[30] Abd El-Monaem EM, Omer AM, Khalifa RE, Eltaweil AS. Floatable cellulose acetate beads embedded with flower-like zwitterionic binary MOF/ PDA for efficient removal of tetracycline. Journal of Colloid and Interface Science. 2022;**620**:333-345. DOI: 10.1016/J.JCIS.2022.04.010

[31] Mao W, Wang D, Wang X, Hu X, Gao F, Su Z. Efficient cobalt-based metal-organic framework derived magnetic Co@C-600 nanoreactor for peroxymonosulfate activation and oxytetracycline degradation. Colloids and Surfaces A: Physicochemical and Engineering Aspects. 2022;**648**:129234. DOI: 10.1016/J.COLSURFA.2022.129234

[32] Lu Y, Yu L, Zhang S, Su P, Li X, Hao X, et al. Dual-functional fluorescent metal-organic framework based beads for visual detection and removal of oxytetracycline in real aqueous solution. Applied Surface Science. 2023;**625**:157202. DOI: 10.1016/J.APSUSC.2023.157202

[33] Vo TK, Kim J. Facile synthesis of magnetic framework composite MgFe2O4@UiO-66(Zr) and its applications in the adsorption-photocatalytic degradation of tetracycline. Environmental Science and Pollution Research International. 2021; **28**:68261-68275. DOI: 10.1007/ S11356-021-15423-Y

[34] Li S, Tang Y, Wang M, Kang J, Jin C, Liu J, et al. NiO/g-C3N4 2D/2D heterojunction catalyst as efficient peroxymonosulfate activators toward tetracycline degradation: Characterization, performance and mechanism. Journal of Alloys and Compounds. 2021;**880**:160547. DOI: 10.1016/J.JALLCOM.2021.160547

[35] Liu X, Zhang T, Li Y, Zhang J, Du Y, Yang Y, et al. Construction of core-shell ZnS@In2S3 rhombic dodecahedron Z-scheme heterojunction structure: Enhanced photocatalytic activity and mechanism insight. Chemical Engineering Journal. 2021;**423**:130138. DOI: 10.1016/J.CEJ.2021.130138

[36] Li Y, Wang J, Huang Z, Qian C, Tian Y, Duan Y. An Eu-doped Zr-metal-organic framework for simultaneous detection and removal of antibiotic tetracycline. Journal of Environmental Chemical Engineering. 2021;**9**:106012. DOI: 10.1016/J.JECE.2021.106012

[37] Hu X, Xie Y, He R, Yao L, Ma S, Bai C. Nano-iron wrapped by graphitic carbon in the carbonaceous matrix for efficient removal of chlortetracycline. Separation and Purification Technology. 2021;**279**:119693. DOI: 10.1016/J.SEPPUR.2021.119693

[38] Van Tran T, Nguyen DTC, Nguyen TT, Pham QT, Vo DVN, Nguyen TD, et al. Linearized and nonlinearized modellings for comparative uptake assessment of metal-organic framework-derived nanocomposite towards sulfonamide antibiotics. Environmental Science and Pollution Research International. 2021;**28**:63448-63463. DOI: 10.1007/S11356-020-09312-Z

[39] Chen Y, Zhao J, Hu L, Tian J, Liu Y. Degradation of sulfamerazine by a novel CuxO@C composite derived from Cu-MOFs under air aeration. Chemosphere. 2021;**280**:130678. DOI: 10.1016/J.CHEMOSPHERE.2021.130678

[40] Jia M, Yang Z, Xiong W, Cao J, Xiang Y, Peng H, et al. Magnetic heterojunction of oxygen-deficient Ti3+-TiO2 and Ar-Fe2O3 derived from metal-organic frameworks for efficient peroxydisulfate (PDS) photo-activation. Applied Catalysis B: Environmental. 2021;**298**:120513. DOI: 10.1016/J.APCATB.2021.120513

[41] Hu M, Zhu P, Liu M, Xu J, Duan M, Lin J. Preparation, performance and mechanism of p-Ag3PO4/n-ZnO/C heterojunction with IRMOF-3 as precursor for efficient photodegradation of norfloxacin. Colloids and Surfaces A: Physicochemical and Engineering Aspects. 2021;**628**:127235. DOI: 10.1016/J.COLSURFA.2021.127235

[42] Hashemzadeh B, Alamgholiloo H, Noroozi Pesyan N, Asgari E, Sheikhmohammadi A, Yeganeh J, et al. Degradation of ciprofloxacin using hematite/MOF nanocomposite as a heterogeneous Fenton-like catalyst: A comparison of composite and core—shell structures. Chemosphere. 2021;**281**:130970. DOI: 10.1016/J.CHEMOSPHERE.2021.130970

[43] Pooresmaeil M, Namazi H. Chitosan coated Fe3O4@Cd-MOF microspheres as an effective adsorbent for the removal of the amoxicillin from aqueous solution. International Journal of Biological Macromolecules. 2021;**191**:108-117. DOI: 10.1016/J.IJBIOMAC.2021.09.071

[44] Cho HJ, Kang E, Kim S, Yang DC, Nam J, Jin E, et al. Impact of Zr6 node in a metal-organic framework for adsorptive removal of antibiotics from water. Inorganic Chemistry. 2021;**60**:16966-16976. DOI: 10.1021/ACS.INORGCHEM.1C01890

[45] Wang H, Liu SY, Lv XJ, Ma R, Zhang ZQ. Assembly of a Fe–pamoate porous complex on magnetic microspheres for extraction of sulfonamide antibiotics from environmental water samples. Analytical Methods. 2015;**7**:4939-4946. DOI: 10.1039/C5AY00539F

[46] Azhar MR, Abid HR, Sun H, Periasamy V, Tadé MO, Wang S. Excellent performance of copper based metal organic framework in adsorptive removal of toxic sulfonamide antibiotics from wastewater. Journal of Colloid and Interface Science. 2016;**478**:344-352. DOI: 10.1016/J.JCIS.2016.06.032

[47] Wang X, Ye N, Hu X, Liu Q, Li J, Peng L, et al. Open-tubular capillary electrochromatographic determination of ten sulfonamides in tap water and milk by a metal-organic framework-coated capillary column. Electrophoresis. 2018;**39**:2236-2245. DOI: 10.1002/ELPS.201800047

[48] Deng ZH, Xu GJ, Wang XL, Wang X, Wang ML, Lin JM, et al. A Zr (IV)-based porphyrinic metal-organic framework as a solid-phase sorbent for extraction of sulfonamides prior to their quantitation by LC-MS. Mikrochimica Acta. 2018;**185**:1-8. DOI: 10.1007/S00604-018-2985-1

[49] Zhang J, Chen Z, Tang S, Luo X, Xi J, He Z, et al. Fabrication of porphyrin-based magnetic covalent organic framework for effective extraction and enrichment of sulfonamides. Analytica Chimica Acta. 2019;**1089**:66-77. DOI: 10.1016/J.ACA.2019.08.066

[50] Wang G, Zhou T, Lei Y. Exploration of a novel triazine-based covalent organic framework for solid-phase extraction of antibiotics. RSC Advances. 2020;**10**:11557-11564. DOI: 10.1039/C9RA10846G

[51] Lu D, Liu C, Qin M, Deng J, Shi G, Zhou T. Functionalized ionic liquids-supported metal organic frameworks for dispersive solid phase extraction of sulfonamide antibiotics in water samples. Analytica Chimica Acta. 2020;**1133**:88-98. DOI: 10.1016/J.ACA.2020.07.074

[52] Velmurugan S, Yang TCK, Chen SW, Chen JN. Metal-organic frameworks derived ZnO-Co3O4 pn heterojunction photocatalyst for the photoelectrochemical detection of sulfadiazine. Journal of Environmental Chemical Engineering. 2021;**9**:106169. DOI: 10.1016/J.JECE.2021.106169

[53] Ouyang J, Chen J, Ma S, Xing X, Zhou L, Liu Z, et al. Adsorption removal of sulfamethoxazole from water using UiO-66 and UiO-66-BC composites. Particuology. 2022;**62**:71-78. DOI: 10.1016/J.PARTIC.2021.04.012

[54] Liu H, He Z, Wang S, Zhao S. CoZn-ZIF and melamine co-derived double carbon layer matrix supported highly dispersed and exposed Co nanoparticles for efficient degradation of sulfamethoxazole. Chemical Engineering Journal. 2023;**469**:144054. DOI: 10.1016/J.CEJ.2023.144054

[55] Tan JX, Chen ZY, Chen CH, Hsieh MF, Lin AYC, Chen SS, et al. Efficient adsorption and photocatalytic degradation of water emerging contaminants through nanoarchitectonics of pore sizes and optical properties of zirconium-based MOFs. Journal of Hazardous Materials. 2023;**451**:131113. DOI: 10.1016/J.JHAZMAT.2023.131113

[56] Han L, Qin P, Li M, Li D, Mu M, Gao Y, et al. Hierarchically porous zirconium-based metal–organic frameworks for rapid adsorption and enrichment of sulfonamide antibiotics.

Chemical Engineering Journal. 2023;**456**: 140969. DOI: 10.1016/J.CEJ.2022.140969

[57] Avval MM, Khani R, Farrokhi A. Aluminium fumarate biological metal–organic framework as an emerging tool for isolation and detection trace amounts of sulfadiazine in food and water samples. Spectrochimica Acta Part A: Molecular and Biomolecular Spectroscopy. 2024;**308**:123716. DOI: 10.1016/J.SAA.2023.123716

[58] Wang B, Lv XL, Feng D, Xie LH, Zhang J, Li M, et al. Highly stable Zr (IV)-based metal-organic frameworks for the detection and removal of antibiotics and organic explosives in water. Journal of the American Chemical Society. 2016;**138**:6204-6216. DOI: 10.1021/JACS.6B01663/ASSET/IMAGES/JA-2016-016633_M004.GIF

[59] Li G, Yan T, Zhao H, Liu D. Effective removal of nitroimidazole antibiotics in aqueous solution by an aluminum-based metal-organic framework: Performance and mechanistic studies. Journal of Solid State Chemistry. 2023;**317**:123659. DOI: 10.1016/J.JSSC.2022.123659

[60] Oni BA, Sanni SE, Agu KC, Tomomewo OS. Green synthesis of Ag nanoparticles from Argemone mexicana L. leaf extract coated with MOF-5 for the removal of metronidazole antibiotics from aqueous solution. Journal of Environmental Management. 2023;**342**:118161. DOI: 10.1016/J.JENVMAN.2023.118161

[61] Escamilla P, Bartella L, Sanz-Navarro S, Percoco RM, Di Donna L, Prejanò M, et al. Degradation of penicillinic antibiotics and β-lactamase enzymatic catalysis in a biomimetic Zn-based metal–organic framework. Chemistry : A European Journal. 2023; **29**:e202301325. DOI: 10.1002/CHEM.202301325

[62] Abazari R, Mahjoub AR, Shariati J. Synthesis of a nanostructured pillar MOF with high adsorption capacity towards antibiotics pollutants from aqueous solution. Journal of Hazardous Materials. 2019;**366**:439-451. DOI: 10.1016/J.JHAZMAT.2018.12.030

[63] Bi W, Wang G, Hu X. Fabrication of Zn-MOF derived graphitic carbon materials with mesoporous structure for adsorptive removal of ceftazidime from aqueous solutions. Colloids and Surfaces A: Physicochemical and Engineering Aspects. 2022;**652**:129758. DOI: 10.1016/J.COLSURFA.2022.129758

[64] Sallam S, Alorabi AQ, Almotairy ARZ, Ibarhiam SF, Aljuhani E, Al-Qahtani SD, et al. Superior and effective adsorption of amoxicillin by using novel metal organic framework and its composite: Thermodynamic, kinetic, and optimization by box–Behnken design. Applied Organometallic Chemistry. 2023;**37**:e7184. DOI: 10.1002/AOC.7184

[65] Yang L, Hu D, Liu H, Wang X, Liu Y, Xia Q, et al. Biodegradation pathway of penicillins by β-lactamase encapsulated in metal-organic frameworks. Journal of Hazardous Materials. 2021;**414**:125549. DOI: 10.1016/J.JHAZMAT.2021.125549

[66] Eskandari H, Amirzehni M, Safavi E, Hassanzadeh J. Synthesis of Zn metal–organic framework doped magnetic graphene oxide for preconcentration and extraction of cefixime followed by its measurement using HPLC. Microchemical Journal. 2021;**169**:106537. DOI: 10.1016/J.MICROC.2021.106537

[67] Bhattacharyya P, Sarma L, Taneja A, Parmar PR, Jain G, Bandyopadhyay D, et al. MOF-Heterostructure-mediated pH-responsive adsorptive removal of ciprofloxacin: A step towards pharmaceutical wastewater treatment.

ChemNanoMat. 2023;**9**:e202200518. DOI: 10.1002/CNMA.202200518

[68] Priyadarshini N, Kumar Das K, Mansingh S, Parida K. Facile fabrication of functionalised Zr co-ordinated MOF: Antibiotic adsorption and insightful physiochemical characterization. Results in Chemistry. 2022;**4**:100450. DOI: 10.1016/J.RECHEM.2022.100450

[69] Kim N, Cha B, Yea Y, Njaramba LK, Vigneshwaran S, Elanchezhiyan SS, et al. Effective sequestration of tetracycline and ciprofloxacin from aqueous solutions by Al-based metal organic framework and reduced graphene oxide immobilized alginate biosorbents. Chemical Engineering Journal. 2022;**450**: 138068. DOI: 10.1016/J.CEJ.2022.138068

[70] Yu F, Bai X, Liang M, Ma J. HKUST-1-derived Cu@Cu(I)@Cu(II)/carbon adsorbents for ciprofloxacin removal with high adsorption performance. Separation and Purification Technology. 2022;**288**:120647. DOI: 10.1016/J.SEPPUR.2022.120647

[71] Bagri F, Hassani A, Jarrah A, Parnianchi F. Highly effective elimination of tetracycline and ciprofloxacin antibiotics from synthetic wastewater by novel magnetic P2W18O62/MIL-101 (Fe)/NiFe2O4 nanocomposite. Separation and Purification Technology. 2024;**329**: 125128. DOI: 10.1016/J.SEPPUR.2023.125128

[72] Chai Y, Zhang Y, Wang L, Du Y, Wang B, Li N, et al. In situ one-pot construction of MOF/hydrogel composite beads with enhanced wastewater treatment performance. Separation and Purification Technology. 2022;**295**:121225. DOI: 10.1016/J.SEPPUR.2022.121225

Chapter 9

Tin Oxide Materials for Industrial Wastewater Treatment: Promising Adsorbents and Catalyst

Nyiko M. Chauke and Mpfunzeni C. Raphulu

Abstract

Tin oxide (SnO) materials have gained significant attention in recent years for their potential application in industrial wastewater treatment. This abstract provides an overview of the use of SnO materials in various treatment processes and their effectiveness in removing pollutants from wastewater. SnO materials exhibit excellent adsorption and catalytic properties, making them suitable for the removal of various contaminants, including heavy metals, organic compounds, and dyes. The unique surface chemistry and high surface area of SnO nanoparticles enable efficient adsorption and reaction with pollutants, leading to their successful removal from wastewater. Additionally, SnO-based photocatalysts have shown promising results in the degradation of organic pollutants under ultraviolet (UV) light irradiation. Moreover, SnO materials can be easily synthesized, and their surface properties can be tailored by controlling the synthesis parameters, such as particle size, morphology, and doping. However, challenges still exist in terms of cost-effectiveness, long-term stability, and large-scale implementation of SnO materials in wastewater treatment processes. Further research and development efforts are necessary to optimize the performance of SnO materials, enhance their stability, and address their limitations for practical application in industrial wastewater treatment systems. Overall, SnO materials hold great potential as effective adsorbents and catalysts for pollutant removal, contributing to the advancement of sustainable wastewater treatment technologies.

Keywords: tin oxide, materials, industrial wastewater treatment, adsorbents, catalysts, pollutants, removal

1. Introduction

The escalating concerns over industrial wastewater pollution necessitate innovative solutions for efficient treatment. Tin oxide (SnO) materials, with their distinct surface chemistry and high surface area, present a viable avenue for addressing these concerns [1, 2]. This chapter provides an overview of the potential of SnO materials in mitigating the adverse effects of industrial wastewater pollutants.

Industrial activities have undoubtedly contributed to human progress and economic growth, however, they have also led to the generation of copious amounts of wastewater laden with a diverse range of pollutants [2–4]. The disposal of untreated or inadequately treated industrial effluents poses a severe threat to ecosystems, human health, and overall environmental sustainability. In this context, the exploration of advanced and efficient materials for wastewater treatment has become imperative [5].

Tin oxide materials have surfaced as highly promising candidates in the arena of industrial wastewater treatment, presenting a multifaceted approach to address the intensifying challenges associated with water pollution [6, 7]. Their distinctive surface properties, coupled with their versatile chemical composition, have propelled them into the limelight of wastewater treatment research [6]. The intrinsic characteristics of SnO materials, including their tunable surface chemistry, high surface area, and crystalline structures, have garnered significant attention due to their potential as effective agents for pollutant removal and degradation in wastewater treatment processes [8, 9].

In the context of adsorption, the exceptional surface reactivity of SnO materials facilitates robust interactions with a diverse array of contaminants present in industrial effluents [10, 11]. Their high surface area offers an expansive platform for adsorption sites, allowing for the efficient uptake of various pollutants, ranging from heavy metals to organic compounds and dyes [11, 12]. This adsorption process is often driven by mechanisms such as chemisorption and physisorption, where pollutant species adhere to the SnO surface through chemical bonding or electrostatic forces [13]. The intricate interplay between the surface properties of SnO materials and the molecular structures of pollutants determines the overall adsorption efficiency, rendering these materials effective adsorbents in wastewater treatment [14, 15].

Furthermore, the catalytic prowess of SnO materials has unveiled a new dimension in industrial wastewater treatment strategies. Employed as catalysts, particularly in the realm of photocatalysis, SnO materials exhibit the capability to initiate and facilitate degradation reactions of organic pollutants under ultraviolet (UV) light irradiation [16, 17]. This catalytic activity stems from the generation of electron-hole pairs upon exposure to UV light, which subsequently triggers redox reactions leading to the breakdown of complex pollutant molecules into simpler and less harmful compounds [18]. This photocatalytic degradation mechanism presents a sustainable and energy-efficient route to neutralize recalcitrant pollutants, showcasing the potential of these materials to transform the landscape of wastewater treatment practices [18, 19].

In essence, the convergence of tin oxide's unique surface properties, versatile chemistry, and reactivity underscore its prominence as an emerging star in the domain of industrial wastewater treatment [20]. The following subsequent sections of this chapter will delve deeper into the intricacies of these mechanisms, exploring the diverse applications, synthesis methodologies, challenges, and future prospects associated with SnO materials in this critical field. As such, the overarching objective of this chapter is to provide a comprehensive overview of the utilization of SnO materials in industrial wastewater treatment. This exploration encompasses their role as potent adsorbents, their catalytic prowess, methods of synthesis and tailoring, as well as the challenges faced in their practical implementation. Through a systematic examination of the current state of research and practical applications, this chapter also aims to shed light on the potential of SnO materials to revolutionize the landscape of industrial wastewater treatment. Therefore, the subsequent sections will delve into the intricacies of SnO materials, elucidating their mechanisms of action, their efficacy in pollutant removal, and the avenues for optimizing their performance. By obtaining a more comprehensive comprehension of the potentials and constraints of these materials, a pathway can be established for formulating sustainable and efficient methodologies to tackle the formidable challenges presented by pollution in industrial wastewater.

2. Tin oxide as an adsorbent in industrial wastewater treatment

Tin oxide materials have garnered significant attention for their remarkable capabilities as adsorbents in industrial wastewater treatment. Their distinct surface chemistry, high surface area, and tunable properties make them effective candidates for the removal of a wide range of contaminants from wastewater. The adsorption process involves the attachment of pollutants onto the surface of the adsorbent [21, 22]. In the case of SnO materials, this interaction is influenced by their unique surface properties, including the presence of oxygen vacancies, hydroxyl groups, and active sites [23]. These surface features provide ample binding sites for pollutants, enabling both physical adsorption through van der Waals forces and chemical adsorption through electrostatic interactions and covalent bonding [24]. The chemisorption mechanism is particularly relevant when dealing with heavy metal ions, where the formation of metal-oxygen bonds on the SnO surface leads to strong and selective adsorption [25].

Table 1 showcases the remarkable adsorption capacities exhibited by various SnO-based materials. These results underscore the immense potential of SnO materials in the field of industrial wastewater treatment. The diverse range of pollutants, from heavy metals like lead and arsenic to organic compounds like phenols and dyes,

Tin oxide nanomaterials	Pollutant(s) removed	Adsorbent dosage	Adsorption capacity	% Removal efficiency	Reference
SnO_2-NP-AC	Murexide dye	0.025 g	~		[26]
SnO_2 nanoparticles	Zn^{2+}		102 mg/g	83.2	[27]
SnO_2–NP–AC	Malachite green-oxalate	0.01 g	142.83 mg/g	95	[28]
Zinc/Tin selenide (ZnSnSe-1 and ZnSnSe-1 K)	Cs^+, $Ba2^+$, Co^{2+}, and Ni^{2+}	2.3 g	253.61 mg/g	97–100 Cs^+	[29]
Tin and Titanium incorporated rice husk Silica	Methylene blue	0.005–0.02 g	~	98	[30]
Polypyrrole/ hydrous Tin oxide nanocomposites	Fluoride ions	0.1 g	28.98 mg/g	88.45	[31]
Nano Tin Ferrous oxide decorated Graphene Oxide	Arsenic (III)	1.0 g	105 mg/g	~	[32]
Tin Titanate@ RhB coated on polymeric sphere	Herbicide Fomesafen	0.02–0.05 g	~	92	[33]

Table 1.
Adsorption capacities of different tin oxide-based materials.

demonstrates the versatility of these materials. The consistently high adsorption capacities highlight the significance of their tailored surface properties and controlled synthesis methods.

The data presented here reinforces the notion that different forms and compositions of SnO-based materials possess unique characteristics that influence their adsorption capabilities. Such findings not only emphasize the potential of tin oxide-based materials to address complex wastewater treatment challenges but also indicate the need for further exploration and optimization [34]. As such, continued research and innovation in this field hold the promise of harnessing the full adsorption potential of SnO materials, ultimately contributing to more effective and sustainable wastewater treatment strategies.

Tin oxide's versatility as an adsorbent is evident in its ability to remove a diverse array of pollutants commonly found in industrial wastewater [26]. Heavy metals, such as lead, cadmium, and mercury, are efficiently captured due to their affinity for binding with oxygen-containing sites on the SnO surface [35, 36]. Organic compounds, including dyes, phenols, and organic solvents, also undergo effective adsorption driven by a combination of electrostatic and hydrophobic interactions. The large surface area of SnO nanoparticles further enhances their capacity to adsorb pollutants, ensuring efficient removal even at low concentrations [37].

The kinetics of adsorption on SnO materials depend on factors like the initial pollutant concentration, adsorbent dosage, and temperature. In most cases, the adsorption process follows pseudo-second-order kinetics, indicating the involvement of multiple steps including surface diffusion and chemisorption [13, 26, 31, 32, 38]. Thermodynamic studies often reveal the spontaneous and endothermic nature of adsorption on SnO surfaces, affirming the affinity between the adsorbate and adsorbent [39].

One of the advantages of SnO as an adsorbent is its potential for regeneration and reusability [40]. Desorption methods, such as pH adjustment or complexation, can release the adsorbed pollutants from the surface, allowing the adsorbent to be regenerated and reused. This characteristic significantly enhances the cost-effectiveness and sustainability of wastewater treatment processes [41]. Therefore, SnO materials, owing to their unique surface properties and versatile chemistry, present a highly effective adsorbent option for the removal of a wide range of contaminants from industrial wastewater. Their capacity for selective adsorption, coupled with the potential for regeneration, positions them as a crucial component in the development of efficient and sustainable wastewater treatment strategies. However, the challenges of scalability, long-term stability, and cost-efficiency must be addressed for the full realization of SnO potential in practical industrial applications.

3. Tin oxide catalysis in wastewater treatment

Tin oxide materials have emerged as promising catalysts in the field of industrial wastewater treatment, particularly in the context of advanced oxidation processes such as photocatalysis [8, 32, 40]. Leveraging their unique properties, SnO-based catalysts facilitate the degradation of organic pollutants through redox reactions, offering a sustainable and efficient approach to wastewater treatment [42].

The photocatalytic activity of SnO materials primarily hinges on their ability to generate electron-hole pairs when exposed to UV light. This phenomenon is due to the bandgap energy of SnO, which corresponds to the energy of UV photons [8, 43]. The excited electrons and holes can initiate a cascade of redox reactions on the catalyst's surface, leading to the formation of reactive oxygen species (ROS) like hydroxyl radicals. These ROS exhibit high oxidative potential and react with organic pollutants adsorbed on the catalyst's surface, breaking down complex organic molecules into smaller, less harmful fragments [43, 44].

Tin oxide photocatalysis has demonstrated impressive efficacy in degrading a range of organic pollutants found in industrial wastewater, including dyes, phenols, pesticides, and pharmaceuticals [44, 45]. The process offers several advantages, including its potential to mineralize pollutants into carbon dioxide (CO_2) and water (H_2O), rendering them harmless. Moreover, the non-selective nature of ROS-driven reactions ensures the breakdown of even recalcitrant compounds, which are traditionally challenging to remove using conventional treatment methods [46]. **Figure 1** illustrates the fundamental photocatalytic process involving the SnO photocatalyst. When SnO is exposed to UV or visible light, its electrons in the valence band (VB) get excited to the higher-energy conduction band (CB), generating electron-hole pairs. Importantly, the design of SnO enables a lower recombination rate of these electron-hole pairs, meaning that they remain separated for a more extended period. This reduced recombination rate is a critical factor in enhancing the efficiency of the semiconductor during the excitation state.

The improved efficiency allows SnO to act as an effective catalyst in the degradation of toxins present in wastewater. As the excited electrons (e^-) and holes (h^+) are available for reactions, they engage with adsorbed water molecules or oxygen in the water, producing highly reactive oxygen species like hydroxyl radicals (•OH) [47]. These ROS possess potent oxidative capabilities and play a pivotal role in breaking down complex organic pollutants and transforming them into simpler, less harmful compounds [47, 48].

The significance of this process lies in the fact that SnO can harness the energy from UV or visible light to drive the degradation of contaminants, making it an energy-efficient and sustainable approach to wastewater treatment [49]. This mechanism of photocatalysis demonstrates SnO's potential to revolutionize the field

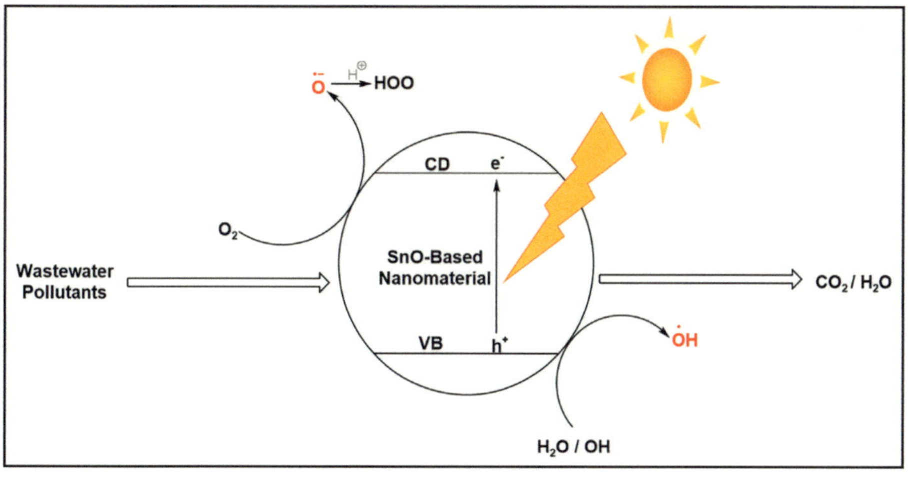

Figure 1.
Photocatalytic process of SnO as a photocatalyst.

of industrial wastewater treatment by offering a pathway to efficiently neutralize pollutants and mitigate their adverse environmental impacts [50].

As such, to optimize SnO catalytic efficiency, researchers have explored strategies such as modifying the material's surface properties and enhancing light absorption. Doping SnO with elements like nitrogen or metal nanoparticles can extend its light absorption range into visible light, thereby broadening its application potential. Additionally, controlling factors such as particle size and morphology can impact the catalyst's surface area and surface reactivity, influencing its overall performance [51, 52].

While SnO photocatalysis holds immense promise, challenges persist. Catalyst stability over prolonged exposure to UV light, catalyst recovery, and the potential toxicity of generated byproducts are areas that warrant further investigation [53]. Moreover, efficient utilization of solar energy by SnO catalysts is an ongoing research focus, aiming to reduce energy consumption and enhance sustainability. Therefore, SnO catalysis, particularly in the domain of photocatalysis, emerges as a promising avenue for industrial wastewater treatment. Its ability to harness light energy to initiate pollutant degradation aligns well with the goals of sustainability and efficiency in wastewater treatment. As research continues to refine the catalyst design, enhance its stability, and mitigate associated challenges, SnO materials could play a pivotal role in transforming the way industrial wastewater is treated, contributing to a cleaner and healthier environment.

4. Synthesis and tailoring of tin oxide materials

The effective utilization of SnO materials in industrial wastewater treatment hinges on the precision of their synthesis and the ability to tailor their properties to match specific treatment requirements [54, 55]. By manipulating synthesis methods and controlling various parameters, researchers can engineer SnO materials with desired characteristics, ensuring optimal performance as adsorbents and catalysts [55, 56].

Several methods are employed to synthesize SnO materials, each yielding distinct morphologies and properties. The sol-gel method, for instance, involves the controlled hydrolysis and condensation of precursor compounds, allowing precise control over particle size and surface properties [56, 57]. Hydrothermal and solvothermal processes utilize elevated temperatures and pressures to induce crystallization, yielding well-defined nanocrystalline structures [58, 59]. Chemical vapor deposition (CVD) and physical vapor deposition (PVD) techniques enable the creation of thin films and coatings with tailored properties [60, 61]. Additionally, hydrolysis of organotin compounds and precipitation reactions offer facile routes for generating SnO nanoparticles [62].

The surface properties of SnO materials play a pivotal role in their adsorption and catalytic performance. Surface modification and doping emerge as effective strategies to tailor these properties [63]. Doping with elements like nitrogen, fluorine, or metals can introduce new functionalities, alter band structures, and enhance catalytic activity. Surface functionalization with organic groups or metal nanoparticles can fine-tune the material's interaction with pollutants, enhancing adsorption selectivity [23, 51, 64, 65].

Particle size and morphology significantly impact the efficacy of SnO materials in wastewater treatment. Smaller nanoparticles offer higher surface area and increased adsorption capacity [66]. As such, controlling particle size through synthesis parameters such as reaction temperature, precursor concentration, and surfactant use is crucial. Morphology control is equally important, as nanotubes, nanowires, and nanoparticles exhibit varying properties. Tailoring morphology can be achieved through the choice of solvents, surfactants, and growth conditions [67].

While synthesis in the laboratory often focuses on producing small quantities, the scalability of SnO synthesis for industrial applications is a critical consideration. Developing cost-effective and efficient synthesis techniques that can be translated to larger scales without compromising the quality of the materials is essential. Process optimization and automation play a pivotal role in ensuring commercial viability. Therefore, the synthesis and tailoring of SnO materials present a gateway to unlocking their potential in industrial wastewater treatment. By manipulating synthesis methods, doping, and controlling particle size and morphology, researchers can fine-tune the materials' properties to meet the demands of specific treatment scenarios. As the field continues to advance, achieving a delicate balance between effective pollutant removal and practical scalability will be instrumental in realizing the widespread application of tailored tin oxide materials for sustainable wastewater treatment.

5. Challenges and future directions in tin oxide materials for industrial wastewater treatment

Despite the significant promise exhibited by SnO materials in industrial wastewater treatment, several challenges persist that need to be addressed to fully realize their potential. As researchers strive to overcome these challenges, new avenues and future directions are emerging, shaping the trajectory of SnO's role in sustainable wastewater treatment.

5.1 Challenges

1. Long-term stability: The stability of SnO materials, particularly under extended exposure to harsh wastewater conditions, remains a concern. Oxidation, dissolution, and aggregation of nanoparticles over time can compromise their effectiveness [68].

2. Scalability: The transition from laboratory-scale synthesis to large-scale production poses challenges in maintaining material quality, uniformity, and cost-effectiveness. Upscaling while retaining the desired properties is crucial.

3. Regeneration and reusability: While the regenerability of SnO adsorbents and catalysts is a valuable feature, practical methods for efficient regeneration without compromising material integrity need further development [69–71].

4. Specificity: Enhancing the selectivity of SnO materials towards specific pollutants is a challenge. Tailoring their surface properties to preferentially adsorb or catalyze certain contaminants while ignoring others is a complex task [72].

5. Energy consumption: In the case of photocatalysis, ensuring that the energy consumed during catalyst activation (e.g., UV light irradiation) is outweighed by the energy savings from pollutant degradation is an ongoing challenge.

5.2 Future prospects and directions

1. Advanced material engineering: Continued research into innovative synthesis techniques and material engineering will lead to the development of more stable and efficient tin oxide-based materials.

2. Multifunctional nanocomposites: Designing nanocomposites that combine SnO with other materials (e.g., graphene, metal nanoparticles) could yield synergistic effects, enhancing both adsorption and catalytic capabilities.

3. Hybrid treatment systems: Integrating SnO-based processes with other treatment methods, such as membrane filtration or biological processes, could lead to more comprehensive and effective wastewater treatment strategies.

4. In-situ regeneration techniques: Developing in-situ regeneration methods that minimize the need for external chemicals and energy can enhance the sustainability and practicality of SnO-based treatment processes.

5. Real-time monitoring: Incorporating sensors and monitoring systems that can assess pollutant levels, catalyst activity, and material integrity in real-time can enable proactive management of treatment processes.

6. Life cycle analysis: Conducting thorough life cycle assessments to evaluate the environmental and economic impacts of SnO-based treatment systems can guide sustainable technology development.

7. Commercialization and industry collaboration: Bridging the gap between research and industrial implementation through partnerships and collaboration will facilitate the translation of promising laboratory findings into practical solutions.

Therefore, while challenges persist, the trajectory of SnO materials in industrial wastewater treatment is undeniably promising. Through a concerted effort to address these challenges and navigate emerging directions, researchers and practitioners can harness the unique properties of SnO materials to develop sustainable, efficient, and economically viable solutions for the pressing problem of industrial wastewater pollution.

6. Case studies and practical applications of tin oxide materials in industrial wastewater

Real-world applications serve as tangible evidence of the efficacy and potential of SnO materials in addressing industrial wastewater treatment challenges. Several case studies highlight the successful integration of these materials into practical treatment strategies, showcasing their diverse applications and contributions to sustainable environmental management [73].

A metal plating facility faced significant challenges in removing trace heavy metals like chromium and cadmium from its wastewater before discharge [74]. Farhan et al., wrote a review on nanocomposites for the removal of hazardous metals from water using advanced remediation techniques and photocatalytic oxidation [75]. They highlighted that wastewater has been found to have high levels of chromium, mercury, and arsenic, posing a threat to human health. Although, for the removal of heavy/hazardous metals, a range of nanoadsorbents and nanophotocatalysts are employed. Tin oxide nanoparticles were synthesized and employed as adsorbents by Kumar et al. [76] due to their high affinity for metal ions [77]. The nanoparticles exhibited exceptional adsorption capacity and selectivity for heavy metals, successfully reducing their concentrations in the effluent to permissible levels. The adsorbent could be regenerated efficiently, enabling its cost-effective long-term use [78].

A textile dye manufacturing plant grappled with the disposal of colored wastewater containing recalcitrant organic dyes [79]. Tin oxide nanoparticles, doped with nitrogen to enhance their photocatalytic activity, can be utilized in dye removal [80–82]. Under UV light irradiation, the nanoparticles exhibited remarkable catalytic degradation of dye molecules, which can be in colorless and less toxic effluents [83].

Pharmaceutical production facilities are encountering challenges in removing emerging contaminants, such as pharmaceutical residues, from their wastewater [3, 84, 85]. SnO materials can be incorporated into an advanced treatment system, combining adsorption and catalysis [85]. The tailored surface properties of the materials will allow efficient adsorption of complex organic molecules, followed by photocatalytic degradation. The treated effluent will meet stringent discharge standards, which will showcase the potential of tin oxide-based hybrid treatment approaches [65].

Industrial landfill sites also generate leachate containing a cocktail of pollutants, including heavy metals and organic compounds. A multifaceted treatment system can be integrated with tin oxide adsorption and photocatalysis [86]. The adsorbent will efficiently remove heavy metals, while the photocatalyst effectively will degrade organic pollutants. This combined approach will ensure comprehensive treatment, transforming the hazardous leachate into environmentally safe effluent.

These case studies exemplify the versatility and effectiveness of tin oxide materials in tackling diverse challenges in industrial wastewater treatment. From heavy metal removal to organic pollutant degradation, tin oxide's distinct surface properties and tailored chemistry offer a range of solutions. By adapting these materials to specific treatment scenarios and optimizing their integration into larger treatment systems, practical applications demonstrate their potential to revolutionize the landscape of industrial wastewater management.

7. Conclusions

In the realm of industrial wastewater treatment, tin oxide materials have emerged as dynamic tools with the potential to reshape the landscape of pollutant removal processes. Their unique surface properties, tunable chemistry, and multifaceted applications have sparked optimism for more effective, sustainable, and efficient treatment strategies. From their role as adsorbents adept at capturing diverse contaminants to their catalytic prowess in dismantling recalcitrant pollutants, tin oxide materials have showcased their versatility and promise in addressing the complex challenges posed by industrial wastewater pollution. However, as this chapter has outlined, challenges such as stability, scalability, and selectivity remain to be surmounted. The journey towards widespread implementation necessitates concerted research efforts and innovative approaches. Advancements in synthesis methods, tailoring of properties, and development of hybrid systems hold the key to overcoming these challenges. The case studies presented underscore the real-world applicability of tin oxide materials, highlighting their transformative impact across diverse industries. Nevertheless, these successes are only the beginning. Continued research and innovation are imperative to refine existing methodologies, optimize performance, and explore novel applications. The journey towards integrating tin oxide materials seamlessly into industrial wastewater treatment practices requires interdisciplinary collaboration, technological adaptation, and a dedication to addressing the evolving demands of sustainability and environmental stewardship. In the ever-evolving landscape of industrial wastewater treatment, tin oxide materials stand as promising agents of change. Their potential to revolutionize pollutant removal processes is undeniable, and their journey from laboratory innovation to widespread industrial application is a testament to human ingenuity. As research continues to unlock its full potential and surmount the challenges that lie ahead, tin oxide materials hold the promise of offering more sustainable and environmentally conscious solutions to the critical challenge of industrial wastewater pollution.

Acknowledgements

The authors would like to express their heartfelt gratitude for the generous financial support provided by the Mintek Science Vote Grant. This funding has been instrumental in facilitating our research endeavors and enabling us to make significant strides in advancing scientific knowledge. We sincerely appreciate Mintek's commitment to promoting research and innovation in the field, and we are honored to be recipients of their support.

Conflict of interest

The authors declare no conflict of interest.

Author details

Nyiko M. Chauke* and Mpfunzeni C. Raphulu
Mintek, Catalysis Group, Advanced Materials Division, Randburg, South Africa

*Address all correspondence to: nyikoc@mintek.co.za

References

[1] Chandran P et al. Biological treatment solutions using bioreactors for environmental contaminants from industrial wastewater. Journal of Umm Al-Qura University for Applied Sciences. 2023:1-23

[2] Iloms E, Ololade OO, Ogola HJO, Selvarajan R. Investigating industrial effluent impact on municipal wastewater treatment plant in Vaal, South Africa. International Journal of Environmental Research and Public Health. 2020;**17**(3):1096

[3] Samal K, Mahapatra S, Ali MH. Pharmaceutical wastewater as emerging contaminants (EC): Treatment technologies, impact on environment and human health. Energy Nexus. 2022;**6**:100076

[4] Edokpayi JN, Odiyo JO, Durowoju OS. Impact of wastewater on surface water quality in developing countries: A case study of South Africa. Water Quality. 2017;**10**(66561):10-5772

[5] Obaideen K, Shehata N, Sayed ET, Abdelkareem MA, Mahmoud MS, Olabi AG. The role of wastewater treatment in achieving sustainable development goals (SDGs) and sustainability guideline. Energy Nexus. 2022;**7**:100112

[6] Kumar S. Smart and innovative nanotechnology applications for water purification. Hybrid Advances. 2023;**3**:100044

[7] Khatami M, Iravani S. MXenes and MXene-based materials for the removal of water pollutants: Challenges and opportunities. Comments on Inorganic Chemistry. 2021;**41**(4):213-248

[8] Al-Hamdi AM, Rinner U, Sillanpää M. Tin dioxide as a photocatalyst for water treatment: A review. Process Safety and Environmental Protection. 2017;**107**:190-205

[9] Nabgan W et al. A review on the design of nanostructure-based materials for photoelectrochemical hydrogen generation from wastewater: Bibliometric analysis, mechanisms, prospective, and challenges. International Journal of Hydrogen Energy. 2023;**52**:622-663

[10] Chakraborty U, Kaur G, Rubahn H-G, Kaushik A, Chaudhary GR, Mishra YK. Advanced metal oxides nanostructures to recognize and eradicate water pollutants. Progress in Materials Science. 2023;**139**:101169

[11] Kumari H et al. A review on photocatalysis used for wastewater treatment: Dye degradation. Water, Air, & Soil Pollution. 2023;**234**(6):349

[12] Xu G-R, An Z-H, Xu K, Liu Q, Das R, Zhao H-L. Metal organic framework (MOF)-based micro/nanoscaled materials for heavy metal ions removal: The cutting-edge study on designs, synthesis, and applications. Coordination Chemistry Reviews. 2021;**427**:213554

[13] Darweesh MA, Elgendy MY, Ayad MI, Ahmed AM, Elsayed NMK, Hammad WA. Adsorption isotherm, kinetic, and optimization studies for copper (II) removal from aqueous solutions by banana leaves and derived activated carbon. South African Journal of Chemical Engineering. 2022;**40**:10-20

[14] Zhang L-L, Zaoui A, Sekkal W, Zheng Y-Y. Interlayer adsorption of cationic dye on cationic surfactant-modified and unmodified

montmorillonite. Journal of Hazardous Materials. 2023;**442**:130107

[15] Zhang Z et al. Functionalized attapulgite for the adsorption of methylene blue: Synthesis, characterization, and adsorption mechanism. ACS Omega. 2021;**6**(30):19586-19595

[16] Xia H, Ahmad MA, Guo J, Yang Y. Unraveling the roles of MW/UV/TiO2 photocatalysis technologies for organic wastewater treatment. Catalysts. 2023;**13**(4):754

[17] Mohapatra L, Park JH. A Triphasic Superwetting Catalyst for Photocatalytic Wastewater Treatment. London, United Kingdom: IntechOpen-Eurospan; 2023

[18] Pavel M, Anastasescu C, State R-N, Vasile A, Papa F, Balint I. Photocatalytic degradation of organic and inorganic pollutants to harmless end products: Assessment of practical application potential for water and air cleaning. Catalysts. 2023;**13**(2):380

[19] Aslam M et al. Photocatalytic degradation of recalcitrant pollutants of greywater. Catalysts. 2022;**12**(5):557

[20] Oghyanous FA. Nanoparticles in wastewater treatment. Water Conservation: Inevitable Strategy. 2022;**107**:107-119

[21] Yang J et al. Nanomaterials for the removal of heavy metals from wastewater. Nanomaterials. 2019;**9**(3):424

[22] Kaur H, Devi N, Siwal SS, Alsanie WF, Thakur MK, Thakur VK. Metal–organicframework-basedmaterials for wastewater treatment: Superior adsorbent materials for the removal of hazardous pollutants. ACS Omega. 2023;**8**(10):9004-9030

[23] Faustino BMM, Foot PJS, Kresinski RA. Synthesis and photoluminescent properties of Sm3+–doped SnO2 nanoparticles. Ceramics International. 2016;**42**(16):18474-18478

[24] Eslamian M, Salehi A, Nadimi E. The role of oxygen vacancies on SnO2 surface in reducing cross-sensitivity between ambient humidity and CO: A first principles investigation. Surface Science. 2021;**708**:121817

[25] Sopiha KV, Malyi OI, Persson C, Wu P. Chemistry of oxygen ionosorption on SnO2 surfaces. ACS Applied Materials & Interfaces. 2021;**13**(28):33664-33676

[26] Davoodi S, Marahel F, Ghaedi M, Roosta M, Hekmati Jah A. Tin oxide nanoparticles loaded on activated carbon as adsorbent for removal of murexide. Desalination and Water Treatment. 2014;**52**(37-39):7282-7292

[27] Ahmed M, Elektorowicz M, Hasan SW. GO, SiO2, and SnO2 nanomaterials as highly efficient adsorbents for Zn2+ from industrial wastewater—A second stage treatment to electrically enhanced membrane bioreactor. Journal of Water Process Engineering. 2019;**31**:100815

[28] Shamsizadeh A, Ghaedi M, Ansari A, Azizian S, Purkait MK. Tin oxide nanoparticle loaded on activated carbon as new adsorbent for efficient removal of malachite green-oxalate: Non-linear kinetics and isotherm study. Journal of Molecular Liquids. 2014;**195**:212-218

[29] Liu Y et al. Open-framework hybrid zinc/tin selenide as an ultrafast adsorbent for Cs+, Ba2+, Co2+, and Ni2+. Journal of Hazardous Materials. 2023;**458**:132038

[30] Adam F, Appaturi JN, Khanam Z, Thankappan R, Nawi MAM. Utilization

of tin and titanium incorporated rice husk silica nanocomposite as photocatalyst and adsorbent for the removal of methylene blue in aqueous medium. Applied Surface Science. 2013;**264**:718-726

[31] Parashar K, Ballav N, Debnath S, Pillay K, Maity A. Rapid and efficient removal of fluoride ions from aqueous solution using a polypyrrole coated hydrous tin oxide nanocomposite. Journal of Colloid and Interface Science. 2016;**476**:103-118

[32] Sharma M, Ramakrishnan S, Remanan S, Madras G, Bose S. Nano tin ferrous oxide decorated graphene oxide sheets for efficient arsenic (III) removal. Nano-Structures & Nano-Objects. 2018;**13**:82-92

[33] Valadez-Renteria E, Oliva J, Mtz-Enriquez AI, Rodriguez-Gonzalez V. A novel floating compound based on a plastic sphere coated with tin titanate for the removal of the herbicide fomesafen. Journal of Cleaner Production. 2022;**379**:134882

[34] Batzill M, Diebold U. The surface and materials science of tin oxide. Progress in Surface Science. 2005;**79**(2-4):47-154

[35] Zaimee MZA, Sarjadi MS, Rahman ML. Heavy metals removal from water by efficient adsorbents. Water. 2021;**13**(19):2659

[36] De Gisi S, Lofrano G, Grassi M, Notarnicola M. Characteristics and adsorption capacities of low-cost sorbents for wastewater treatment: A review. Sustainable Materials and Technologies. 2016;**9**:10-40

[37] Ahmed M et al. Recent developments in hazardous pollutants removal from wastewater and water reuse within a circular economy. NPJ Clean Water. 2022;**5**(1):12

[38] Kajjumba GW, Emik S, Öngen A, Özcan HK, Aydın S. Modelling of adsorption kinetic processes—Errors, theory and application. Advanced Sorption Process Applications. 2018:1-19

[39] Mahmud N, Benamor A. Magnetic Iron oxide kaolinite nanocomposite for effective removal of Congo red dye: Adsorption, kinetics, and thermodynamics studies. Water Conservation Science and Engineering. 2023;**8**(1):35

[40] Sharif F, Roberts EPL. Electrochemical oxidation of an organic dye adsorbed on tin oxide and antimony doped tin oxide graphene composites. Catalysts. 2020;**10**(2):263

[41] Saha B, Vedachalam S, Dalai AK. Review on recent advances in adsorptive desulfurization. Fuel Processing Technology. 2021;**214**:106685

[42] Martínez-Huitle CA, Rodrigo MA, Sirés I, Scialdone O. A critical review on latest innovations and future challenges of electrochemical technology for the abatement of organics in water. Applied Catalysis B: Environmental. 2023;**328**:122430

[43] Arumugam M, Natarajan TS, Saelee T, Praserthdam S, Ashokkumar M, Praserthdam P. Recent developments on bismuth oxyhalides (BiOX; X= Cl, Br, I) based ternary nanocomposite photocatalysts for environmental applications. Chemosphere. 2021;**282**:131054

[44] Anpo M, Costentin G, Giamello E, Lauron-Pernot H, Sojka Z. Characterisation and reactivity of oxygen species at the surface of metal oxides. Journal of Catalysis. 2021;**393**:259-280

[45] Li S et al. Photocatalytic degradation of hazardous organic pollutants in water by Fe-MOFs and their composites: A review. Journal of Environmental Chemical Engineering. 2021;**9**(5):105967

[46] Kumar V, Shah MP. Advanced oxidation processes for complex wastewater treatment. In: Advanced Oxidation Processes for Effluent Treatment Plants. Elsevier; 2021. pp. 1-31

[47] Kumari H, Chahal S, Kumar P, Kumar A, Parmar R. Photocatalytic degradation of RB dye via cerium substituted SnO2 photocatalysts. Materials Science and Engineering: B. 2023;**296**:116654

[48] Oluwole AO, Olatunji OS. Photocatalytic degradation of tetracycline in aqueous systems under visible light irridiation using needle-like SnO 2 nanoparticles anchored on exfoliated gC 3 N 4. Environmental Sciences Europe. 2022;**34**:1-14

[49] Peleyeju MG, Arotiba OA. Recent trend in visible-light photoelectrocatalytic systems for degradation of organic contaminants in water/wastewater. Environmental Science: Water Research & Technology. 2018;**4**(10):1389-1411

[50] Hojamberdiev M, Czech B, Göktaş AC, Yubuta K, Kadirova ZC. SnO2@ ZnS photocatalyst with enhanced photocatalytic activity for the degradation of selected pharmaceuticals and personal care products in model wastewater. Journal of Alloys and Compounds. 2020;**827**:154339

[51] Akhter P, Arshad A, Saleem A, Hussain M. Recent development in non-metal-doped titanium dioxide photocatalysts for different dyes degradation and the study of their strategic factors: A review. Catalysts. 2022;**12**(11):1331

[52] Yang Y, Niu S, Han D, Liu T, Wang G, Li Y. Progress in developing metal oxide nanomaterials for photoelectrochemical water splitting. Advanced Energy Materials. 2017;7(19):1700555

[53] Ren G et al. Recent advances of photocatalytic application in water treatment: A review. Nanomaterials. 2021;**11**(7):1804

[54] Teh CY, Budiman PM, Shak KPY, Wu TY. Recent advancement of coagulation–flocculation and its application in wastewater treatment. Industrial & Engineering Chemistry Research. 2016;**55**(16):4363-4389

[55] Uday YS, Vidya YS, Manjunatha HC, Manjunath S, Sridhar KN, Nagashree KL. Synthesis and characterization and X-ray/gamma ray absorption properties of tin oxide synthesized via solution combustion method. Applied Surface Science Advances. 2023;**13**:100367

[56] Chen Y, Yuan M, Zhang Y, Wang X, Ke F, Wang H. One-pot synthesis of tin oxide/reduced graphene oxide composite coated fabric for wearable ammonia sensor with fast response/recovery rate. Journal of Alloys and Compounds. 2023;**931**:167585

[57] Ali SI, Dutta D, Das A, Mandal S, Mandal AC. Understanding the structure-property correlation of tin oxide nanoparticles synthesized through the sol-gel technique. Journal of Luminescence. 2023;**253**:119465

[58] Amutha E, Rajaduraipandian S, Sivakavinesan M, Annadurai G. Hydrothermal synthesis and characterization of the antimony–tin oxide nanomaterial and its application as a high-performance asymmetric

supercapacitor, photocatalyst, and antibacterial agent. Nanoscale Advances. 2023;**5**(1):255-267

[59] Furukawa M, Iwamoto D, Inamori K, Tateishi I, Katsumata H, Kaneco S. Synthesis of tungsten-modified Sn3O4 through the cetyltrimethylammonium bromide-assisted solvothermal method for dye decolorization under visible light irradiation. Catalysts. 2023;**13**(8):1179

[60] Liu J et al. Black phosphorus nanodot incorporated tin oxide hollow-spherical heterojunction for enhanced properties of room-temperature gas sensors. Ceramics International. 2023;**49**(5):8248-8258

[61] Kim D-S, Jung J-Y, Seo S, Kim J-H. Facile fabrication of multifunctional transparent electrodes via spray deposition of indium-tin-oxide nanoparticles. Applied Surface Science. 2023;**611**:155756

[62] Kamali AR, Li S. Molten salt-assisted valorization of waste PET plastics into nanostructured SnO2@ terephthalic acid with excellent Li-ion storage performance. Applied Energy. 2023;**334**:120692

[63] Roy H et al. Toxic dye removal, remediation, and mechanism with doped SnO2-based nanocomposite photocatalysts: A critical review. Journal of Water Process Engineering. 2023;**54**:104069

[64] Al-Naggar AH, Shinde NM, Kim J-S, Mane RS. Water splitting performance of metal and non-metal-doped transition metal oxide electrocatalysts. Coordination Chemistry Reviews. 2023;**474**:214864

[65] Jin C et al. Efficient photocatalytic degradation and adsorption of tetracycline over type-II heterojunctions consisting of ZnO nanorods and K-doped exfoliated g-C3N4 nanosheets. Industrial & Engineering Chemistry Research. 2020;**59**(7):2860-2873

[66] Haruna A, Chong F-K, Ho Y-C, Merican ZMA. Preparation and modification methods of defective titanium dioxide-based nanoparticles for photocatalytic wastewater treatment—A comprehensive review. Environmental Science and Pollution Research. 2022;**29**(47):70706-70745

[67] Kaur A, Bajaj B, Kaushik A, Saini A, Sud D. A review on template assisted synthesis of multi-functional metal oxide nanostructures: Status and prospects. Materials Science and Engineering: B. 2022;**286**:116005

[68] Rizwan M et al. Effect of metal and metal oxide nanoparticles on growth and physiology of globally important food crops: A critical review. Journal of Hazardous Materials. 2017;**322**:2-16

[69] Notthakun S, Crittenden JC, Hand DW, Perram DL, Mullins ME. Regeneration of adsorbents using heterogeneous advanced oxidation. Journal of Environmental Engineering. 1993;**119**(4):695-714

[70] Campisi S et al. Multifunctional interfaces for multiple uses: Tin (II)-hydroxyapatite for reductive adsorption of Cr (VI) and its upcycling into catalyst for air protection reactions. Journal of Colloid and Interface Science. 2023;**630**:473-486

[71] Tan Z et al. SnO2/ATP catalyst enabling energy-efficient and green amine-based CO2 capture. Chemical Engineering Journal. 2023;**453**:139801

[72] Kim T et al. Tuning the oxidation state of SnO x and mass transport to

enhance catholyte-free CO 2-to-formate electrolysis. Sustainable Energy & Fuels. 2023;**7**(14):3395-3403

[73] Anish M et al. Utilization of nano materials in hydrogen production-emerging technologies and its advancements: An overview. International Journal of Hydrogen Energy. 2023;**52**:140-158

[74] Rahman ML et al. Heavy metals removal from electroplating wastewater by waste fiber-based poly (amidoxime) ligand. Water. 2021;**13**(9):1260

[75] Farhan A et al. Removal of toxic metals from water by nanocomposites through advanced remediation processes and photocatalytic oxidation. Current Pollution Reports. 2023;**9**:1-21

[76] Kumar KY, Raj TNV, Archana S, Prasad SBB, Olivera S, Muralidhara HB. SnO2 nanoparticles as effective adsorbents for the removal of cadmium and lead from aqueous solution: Adsorption mechanism and kinetic studies. Journal of Water Process Engineering. 2016;**13**:44-52

[77] Shrestha R et al. Technological trends in heavy metals removal from industrial wastewater: A review. Journal of Environmental Chemical Engineering. 2021;**9**(4):105688

[78] Kumar R, Rauwel P, Rauwel E. Nanoadsorbants for the removal of heavy metals from contaminated water: Current scenario and future directions. PRO. 2021;**9**(8):1379

[79] Gayathiri E et al. Plant microbe based remediation approaches in dye removal: A review. Bioengineered. 2022;**13**(3):7798-7828

[80] Javed M et al. Fabrication of effective Co-SnO2/sgcn photocatalysts for the removal of organic pollutants and pathogen inactivation. Crystals. 2023;**13**(2):163

[81] Šuligoj A, Pavlović J, Arčon I, Rajić N, Novak Tušar N. SnO2-containing clinoptilolite as a composite photocatalyst for dyes removal from wastewater under solar light. Catalysts. 2020;**10**(2):253

[82] Wang L, Shen L, Zhu L, Jin H, Bing N, Wang L. Preparation and photocatalytic properties of SnO2 coated on nitrogen-doped carbon nanotubes. Journal of Nanomaterials. 2012;**2012**:32-32

[83] Sharma M, Jain T, Singh S, Pandey OP. Photocatalytic degradation of organic dyes under UV–visible light using capped ZnS nanoparticles. Solar Energy. 2012;**86**(1):626-633

[84] Mahmood T, Momin S, Ali R, Naeem A, Khan A, Technologies for Removal of Emerging Contaminants from Wastewater. London, United Kingdom: IntechOpen-Eurospan; 2022

[85] Kanakaraju D, Glass BD, Oelgemöller M. Titanium dioxide photocatalysis for pharmaceutical wastewater treatment. Environmental Chemistry Letters. 2014;**12**:27-47

[86] Pisharody L, Gopinath A, Malhotra M, Nidheesh PV, Kumar MS. Occurrence of organic micropollutants in municipal landfill leachate and its effective treatment by advanced oxidation processes. Chemosphere. 2022;**287**:132216

Chapter 10

Wastewater Treatment Approaches for the Removal of Antidepressant Residues

Radu C. Racovita and Maria D. Ciuca

Abstract

Pharmaceuticals are a major class of contaminants of emerging concern for wastewaters and natural waters alike. Among pharmaceuticals, antidepressants are the most rapidly increasing subclass, as more and more people are dealing with depression in their fast-paced and challenging everyday lives. As psychotropic medications, residual antidepressants in water must be carefully monitored and maintained below levels, where human health may be endangered. Moreover, aquatic life forms were shown to be seriously affected when such pollutants entered their natural habitat, in terms of locomotory, reproductive functions, or social behavior. Therefore, modern wastewater treatment plant technologies should incorporate solutions for removing antidepressant residues. This chapter summarizes recent efforts toward this goal and covers a wide range of proposed treatment approaches. Firstly, adsorptive methods are presented, whether based on classic, yet effective adsorbents like activated carbon or silicates, or modern alternatives such as ion-exchange resins or molecularly imprinted polymers. Secondly, extractive methods are considered, although currently impractical due to lack of both cheap and environmentally-benign solvents. Thirdly, advanced oxidation processes are surveyed, including ozone treatment, ultraviolet, gamma radiation, and electrochemical options, some of which, alone or in combination, may yield complete mineralization of antidepressant toxicants. Lastly, biological treatment with microorganisms is discussed, which may be highly specific, but usually does not enable a complete mineralization.

Keywords: contaminants of emerging concern, antidepressants, wastewater, adsorption, extractive methods, advanced oxidation processes, biological treatment

1. Introduction

Contaminants of Emerging Concern (CECs) are pollutants whose occurrence in the environment has been investigated more only in the last 20 years due to the lack of readily available, sensitive analytical methods that can detect their relatively low levels (typically in the order of μg/L) in environmental samples. These contaminants are widespread in both aquatic and terrestrial environments and include anthropogenic and naturally occurring chemicals, food additives, pharmaceuticals and personal care products (PPCPs), metabolites and other transformation products of PPCPs, illicit drugs, and engineered nanomaterials. CECs are not currently regulated in drinking water supplies and, therefore, are not regularly monitored in the environment. However, many of these contaminants have the potential to cause adverse ecological and/or human health effects even at low concentration levels [1].

In the European Union (EU) member states, a watch list of CECs is elaborated within national monitoring programs, as required by the Water Framework Directive (WFD) [2]. This list presents emerging substances requiring further attention due to their high frequency of occurrence, the expected risk for human health and/or aquatic life, and/or for the lack of available monitoring techniques. To date, the WFD has designated, for the EU, 45 substances or compound classes as priority pollutants [3]. However, the means for managing these substances remain a challenge. The sources of PPCPs in the environment are mostly represented by anthropic activities, coming from municipal wastewater treatment plants (WWTPs) and hospital wastewater discharges, illicit drugs, and municipal sewage [4]. In addition, animal agriculture and aquaculture are also important sources, especially for antibiotics that are administered both therapeutically and sub-therapeutically, and for hormones that are excreted naturally or synthetic hormones that are used to regulate the reproductive system and animal growth. These compounds have been detected worldwide in WWTP effluents [5, 6], surface water [7–9], drinking water [10, 11], and groundwater [12].

Over the past two decades, due to continuously growing demands, challenges, and pressures experienced in their professional and daily lives by modern people, a particular class of pharmaceuticals started to be prescribed and consumed more and more across European countries and other developed countries, such as Canada, Australia, New Zealand, South Korea, or Israel [13]. This class is represented by antidepressants. The graph below (**Figure 1**) shows how the defined daily dosage (DDD) per 1000 inhabitants has increased continuously over the past years in several EU and European Economic Area (EEA) countries according to the latest statistics released

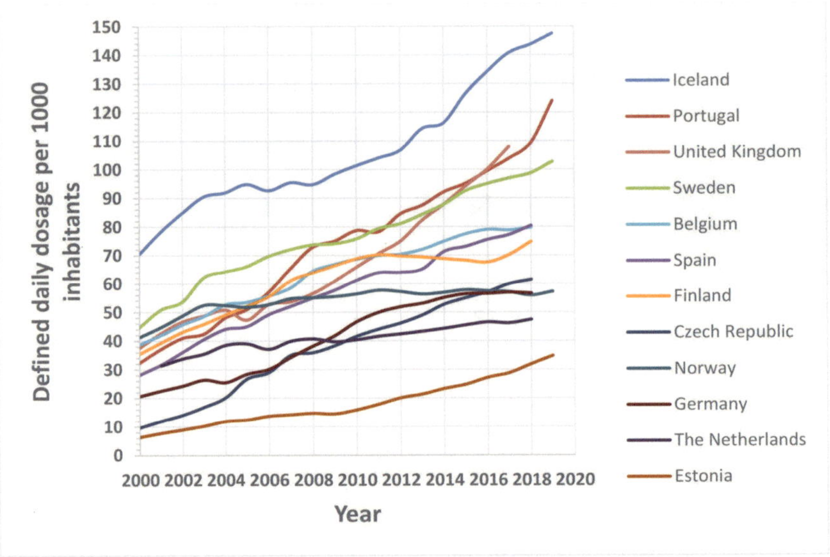

Figure 1.
Trends in consumption of antidepressant drugs in twelve EU/EEA countries over the past two decades according to statistics released by OECD.

by the Organization for Economic Co-operation and Development (OECD). In this plot, the DDD is defined as "the assumed average maintenance dose per day for a drug used on its main indication in adults" [13]. While countries such as Iceland, Portugal, Sweden, and the UK appear to be the highest European consumers of antidepressants, in fact, all European societies have experienced some increase in the prescription and consumption of these pharmaceuticals, somewhat commensurate with their annual healthcare spending per capita [14]. A study from 2015 concluded that the average percentage of the general population of the 27 EU member states who reported using antidepressants regularly was 7.2%, ranging from a high 15.7% in Portugal to a low 2.7% in Greece [14]. These percentages have increased since and will only continue to grow, especially after the recent COVID-19 pandemic measures enforced by governments worldwide, which only augmented depression symptoms through human isolation and limited social interaction [15–17].

Along with high prescription and consumption rates comes a correlated increase in contamination levels with such drugs and their metabolites of municipal waters or wastewaters from pharmaceutical companies or hospitals [18, 19]. Although it has been estimated that, overall, these types of drugs make up only 4% of all pharmaceuticals detected in the environment [20], the reported levels of contamination of wastewaters with antidepressant residues range from the limit of detection (~0.1 ng/L) up to about 3000 ng/L in WWTP raw influents and a few tens or hundreds of ng/L in final effluents [19], which are concerning levels given the toxic effects of these chemical compounds for both humans and aquatic species.

2. Antidepressants and their impact on human health and aquatic life

Antidepressant drugs are typically classified based on their biochemical mode of action and on some structural features (**Figure 2**) into (a) selective serotonin reuptake inhibitors (SSRI), which are the most highly prescribed and encountered, for example, fluoxetine (FLU), sertraline (SER), paroxetine (PAR), fluvoxamine (FLA), and citalopram (CIT); (b) serotonin-noradrenaline reuptake inhibitors (SNRI), like venlafaxine (VEN) and duloxetine (DUL); (c) tricyclic antidepressants (TCA), including imipramine (IMI), clomipramine (CLO), trimipramine (TRI), amitriptyline (AMI), and doxepine (DOX), which are usually employed in major depressive disorders [21]; (d) monoamine oxidase inhibitors (MAOI), such as phenelzine, isocarboxazid, or selegiline; and (e) atypical antidepressants, like bupropion (BUP), mianserin (MIA), and trazodone, among others [22]. Regardless of their particular class, the administration

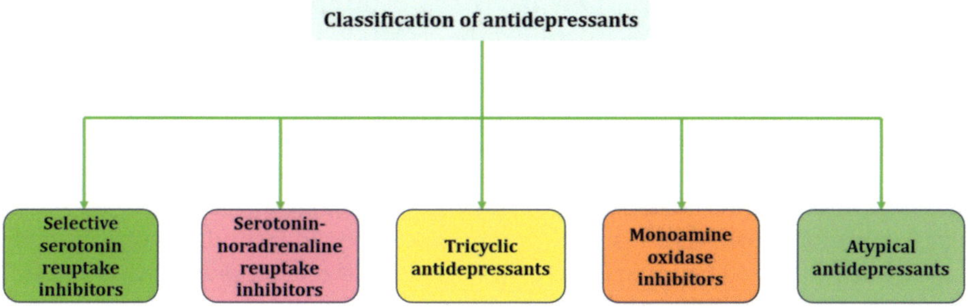

Figure 2.
Classes of antidepressants.

of antidepressants to patients is always done under careful supervision by medical practitioners, often in combination with psychotherapy and with regular monitoring of blood levels for an informed adjustment of dosages in the long term [23].

Therefore, intoxications may occur upon ingestion of uncontrolled concentrations of antidepressants by people. Many studies have shown that antidepressants can cause health problems ranging from mild, such as dry mouth, fatigue, nausea, headaches, and vomiting, to very serious, for example, cardiac dysrhythmias, sexual dysfunctions, seizures, and even death [24]. The fatal toxicity index (FTI), *i.e.* the number of deaths per million prescriptions, varies from a low 2.0 to 6.2 in the case of SSRI and atypical antidepressants to around 13.5 for MAOI and a high 34.1 in the case of TCAs, which are the most concerning antidepressant class [24].

When WWTP effluents arrive eventually in natural water reservoirs, remnant antidepressant residues can affect fish and other forms of aquatic life. Several studies revealed various negative consequences of antidepressant ingestion by fish: reduction of food intake, reduction of swimming velocity and distance covered; activation of locomotory organs irrespective of the presence or absence of a predator, modification of courtship behavior, which in turn affects population dynamics, weight loss, and reduction of hatching period [19, 25]. Aquatic invertebrates were also shown to be affected even by trace amounts of antidepressants. Spawning and larval release in bivalves, as well as locomotion and fecundity in snails, were reduced. Among crustaceans, antidepressants affected the photo- and geotactic behavior of amphipods, the aggression of crayfish, and the reproduction and development of daphnids [26].

As such, effective technologies for minimizing antidepressant residues in wastewaters must be developed and implemented in the immediate future.

3. Overview of water treatment approaches for the removal of antidepressant residues

The current approaches for water treatment to remove residual antidepressant drugs are, broadly, the same as for other types of CECs. Treatment options for the removal of CECs from drinking water as well as wastewater include physical-chemical separation [27], coagulation and sorption processes [28], ozone-based and other advanced oxidation processes (AOPs) [29, 30], including ultraviolet (UV) [31–33] or gamma radiation treatment [34], biological and enzymatic treatment [35, 36], ultrasonic degradation [37], ultrafiltration (UF) and nanofiltration (NF) through membranes [37], electrochemical oxidation [38], and reverse osmosis (RO) using special membranes [34, 39].

Each of these methods has its pros and cons when applied to PPCPs. Physical-chemical treatments such as coagulation or flocculation were generally found to be unable to effectively remove PPCPs. On the other hand, although AOPs can be effective for the removal of some pharmaceuticals, these processes can lead to the formation of oxidation intermediates that have occasionally been found to be even more toxic than the parent compounds, thus raising concerns about their presence in the environment and the necessity to assess their ecological risks along with their parent compounds [40, 41]. Adsorption processes do not add undesirable by-products and have been found to be superior to other techniques for wastewater treatment in terms of simplicity of design and operation, and insensitivity to toxic substances [42]. Among several materials used as adsorbents, activated carbon (AC) has been used for the removal of different types of PPCPs, but its use is sometimes restricted due to high costs [43]. Furthermore, once the AC has been exhausted, it can be regenerated for further use, but regeneration processes have additional costs, may result in a loss of carbon, and the regenerated product may have a lower adsorption capacity after a few rounds of regeneration in comparison to the non-recycled AC [44]. This has triggered attempts by various scientists and engineers to prepare low-cost alternative adsorbents, which may replace AC in pollution control by adsorption and may overcome its economic disadvantages [42]. Recently, natural materials that are available in large quantities from agricultural activities have been evaluated as low-cost and environmentally-friendly sorbents. Moreover, the utilization of these waste materials as such or after some minor treatment as adsorbents is receiving attention because they represent unused secondary resources and would otherwise cause serious disposal problems [45]. When comparing biological treatment technologies, waste stabilization ponds, trickling filters, and activated sludge supplemented with a biofilter system for estrogenic and aryl hydrocarbon receptor, the waste stabilization ponds showed greater or similar removal efficiency as activated sludge with biofilters, and both were much more efficient than the trickling filter, which had the lowest removal efficiency [46]. For enhanced performance, biological treatment has been combined sometimes with ultrasound and membrane filtration [47]. Electro-oxidation (anodic oxidation) performed using high cell voltages in wastewater treatment can involve either only reactive oxygen species (like HO· or H_2O_2) or, for chloride-rich waters, also chlorine species (Cl_2, $HClO/ClO^-$, ClO_2^-), which further improve the oxidative action against pharmaceutical contaminants [34].

For removal of antidepressant drug residues from wastewater and other contaminated waters, treatment options considered to date can be classified into adsorptive, extractive, oxidative (O_3 and/or other AOP), and biological degradation processes (**Figure 3**).

3.1 Adsorptive methods

Adsorbents used to date for the removal of antidepressants from aqueous media include cation-exchange resins [48], activated carbon, silicates, and titanium dioxide nanoparticles [49], or molecularly imprinted polymers [50].

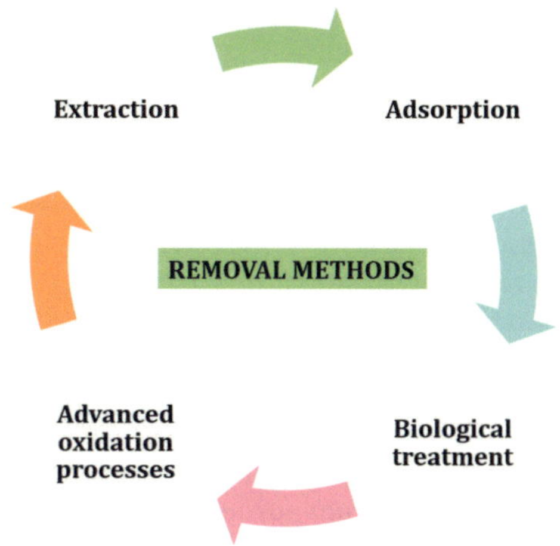

Figure 3.
Methods for removal of antidepressant residues from wastewater.

Choi et al. [48] screened six different cation-exchange resins for their adsorption efficiencies with respect to tricyclic antidepressants. Three of these had sulfone functional groups (Amberlite IR-120, Dowex 50WX4-200, and Dowex HCR W2) and the other three had carboxyl functionalities (Amberlite IRC-86, Dowex MAC-3, and Amberlite IRP-64). Using one-point check tests with AMI as a model pharmaceutical contaminant, they showed that Dowex 50WX4-200 had the highest uptake of AMI. Aqueous pH was shown to have an important effect on adsorption efficacy, specifically a negative effect above 7, where TCAs, having pKa values around 9–10, tend to partially deprotonate and become insoluble in water. However, pH values between 3 and 6.5 did not inflict any significant changes on adsorption capacity, therefore, an almost neutral pH of 6.5 was chosen for further measurements. Using adsorption isotherm models for the experimental results, the maximum adsorption capacities of the resin were determined for a series of four TCAs (AMI, CLO, IMI, and its demethylated derivative desipramine DES) and ranged from 2.53 to 3.76 mmol/g. The removal efficiency of DES from real wastewater samples containing other miscellaneous cations was also probed and, interestingly, it was weaker in the case of filtered wastewater (56.66%) as compared to unfiltered (77.99%). The latter was explained based on an additional adsorptive contribution from the activated sludge found in unfiltered wastewater [48].

Activated charcoal was also used successfully as an adsorbent for IMI, exhibiting a maximum adsorptive capacity of 610 mg/g (2.17 mmol/g) at pH 8.5 and 372 mg/g (1.33 mmol/g) at pH 6.8 according to the Langmuir adsorption model. Unlike with cation-exchange resins, the deprotonated form (electrically neutral) of IMI adsorbed better on the carbonaceous substrate than the protonated one [51]. In a different study, AMI and its demethylated derivative, nortriptyline (NOR), were removed from water using activated carbon, diosmectite (a natural silicate), and TiO_2 nanoparticles [49]. Activated carbon performed best, with specific adsorption capacities of 3.75 and 1.87 mmol/g for AMI and NOR, respectively. Nonetheless, diosmectite performed very well also, with specific capacities of 3.30 and 1.82 mmol/g, respectively, while TiO_2 nanoparticles showed only 2.97 and 0.79 mmol/g maximum adsorption capacities for AMI and NOR. The mechanism of adsorption was also investigated and concluded to be a combination of (1) electrostatic interaction between positively charged antidepressant molecules and the negative surface charge of the adsorbents and (2) π-π stacking interactions between the delocalized aromatic rings of antidepressants and carbon structures on the surface of activated carbon, thus explaining why this particular material performed better. Nevertheless, the cost of production of activated carbon is quite high [52], however, several options exist for its cost-effective regeneration such as chemical [53], biological [54], and electrochemical regeneration [44].

One of the most recent developments in adsorptive materials for environmental remediation is represented by molecularly imprinted polymers (MIPs), which are polymers that had been imprinted during polymerization with a desired template to enable highly selective recognition sites onto their surface, which is why they are sometimes referred to as artificial antibodies or synthetic receptors [50]. Gornik et al. prepared such MIPs using methacrylic acid as monomer, either alone or co-polymerized with methyl methacrylate or 2-hydroxymethyl methacrylate. Although adsorptive capacity was lower than for activated carbon (e.g. 72.6 mg/g or 0.24 mmol/g for SER), primarily due to much smaller specific surface area of the polymer materials (27.4–193.8 m^2/g) compared to the highly porous AC (1400 m^2/g), the main advantage of these MIPs was represented by their higher selectivities and affinities for their target contaminants. The MIPs selectively removed SSRI antidepressants from water, like SER, FLU, PAR, and CIT, leaving behind interfering related molecules such as

BUP and the structurally related drug, bupivacaine. Furthermore, the polymers required regeneration after longer intervals compared to AC, which is advantageous in terms of cost of operation.

Although the best-performing adsorbents can have high initial costs of production, the available technologies for their regeneration and reuse make adsorptive methods a highly relevant option for the removal of antidepressant residues from contaminated water.

3.2 Extractive methods

In general, liquid-liquid extraction methods are only of historical significance as most organic immiscible solvents that were tried over the years are not green solvents and would actually augment the contamination of wastewaters with their own residues, of even greater concern than antidepressants themselves. Kocoglu et al. [55] tested the extraction efficacy of halogenated and non-halogenated organic solvents, such as dichloromethane, chloroform, and ethyl acetate, using SER as a model antidepressant molecule. Ethyl acetate yielded the best results, especially when 2% (v/v) ethanol was added to it, however, ethyl acetate presents a plethora of toxic effects [56] and this method is not suitable for wastewater treatment.

Alternative, greener solvents were tested in other studies for antidepressant extraction from water. Thus, water-in-oil microemulsions, prepared using various homologous carboxylic esters and Brij 30, a non-ionic, environmentally-benign surfactant, were used to extract four antidepressants (AMI, DOX, IMI, and CLO) from contaminated water with extraction efficiencies ranging from 67.2 to 99.5% at nearly neutral pH and low ionic strength [57]. Propyl acetate-based microemulsions performed best, maintaining extraction yields between 88 and 99.3% even at high concentrations of antidepressant contaminants (up to 100 mg/L). The presence of other ions in the water (high ionic strength) or a more basic pH of 9 produced further increases in extraction yields, from 93 to nearly 100%, revealing that these extraction solvents remained very promising even with more diverse contaminant loadings of wastewaters. Still, the use of organic ester solvents in the composition of the microemulsions, even though in smaller proportions, remains an inconvenient factor, as is also the cost of the non-ionic surfactant.

Ionic liquids, with their well-known advantages of negligible volatility, high chemical and thermal stability, excellent solvation ability, and non-corrosive nature, were also proposed as green extraction solvents for antidepressants [58]. Zawadski et al. [59] used aqueous biphasic systems based on either tetrabutylammonium or tetrabutylphosphonium salts and phosphate buffers to extract AMI from pharmaceutical waste. The observed extraction efficiencies ranged from 94 to 100%, and they were improved when the pH was raised from 6.6 to 9.6, however, they remained virtually unchanged when further raised to 13.2, proving that advanced alkalization is not necessary in this case. Nevertheless, the high costs currently associated with ionic liquids make these solvents only promising candidates for future technologies.

Given the toxicity associated with classic organic solvents and the high costs of modern green solvents, extractive methods do not represent currently a viable method of decontamination for wastewaters containing antidepressant residues.

3.3 Advanced oxidation processes (AOPs)

UV irradiation alone, whether UVA, UVB, or UVC, has proven itself to be a suitable technique for water decontamination when polluted by antidepressants or their derivatives [60]. UVC treatment has been most effective, as shown by recent studies of photo-degradation of tricyclic antidepressants like AMI and CLO [61] or TRI [62] in water with efficiencies above 90%, while the combination of UVA + UVB applied to DES in water yielded a very long half-life of 36 hours [63] and UVA irradiation alone was only able to degrade less than 30–40% of both DOX and VEN even after 114 days of irradiation [64]. The performance of the latter was only improved significantly when used in combination with a photocatalyst, TiO_2, or with Fe^{2+} and H_2O_2, *i.e.* a photo-Fenton process, which led to 75% degradation of IMI in 4 hours and complete degradation in 24 hours [65]. In general, for increased effectiveness, UV treatment tends to be combined with some other type of advanced oxidation process.

Ozone (O_3) has proven to be highly effective at degrading antidepressant residues from wastewater. Mendez-Arriaga et al. [66] performed a study of oxidative disposal of FLU from aqueous environment using O_3, $O_3 + TiO_2$, $O_3 + H_2O_2$ (PEROXONE process), and $O_3 + TiO_2 + H_2O_2$, with and without simultaneous UV irradiation. The effectiveness varied, leading to partial or complete photo-mineralization of FLU to yield CO_2 and inorganic ions like NO_2^-, NO_3^-, and F^-. Under dark conditions, no depletion of FLU occurred at either acid or basic pH. Methods involving TiO_2 were only effective in alkaline medium, because adsorption onto the TiO_2 surface was a mandatory pre-condition for degradation and cationic antidepressant species in acid environment were repelled by the positively charged surface of the oxide catalyst. UVA (360 nm) irradiation was also not able to degrade FLU in acidic medium, but after alkalization, the concentration of FLU decayed by ca. 15% after 60 min of illumination. The extent of photo-degradation doubled when combining UVA and O_3 treatment with a small amount of H_2O_2. The heterogeneous photo-assisted process UVA + O_3 + TiO_2 led to 50% mineralization in 60 minutes. The presence of H_2O_2 enhanced photo-mineralization further, such that the hybrid configuration UVA + O_3 + H_2O_2 + TiO_2 attained 97% mineralization, suggesting it to be the best combination for wastewater treatment against such recalcitrant drug pollutants.

An alternative to UV radiation is represented by ionizing radiation, namely gamma (γ) radiation. Bojanowska-Czajka et al. investigated the γ-ray-inflicted degradation of SER and CIT in surface water, including in the presence of commonly encountered radical scavengers such as CO_3^{2-}, NO_3^-, and humic acid [67]. As γ radiation source, ^{60}Co was used, with radiation doses between zero and 500 Gy. At concentration levels of 1 mg/L, both antidepressants were fully degraded by a dose of 100 Gy, while at 10 mg/L, SER needed 200 Gy and CIT 400 Gy for complete mineralization, which for a dose rate of 2.15 kGy of the radiation source translated into only 11 minutes of irradiation, a much shorter time than in the case of UV radiation. When 10 mg/L each of CO_3^{2-}, NO_3^-, and humic acid were added to simulate radical scavengers that may be commonly encountered in such media, the degradation of SER was almost unaffected at a concentration 1 mg/L and 100 Gy absorbed dose. The degradation of CIT at the same level of concentration was unaffected by nitrates, but carbonates and humic acid did have an impact. For an absorbed dose of 50 Gy, the degradation efficiency was 70% in the presence of CO_3^{2-} and 50% with humic acid. Carbonates continued to inhibit somewhat the degradation of CIT even at 200 Gy, where an 80% efficiency was attained. Interestingly, 100% degradation of CIT was achieved with humic acid additive at a dose of only 150 Gy, better than in the absence of added radical scavengers.

Electro-oxidation (anodic oxidation) has also been reported as a suitable technique for destruction of antidepressant residues found in water. Depending on the simultaneous presence or absence of chloride anions in the same aqueous environment, electro-oxidation processes can be with or without participation of active chlorine species [34]. Various reactive oxygen species (ROS) are generated from water discharge at the anode surface, *e.g.* H_2O_2, O_3, and physisorbed OH radical, which is the strongest oxidant. Anodes that are typically used in this type of water treatment are Pt, IrO_2, or RuO_2, known as "active" electrodes, which tend to oxidize water molecules to strongly bound, chemisorbed "active oxygen" or metal superoxide species, as well as PbO_2, SnO_2, and boron-doped diamond (BDD), considered "non-active" electrodes, where water splitting only leads to physisorbed OH radicals, which interact so weakly with the anode surface that they easily oxidize organics to their complete mineralization. BDD is usually the most potent among them [34]. Melin et al. investigated the degradation of 130 mg/L AMI by three different electrochemical AOPs, where active OH radicals were generated by simple anodic oxidation (AO), an electro-Fenton (EF), and a photo-electro-Fenton (PEF) process, respectively [68]. For AO, at a fixed current density of 100 mA/cm^2, they also compared the efficiency of different electrodes, such as Pt, RuO_2, and BDD, the latter yielding the best results, namely 76% mineralization after 360 minutes, as revealed by total organic carbon (TOC) analysis. Only 21 and 27%, respectively, mineralization was achieved with Pt and RuO_2, and even this was mainly the outcome of the presence of Cl^- in the system, as the AMI hydrochloride salt was employed. Under these conditions, at pH less than 3.3, the main oxidant is Cl_2, while at higher pHs, hypochlorite ClO^- takes over [68]. Using the BDD anode, Melin et al. also studied the influence of current density and obtained within 360 minutes degrees of mineralization of 50, 64, 76, and 83% with applied currents of 33.3, 66.7, 100, and 150 mA/cm^2, respectively. By adding 0.5 mmol/L Fe^{2+} catalyst to the medium, conditions for an electro-Fenton (EF) process were created, as H_2O_2 is formed anyways by OH radical coupling during AO. Indeed, the EF process yielded 78% mineralization faster, within 240 min of electrolysis at 100 mA/cm^2 using the BDD anode. Interestingly, doubling the concentration of catalyst did not improve mineralization, because iron hydroxide species are formed, limiting the generation of OH radicals in the bulk and thus causing a decrease of the efficiency of the EF process [68]. It was, however, the PEF process which allowed the best performance for mineralization of AMI due to the synergistic action of highly reactive physisorbed OH onto BDD, homogeneous OH in the bulk, and UVA (360 nm) radiation. After 360 min of treatment, maintaining 100 mA/cm^2 current density and 0.5 mmol/L Fe^{2+}, 95% mineralization was achieved. Using gas chromatography and high-performance liquid chromatography coupled with mass spectrometry analysis, the authors of the study were also able to evidence some of the intermediates formed from AMI along its mineralization pathway, namely polyaromatic hydrocarbons and their oxidized derivatives, such as dibenzosuberone and 5-dibenzosuberenone, and carboxylic acids like succinic, malic, oxalic, and formic acid [68].

3.4 Biological treatment

The biological treatment of wastewaters containing antidepressant drug residues involves the employment of microorganisms, which typically produce enzymes that are responsible for the breakdown of the structurally complex, toxic pharmaceutical molecules into simpler, less toxic products. The biological transformation of pharmaceutical residues in WWTP occurs via two principal mechanisms: co-metabolism, where the drug pollutant is degraded by some of the enzymes secreted by the microbial communities found in the sewage sludge, or by sole substrate degradation, in which the target compound constitutes the sole source of carbon and energy for the microbes [69]. Depending on the availability of oxygen and its actual usage by microorganisms, drug biotransformation mechanisms can also be classified as aerobic or anaerobic.

The efficacy of biodegradation of pharmaceutical pollutants depends primarily on their solubility in the wastewater. If the solubility is low (hydrophobic drug molecules), then they will be retained in the sewage sludge, providing more time for microbial degradation, whether by action of catabolic enzymes or for utilization as single carbon source. Highly soluble drug micropollutants, on the other hand, may evade the biodegradation process, leaving the WWTP after a very short period of residence [69].

Gornik et al. performed a biotransformation study of the most commonly prescribed antidepressant, SER, using two different approaches: (1) batch biodegradation and sorption and (2) flow-through pilot wastewater treatment bioreactor [70]. For the batch experiments, sorption to the activated sludge was the true leading removal process for SER, around 90%, and only secondary was the biodegradation, highly influenced also by the presence of other carbon sources easily biodegradable by the microbial population. Identified products of biodegradation included norsertraline, sertraline ketone, hydroxyl-sertraline, and traces of a few others, *i.e.* biological degradation did not yield complete breakdown to inorganic reaction products. The highest removal effectiveness achieved with the flow-through bioreactor was 94%. When real wastewaters from two WWTP and one psychiatric hospital were subjected to the flow-through bioreactor treatment process, SER removal efficiencies ranged from 77 to 81%, but, once again, in addition to the small percentage of leftover SER, eight other structurally complex transformation products could still be identified in the resulting effluents [70].

Mycoremediation of wastewater containing 2.5 mg/L each of SER, PAR, FLU, CIT, CLO, VEN, and MIA was attempted in a different study using *Pleurotus ostreatus* [71]. To separate fungal biodegradation from mere adsorption on fungal biomass, control experiments were also performed, where the fungus was autoclaved for 15 min at 120°C to inactivate all its enzymes. Based on combined adsorption+enzymatic degradation, removal efficiencies above 90% were attained after 96 hours of fungal biomass exposure for SER, PAR, CLO, and MIA, about 85% was attained in the case of FLU, and 50% and 22% only for CIT and VEN, respectively. When considering only the enzymatic contribution from the fungus, CIT and VEN were essentially unaffected, MIA and CLO exceeded 80% degradation, while PAR, SER, and FLU were degraded in proportion of 50, 48, and 24%, respectively [71]. The authors of the study interpreted these findings in terms of the degree of structural similarity between each antidepressant and lignin monomeric structures, which represent the natural substrates of the fungal enzymes. Thus, lignin-like multi-aromatics such as SER, CLO, MIA, PAR, and FLU showed the highest degradation efficiencies, slightly lower in the case of FLU because of its three electron-withdrawing fluorine atoms causing a reduced aromaticity of one of the benzene rings. A similar reason may have caused the increased resistance of CIT, while VEN may have not been degraded significantly because it contains a single aromatic ring and was thus not recognized as a substrate by the specific enzymes of the fungus [71]. Overall, the results suggested that certain microorganisms will perform better when dealing with a specific type of environmental drug pollutant.

Table 1 summarizes the comparative efficiencies of the various methods of removal of antidepressant residues from wastewater discussed in this chapter.

Wastewater treatment method	Specific conditions	Antidepressant removed	Efficiency of removal (%)	Reference
Adsorption	Cation-exchange resins	AMI, CLO, IMI, DES	57–78	[48]
	Activated carbon	AMI, NOR	17–37	[49]
		IMI	37–61	[51]
	Diosmectite	AMI, NOR	16–31	[51]
	TiO_2 nanoparticles	AMI, NOR	7–22	[51]
	Molecularly imprinted polymers	SER, FLU, CIT, PAR	70–100	[50]
		BUP	55–90	[50]
Extraction	Ethyl acetate	SER	86–95	[55]
	Propyl acetate+Brij 30 + Water microemulsion	AMI, DOX, IMI, CLO	88–100	[57]
	Tetrabutyl-ammonium/ phosphonium salt+phosphate buffer	AMI	94–100	[59]
Advanced oxidation processes	UVA irradiation	DOX, VEN	30–40	[64]
		FLU	15	[66]
	UVA + UVB irradiation	DES	50	[63]
	UVC irradiation	AMI, CLO	88–100	[61]
		TRI	92	[62]
	O_3 + TiO_2 + UVA	FLU	50	[66]
	O_3 + H_2O_2 (PEROXONE) + UVA	FLU	70	[66]
	O_3 + TiO_2 + H_2O_2 + UVA	FLU	97	[66]
	Gamma (γ) radiation	SER, CIT	80–100	[67]
	Electro-oxidation using BDD anode	AMI	76	[68]
	Electro-Fenton (EF)	AMI	78	[68]
	Photo-electro-Fenton (PEF)	AMI	95	[68]
Biological degradation	Activated sludge	SER	77–81	[70]
	Fungal adsorption+ enzymatic degradation	SER, PAR, CLO, MIA, FLU, CIT, VEN	22–98	[71]

Table 1.
Efficiencies of removal of various antidepressant residues from wastewater by currently available treatment methods.

4. Conclusions

In conclusion, antidepressant drug residues have become of greater concern every year when considering wastewater treatment technologies, because of their continuously increasing presence in the society and implicitly in the environment. Their occurrence should not be neglected in light of recent evidence suggesting they can severely affect aquatic ecosystems and, at higher levels, even human health.

Although they have only started to emerge, several wastewater treatment options are available nowadays to deal with this new class of drug pollutants, ranging from classic methods like sorption onto activated carbon, silicates, or ion-exchange resins, to advanced oxidation processes using ozone and other reactive oxygen species generated either photo-, electro-, or radio-chemically, and ending with modern technologies such as biological degradation by fungi or other microorganisms with high affinity for such organic contaminants.

Acknowledgements

This work was supported by a grant from the Ministry of Research, Innovation and Digitization, CNCS-UEFISCDI, project number PN-III-P1-1.1-TE-2021-1216, within PNCDI III.

Conflict of interest

The authors declare no conflict of interest.

Author details

Radu C. Racovita* and Maria D. Ciuca
Department of Inorganic Chemistry, Physical Chemistry and Electrochemistry, Faculty of Chemical Engineering and Biotechnologies, National University of Science and Technology POLITEHNICA Bucharest, Bucharest, Romania

*Address all correspondence to: radu.racovita@upb.ro

References

[1] Noguera-Oviedo K, Aga DS. Lessons learned from more than two decades of research on emerging contaminants in the environment. Journal of Hazardous Materials. 2016;**316**:242-251

[2] European Commission. Directive 2000/60/EC of the European Parliament and of the council of 23 October 2000 establishing a framework for community action in the field of water policy. Official Journal of the European Union. 2000;**L327**(43):1-73

[3] European Commission. Directive 2013/39/EU of the European Parliament and of the council amending directives 2000/60/EC and 2008/105/EC as regards priority substances in the field of water policy. Official Journal of the European Union. 2013;**L226**:1-17

[4] Bell KY, Wells MJM, Traexler KA, Pellegrin ML, Morse A, Bandy J. Emerging pollutants. Water Environment Research. 2011;**83**(10):1906-1984

[5] Samaras VG, Stasinakis AS, Mamais D, Thomaidis NS, Lekkas TD. Fate of selected pharmaceuticals and synthetic endocrine disrupting compounds during wastewater treatment and sludge anaerobic digestion. Journal of Hazardous Materials. 2013;**244-245**:259-267

[6] Sun Q, Lv M, Hu A, Yang X, Yu CP. Seasonal variation in the occurrence and removal of pharmaceuticals and personal care products in a wastewater treatment plant in Xiamen, China. Journal of Hazardous Materials. 2014;**277**:69-75

[7] Batt AL, Kincaid TM, Kostich MS, Lazorchak JM, Olsen AR. Evaluating the extent of pharmaceuticals in surface waters of the United States using a national-scale Rivers and streams assessment survey. Environmental Toxicology and Chemistry. 2016;**35**(4):874-881

[8] Bayen S, Zhang H, Desai MM, Ooi SK, Kelly BC. Occurrence and distribution of pharmaceutically active and endocrine disrupting compounds in Singapore's marine environment: Influence of hydrodynamics and physical-chemical properties. Environmental Pollution. 2013;**182**:1-8

[9] Kong L, Kadokami K, Wang S, Duong HT, Chau HTC. Monitoring of 1300 organic micro-pollutants in surface waters from Tianjin, North China. Chemosphere. 2015;**122**:125-130

[10] Padhye LP, Yao H, Kung'u FT, Huang CH. Year-long evaluation on the occurrence and fate of pharmaceuticals, personal care products, and endocrine disrupting chemicals in an urban drinking water treatment plant. Water Research. 2014;**51**:266-276

[11] Yang GC, Yen CH, Wang CL. Monitoring and removal of residual phthalate esters and pharmaceuticals in the drinking water of Kaohsiung City, Taiwan. Journal of Hazardous Materials. 2014;**277**:53-61

[12] Cabeza Y, Candela L, Ronen D, Teijon G. Monitoring the occurrence of emerging contaminants in treated wastewater and groundwater between 2008 and 2010. The Baix Llobregat (Barcelona, Spain). Journal of Hazardous Materials. 2012;**239-240**:32-39

[13] OECD Health Statistics 2020: OECD. 2020. Available from: https://www.oecd.org/els/health-systems/health-data.htm

[14] Lewer D, O'Reilly C, Mojtabai R, Evans-Lacko S. Antidepressant use in 27 European countries: Associations with sociodemographic, cultural and economic factors. The British Journal of Psychiatry. 2015;**207**(3):221-226

[15] Benke C, Autenrieth LK, Asselmann E, Pane-Farre CA. Lockdown, quarantine measures, and social distancing: Associations with depression, anxiety and distress at the beginning of the COVID-19 pandemic among adults from Germany. Psychiatry Research. 2020;**293**:113462

[16] Brenner MH, Bhugra D. Acceleration of anxiety, depression, and suicide: Secondary effects of economic disruption related to COVID-19. Frontiers in Psychiatry. 2020;**11**:592467

[17] Thoresen S, Blix I, Wentzel-Larsen T, Birkeland MS. Trusting others during a pandemic: Investigating potential changes in generalized trust and its relationship with pandemic-related experiences and worry. Frontiers in Psychology. 2021;**12**:698519

[18] Castillo-Zacarías C, Barocio ME, Hidalgo-Vázquez E, Sosa-Hernández JE, Parra-Arroyo L, López-Pacheco IY, et al. Antidepressant drugs as emerging contaminants: Occurrence in urban and non-urban waters and analytical methods for their detection. Science of the Total Environment. 2021;**757**:143722

[19] Melchor-Martínez EM, Jiménez-Rodríguez MG, Martínez-Ruiz M, Peña-Benavides SA, Iqbal HMN, Parra-Saldívar R, et al. Antidepressants surveillance in wastewater: Overview extraction and detection. Case Studies in Chemical and Environmental Engineering. 2021;**3**:100074

[20] Santos LH, Araujo AN, Fachini A, Pena A, Delerue-Matos C, Montenegro MC. Ecotoxicological aspects related to the presence of pharmaceuticals in the aquatic environment. Journal of Hazardous Materials. 2010;**175**(1-3):45-95

[21] Ciuca MD, Racovita RC. Development of visible spectrophotometric methods for the determination of tricyclic antidepressants based on formation of molecular complexes with p-benzoquinones. International Journal of Molecular Sciences. 2023;**24**(23):16744

[22] Alvano SA, Zieher LM. An updated classification of antidepressants: A proposal to simplify treatment. Personalized Medicine in Psychiatry. 2020;**19-20**:100042

[23] Mann K, Hiemke C, Schmidt LG, Bates DW. Appropriateness of therapeutic drug monitoring for antidepressants in routine psychiatric inpatient care. Therapeutic Drug Monitoring. 2006;**28**(1):83-88

[24] Sarko J. Antidepressants, old and new: A review of their adverse effects and toxicity in overdose. Pharmacologic Advances in Emergency Medicine. 2000;**18**(4):637-654

[25] Magnuson JT, Longenecker-Wright Z, Havranek I, Monticelli G, Brekken HK, Kallenborn R, et al. Bioaccumulation potential of the tricyclic antidepressant amitriptyline in a marine polychaete, Nereis virens. Science of the Total Environment. 2022;**851**(Pt 1):158193

[26] Fong PP, Ford AT. The biological effects of antidepressants on the molluscs and crustaceans: A review. Aquatic Toxicology. 2014;**151**:4-13

[27] Petrie B, McAdam EJ, Hassard F, Stephenson T, Lester JN, Cartmell E. Diagnostic investigation of steroid

estrogen removal by activated sludge at varying solids retention time. Chemosphere. 2014;**113**:101-108

[28] Hassan SSM, Abdel-Shafy HI, Mansour MSM. Removal of pharmaceutical compounds from urine via chemical coagulation by green synthesized ZnO-nanoparticles followed by microfiltration for safe reuse. Arabian Journal of Chemistry. 2019;**12**(8):4074-4083

[29] Ibanez M, Gracia-Lor E, Bijlsma L, Morales E, Pastor L, Hernandez F. Removal of emerging contaminants in sewage water subjected to advanced oxidation with ozone. Journal of Hazardous Materials. 2013;**260**:389-398

[30] Rosal R, Rodriguez A, Perdigon-Melon JA, Petre A, Garcia-Calvo E, Gomez MJ, et al. Occurrence of emerging pollutants in urban wastewater and their removal through biological treatment followed by ozonation. Water Research. 2010;**44**(2):578-588

[31] Jung YJ, Kim WG, Yoon Y, Kang JW, Hong YM, Kim HW. Removal of amoxicillin by UV and UV/H2O2 processes. Science of the Total Environment. 2012;**420**:160-167

[32] Keen OS, Baik S, Linden KG, Aga DS, Love NG. Enhanced biodegradation of carbamazepine after UV/H2O2 advanced oxidation. Environmental Science & Technology. 2012;**46**(11):6222-6227

[33] Yonar T, Kestioglu K, Azbar N. Treatability studies on domestic wastewater using UV/H2O2 process. Applied Catalysis B: Environmental. 2006;**67**(3-4):223-228

[34] Rivera-Utrilla J, Sanchez-Polo M, Ferro-Garcia MA, Prados-Joya G, Ocampo-Perez R. Pharmaceuticals as emerging contaminants and their removal from water. A review. Chemosphere. 2013;**93**(7):1268-1287

[35] Matamoros V, Gutierrez R, Ferrer I, Garcia J, Bayona JM. Capability of microalgae-based wastewater treatment systems to remove emerging organic contaminants: A pilot-scale study. Journal of Hazardous Materials. 2015;**288**:34-42

[36] Kilic Taseli B, Gokcay CF. Biological treatment of paper pulping effluents by using a fungal reactor. Water Science and Technology. 1999;**40**(11-12):93-99

[37] Secondes MF, Naddeo V, Belgiorno V, Ballesteros F Jr. Removal of emerging contaminants by simultaneous application of membrane ultrafiltration, activated carbon adsorption, and ultrasound irradiation. Journal of Hazardous Materials. 2014;**264**:342-349

[38] Christensen PA, Yonar T, Zakaria K. The electrochemical generation of ozone: A review. Ozone: Science & Engineering. 2013;**35**(3):149-167

[39] Rodriguez-Mozaz S, Ricart M, Kock-Schulmeyer M, Guasch H, Bonnineau C, Proia L, et al. Pharmaceuticals and pesticides in reclaimed water: Efficiency assessment of a microfiltration-reverse osmosis (MF-RO) pilot plant. Journal of Hazardous Materials. 2015;**282**:165-173

[40] Celiz MD, Tso J, Aga DS. Pharmaceutical metabolites in the environment: Analytical challenges and ecological risks. Environmental Toxicology and Chemistry. 2009;**28**(12):2473-2484

[41] Escher BI, Fenner K. Recent advances in environmental risk assessment of transformation products.

Environmental Science & Technology. 2011;**45**(9):3835-3847

[42] Tong DS, Zhou CH, Lu Y, Yu H, Zhang GF, Yu WH. Adsorption of acid red G dye on octadecyl trimethylammonium montmorillonite. Applied Clay Science. 2010;**50**(3): 427-431

[43] Katsigiannis A, Noutsopoulos C, Mantziaras J, Gioldasi M. Removal of emerging pollutants through granular activated carbon. Chemical Engineering Journal. 2015;**280**:49-57

[44] Gazigil L, Er E, Yonar T. Determination of the optimum conditions for electrochemical regeneration of exhausted activated carbon. Diamond and Related Materials. 2023;**133**:109741

[45] Bhatnagar A, Sillanpää M. Utilization of agro-industrial and municipal waste materials as potential adsorbents for water treatment—A review. Chemical Engineering Journal. 2010;**157**(2-3):277-296

[46] Dagnino S, Gomez E, Picot B, Cavailles V, Casellas C, Balaguer P, et al. Estrogenic and AhR activities in dissolved phase and suspended solids from wastewater treatment plants. Science of the Total Environment. 2010;**408**(12):2608-2615

[47] Alfonso-Muniozguren P, Serna-Galvis EA, Bussemaker M, Torres-Palma RA, Lee J. A review on pharmaceuticals removal from waters by single and combined biological, membrane filtration and ultrasound systems. Ultrasonics Sonochemistry. 2021;**76**:105656

[48] Choi JW, Bediako JK, Zhao Y, Lin S, Sarkar AK, Han M, et al. Adsorptive removal of cationic tricyclic antidepressants using cation-exchange resin. Environmental Science and Pollution Research International. 2020;**27**(20):24760-24771

[49] Maršálek R, Švidrnoch M. The adsorption of amitriptyline and nortriptyline on activated carbon, diosmectite and titanium dioxide. Environmental Challenges. 2020;**1**:100005

[50] Gornik T, Shinde S, Lamovsek L, Koblar M, Heath E, Sellergren B, et al. Molecularly imprinted polymers for the removal of antidepressants from contaminated wastewater. Polymers (Basel). 2020;**13**:120

[51] Arimori K, Furukawa E, Nakano M. Adsorption of imipramine onto activated charcoal and a cation exchange resin in macrogol-electrolyte solution. Chemical & Pharmaceutical Bulletin. 1992;**40**(11):3105-3107

[52] Santadkha T, Skolpap W. Economic comparative evaluation of combination of activated carbon generation and spent activated carbon regeneration plants. Journal of Engineering Science and Technology. 2017;**12**(12):3329-3343

[53] Larasati A, Fowler GD, Graham NJD. Insights into chemical regeneration of activated carbon for water treatment. Journal of Environmental Chemical Engineering. 2021;**9**(4):105555

[54] El Gamal M, Mousa HA, El-Naas MH, Zacharia R, Judd S. Bio-regeneration of activated carbon: A comprehensive review. Separation and Purification Technology. 2018;**197**:345-359

[55] Kocoglu ES, Bakirdere S, Keyf S. A novel liquid-liquid extraction for the determination of sertraline in tap water and waste water at trace levels

by GC-MS. Bulletin of Environmental Contamination and Toxicology. 2017;**99**(3):354-359

[56] Ikeda M. Public health problems of organic solvents. Toxicology Letters. 1992;**64-65**:191-201

[57] Racovita RC, Ciuca MD, Catana D, Comanescu C, Ciocirlan O. Microemulsions of nonionic surfactant with water and various homologous esters: Preparation, phase transitions, physical property measurements, and application for extraction of tricyclic antidepressant drugs from aqueous media. Nanomaterials (Basel). 2023;**13**(16):2311

[58] Ge D, Lee HK. Ionic liquid based dispersive liquid-liquid microextraction coupled with micro-solid phase extraction of antidepressant drugs from environmental water samples. Journal of Chromatography. A. 2013;**1317**:217-222

[59] Zawadzki M, e Silva FA, Domańska U, Coutinho JAP, Ventura SPM. Recovery of an antidepressant from pharmaceutical wastes using ionic liquid-based aqueous biphasic systems. Green Chemistry. 2016;**18**(12):3527-3536

[60] Trawinski J, Skibinski R. Studies on photodegradation process of psychotropic drugs: A review. Environmental Science and Pollution Research International. 2017;**24**(2):1152-1199

[61] Nassar R, Trivella A, Mokh S, Al-Iskandarani M, Budzinski H, Mazellier P. Photodegradation of sulfamethazine, sulfamethoxypiridazine, amitriptyline, and clomipramine drugs in aqueous media. Journal of Photochemistry and Photobiology A: Chemistry. 2017;**336**:176-182

[62] Khaleel NDH, Mahmoud WMM, Olsson O, Kummerer K. Initial fate assessment of teratogenic drug trimipramine and its photo-transformation products - role of pH, concentration and temperature. Water Research. 2017;**108**:197-211

[63] Gros M, Williams M, Llorca M, Rodriguez-Mozaz S, Barcelo D, Kookana RS. Photolysis of the antidepressants amisulpride and desipramine in wastewaters: Identification of transformation products formed and their fate. Science of the Total Environment. 2015;**530-531**:434-444

[64] Maślanka A, Żmudzki P, Szlósarczyk M, Talik P, Hubicka U. Photodegradation assessment of amisulpride, doxepin, haloperidol, risperidone, venlafaxine, and zopiclone in bulk drug and in the presence of excipients. Monatshefte für Chemie – Chemical Monthly. 2020;**151**(4):483-493

[65] Calza P, Sakkas VA, Villioti A, Massolino C, Boti V, Pelizzetti E, et al. Multivariate experimental design for the photocatalytic degradation of imipramine. Applied Catalysis B: Environmental. 2008;**84**(3-4):379-388

[66] Mendez-Arriaga F, Otsu T, Oyama T, Gimenez J, Esplugas S, Hidaka H, et al. Photooxidation of the antidepressant drug fluoxetine (Prozac(R)) in aqueous media by hybrid catalytic/ozonation processes. Water Research. 2011;**45**(9):2782-2794

[67] Bojanowska-Czajka A, Pyszynska M, Majkowska-Pilip A, Wawrowicz K. Degradation of selected antidepressants sertraline and citalopram in ultrapure water and surface water using gamma radiation. PRO. 2021;**10**:63

[68] Melin V, Salgado P, Thiam A, Henriquez A, Mansilla HD, Yanez J, et al. Study of degradation of amitriptyline

antidepressant by different electrochemical advanced oxidation processes. Chemosphere. 2021;**274**:129683

[69] Tiwari B, Sellamuthu B, Ouarda Y, Drogui P, Tyagi RD, Buelna G. Review on fate and mechanism of removal of pharmaceutical pollutants from wastewater using biological approach. Bioresource Technology. 2017;**224**:1-12

[70] Gornik T, Kovacic A, Heath E, Hollender J, Kosjek T. Biotransformation study of antidepressant sertraline and its removal during biological wastewater treatment. Water Research. 2020;**181**:115864

[71] Kozka B, Nalecz-Jawecki G, Turlo J, Giebultowicz J. Application of Pleurotus ostreatus to efficient removal of selected antidepressants and immunosuppressant. Journal of Environmental Management. 2020;**273**:111131

Chapter 11

Review of Anaerobic Baffled Reactor for Treatment of Industrial Wastewater in Nigeria

Opeyemi K. Olayanju, Kamar T. Oladepo, Oluwanishola A. Azeez, Adedayo O. Adebusuyi and Oluwademilade Obisakin

Abstract

Industrialization, although with its gain has also led to environmental degradation. One of such degradation is pollution. Water we know is essential to life but with industrialization, the existence is being greatly threatened. The need to treat wastewater has become incontrovertibly important as man's activities in the environment has depleted the limited available water sources daily. These activities include indiscriminate discharge of untreated wastewater into neighboring water bodies thereby causing pollution to the environment and aquatic life. For years, conventional wastewater treatment processes have been adopted and found reasonably successful in treating effluents from wastewater in order to meet the required standard before discharge. The review of the anaerobic baffled reactor (ABR) in the management of industrial wastewater is done in this chapter. The review considers the use of ABR in Nigeria for the treatment of industrial wastewater. The ABR's basic principle, applications, merits, demerits and challenges (in Nigeria) is addressed.

Keywords: industrial wastewater, anaerobic baffled reactor, Nigeria, wastewater treatment, environmental standards

1. Introduction

Water is classified among the requisite renewable resources for food production, economic advancement, maintenance and general well-being of all living things. Most of the uses of water are impractical to replace, difficult to clean and expensive to transport. Also, water is one of the most manageable natural resources as it can be diverted, stored, recycled and stored [1]. The accessibility of freshwater that is free from pollution is a simple necessity for any growth of humans, however, the obtainability of such water that is unpolluted and usable is limited. In various rural areas of developing nations, the predominant factor contributing to water pollution is the unrestricted discharge of untreated sewage directly into nearby bodies of water [2].

With economic growth and industrialization, water resources are growing progressively scarce. The condition of the world environment is frequently becoming of poorer quality in most areas. Treatment, management and disposal of industrial effluent is currently one of the most hazardous sources of environmental problems. This reason brings an urgent need to create reliable and efficient technologies for wastewater treatment.

Wastewater treatment can be described as the planning and conversion of wastewater and associated outputs for reuse (secure and free from danger to the environment) or discarding in-order to mitigate health hazards and protect the environment from the detrimental impacts of pollution, it is essential to address and eliminate significant variables such as solids in wastewater, biological oxygen demand, chemical oxygen demand, pathogenic microorganisms, and nutrients. Other impurities that are of importance to be treated to mitigate their harmful effects include heavy metals or persistent organic compounds (POPs).

This dire need to treat wastewater has led to different researches about different methods of treating wastewater. Of the many methods discovered for the treatment of wastewater, the anaerobic process for wastewater treatment has continued stand out in all the various methods of waste water treatment. This method of treatment has advantages such as simplicity of design, high treatment efficiency, use of non-sophisticated equipment, little excess sludge is generated and capital and low operating cost [3–5].

The Anaerobic baffled reactor (ABR) is described as one of the high-level anaerobic reactors, it has been largely used in the treatment analysis of wastewater (both industrial and domestic). According to [5–7], the ABR was originally developed at Stanford University. The ABR functions much like anaerobic sludge blanket reactors (UASBs), utilizing a system of vertically positioned baffles that guide the wastewater to move both below and above them from the inlet to the outlet. This design ensures that the effluent comes into close contact with a substantial quantity of active biomass, whilst maintaining a comparatively low concentration of biological solids in the effluent. With regards to the advantage, an essential benefit of the ABR lies in its capacity to segregate acidogenesis and methanogenesis along the length of the reactor [5, 8]. This enables distinct bacterial populations to thrive in their respective compartments, with acidification prevailing in the initial section and methanogenesis dominating in the subsequent section [5, 8]. As reported [6, 9, 10], either the industrial or domestic wastewater can be effectively treated using the ABR. The rate of chemical oxygen demand (COD) elimination in ABR is stated to be between the scale of 70–90%. While the rate of elimination of the biological oxygen demand (BOD) is stated between 60–85%. The rate of removal for total Suspended Solids (TSS) is up to 90%.

Nigeria is contending with significant challenges when it comes to management of wastewater management, some of the factors responsible for this include rapid urbanization, population growth, and inadequate infrastructure which has culminated into pressing wastewater crisis [11, 12]. Sustainable ecological development and public health is endangered by contaminated water sources, indiscriminate and unrestricted dumping of untreated wastewater in the environment, and scarce sanitation facilities have all contributed to environmental diseases and ecological deterioration [12]. As stated by [12], some of the identified steps to be taken to mitigate against the challenge of wastewater management in Nigeria include investment in infrastructural development which will include modern treatment plants, sanitation facilities and functional and well-maintained sewer system; also, engaging the communities will be beneficial to the management of wastewater as each community can monitor her environment. Environmental policies and regulations should be developed and stringent measures put in place to check defaulters. Other identified measures are capacity development through education and training for wastewater management professionals, innovation and technology which has become the main hub of wastewater treatment in the advanced countries.

2. Sources and characteristics of wastewater in Nigeria

Like many countries of the world, the sources of wastewater in Nigeria are from domestic usage (cleaning, washing, bathing, flushing); industrial manufacturing processing (cleaning, washing, extraction) and agricultural processes (washing and cleaning). The wastewater from industrial manufacturing processes can either be large scale manufacturing processes which in most cases are automated or non-large-scale industries which are most times cottage industries like the starch processing industry, slaughter houses, garri processing factories, etc. These spent waters from sewage (domestics use) and other wastewater effluents (industrial manufacturing processes) have different and varying characteristics. The characterization is done according to the physical, chemical and biological constituent (composition) of the wastewater. These characteristics include high BOD and COD (depending of the source of the wastewater and the pollutant load), pH level which usually vary from acidic to basic (depending on the source). Others are total solid, dissolved solid, suspended solid, nitrogen, phosphorus, chloride, alkalinity which in all cases usually vary from strong concentration to low concentration depending on the processing resulting to the wastewater. It has been observed that in most cases, these characteristics usually exceed the limits set by standard organizations like World Health Organization, Federal Environmental Protection Agency, etc. and therefore, before discharge or disposal of the wastewater into the environment, there is need to treat the it which in most cases is usually a water body. **Table 1** shows the characteristics of raw industrial wastewater as reported by [13]; the values exceed the standards given for safe discharge into the environment.

Parameters	BOD_5, mg/L	COD, mg/L	TSS	pH
Brewery	1609.34–3980.61	1096.41–8926.08	530.67–3728.02	4.6–7.3
Abattoir	476–3850	935–6600	750–4400	6.85–8.19
Cane Sugar	350–2750	1000–4340	760–800	5–6.5
Oil refinery	100–500	150–800	130–600	2–6
Coke oven	510–1360	930–3120	19–3330	6.8–8.2
Tannery	1000–2000	2000–4000	2000–3000	11–12
Textile	50–500	250–8000	100–700	5.6–9

Table 1.
Characteristics of raw industrial wastewater [13].

3. An overview of wastewater treatment practices in Nigeria

The treatment of wastewater in Nigeria is vital to environmental sciences and public health. Common practices vary depending on the place regulatory bodies available, finances available, etc. Wastewater is something virtually every industry or household releases. Over the years in Nigeria, most industries and households have been making use of some very poor treatment practices for wastewater due to ignorance of the dangers it poses to our ecosystem. The Government has been trying to put things and agencies in place to monitor and ensure compliance by industries, households, and offices. The agencies have only been up and doing in terms of ensuring large-scale industries comply, most small and medium-scale industries end up finding their way by evasion.

Some households just channel their wastewater into the drainage system in front of their house or on the road when a drainage system is absent. This is the most familiar wastewater disposal technique used in most rural areas and some urban areas. This is because of its cost-efficiency. It is virtually the cheapest wastewater disposal method. Most small and medium-scale industries make use of this wastewater disposal method. It poses a great risk to the environment because it has a net effect on global warming also this wastewater is eventually disposed of in oceans and water bodies which causes a great risk to aquatic lives and the ocean's sanity. One of the major reasons why this method is used is because of the absence of a central sewage system as present in other developed nations of the world.

The most important aim of treating wastewater is to eliminate from the wastewater (to permissible limit) the solids (suspended as well as dissolved) present therein. The purpose of this process is to enable the discharge of the remaining treated wastewater into the receiving water without compromising its optimal or intended use. The solids extracted from the wastewater primarily consist of organic matter, although they may also encompass inorganic solids. The sludge removed from the wastewater (in the form of solids and liquids) are also treated. Also, the odor is controlled and pathogenic organisms in the wastewater are expelled through relevant methods of treatment. The above reasons inform the main aim of waste-water treatment [14]. Some of the wastewater treatment practices in Nigeria include.

3.1 Aerobic treatment of wastewater

Wastewater undergoes aerobic method of treatment in the presence of oxygen. Within aerobic wastewater treatment systems, aerobic organisms utilize oxygen to convert organic compounds in the wastewater into carbon dioxide and generate new biomass. This method exclusively relies on the use of air for wastewater treatment. In contrast to conventional septic tanks, aerobic treatment systems demonstrate enhanced efficiency in breaking down organic matter, accelerating the decomposition of organic solids, and diminishing the concentration of pollutants [15].

3.2 Physio-chemical treatment of wastewater

Physicochemical treatment is a commonly used method in the treatment of wastewater. It is applied to remove heavy metals, oils, suspended matter, greases, toxic pollutants, high salt concentrations, dissolved organic substances, as well as organic and inorganic components that are difficult to decompose, and phosphorus.

Physicochemical treatment is used as pre-treatment, final treatment, and specific treatment for wastewater recycle as process water. The techniques used in physicochemical treatment include dilution, sedimentation, and other methods [15].

3.3 Physical treatment of wastewater

This is a common method used in treating wastewater in Nigeria. It involves processes like sedimentation, screening, and skimming which are used to remove solid particles. This method involves no chemicals; hence it is a chemical-free process.

Sedimentation is among the most common physical techniques used for treating wastewater. It involves suspending the insoluble or heavy particles in the water, and allowing them to settle at the bottom over a period of time, and then separating the pure water from the sediment. Aeration is another effective physical water treatment technique that involves circulating air through the water to make available oxygen to it. This helps to break down organic matter and remove unpleasant odors. Finally, filtration is used to filter out impurities and particles that are not soluble present in the wastewater. Sand filter which is the most commonly filter used for this purpose. Additionally, grease found on the surface of some wastewater can be easily removed through this filtration method [16].

3.4 Anaerobic treatment of wastewater

Unlike aerobic treatment of wastewater, anaerobic wastewater is treatment of wastewater without the presence of oxygen. The whole process of anaerobic treatment occurs in the absence (devoid) of oxygen. The whole process is very effective in removing biodegradable organic compounds leaving mineralized compounds in the solution [17]. Different industries majorly use anaerobic method of treating wastewater because of its suitability and the basic requirement. **Table 2** shows the summary of the advantages and disadvantages of the anaerobic treatment of different types wastewater. In anaerobic environments, the decomposition of organic matter occurs through the successive and cooperative metabolic interactions among different trophic groups of prokaryotes. These groups include fermenters, acetogens, methanogens, and sulfate-reducing bacteria (SRB) [18]. The metabolic interplay among these microbial groups results in the conversion of intricate organic compounds into simpler substances, including ammonia, hydrogen sulfide carbon dioxide, and methane [18].

3.5 Reverse osmosis treatment

Reverse osmosis (RO), a filtration technique that makes use of a special membrane to remove significant molecules and ions from solutions. To achieve this, it is done by the application pressure to the mixture on one side of the membrane. The membrane allows smaller parts, such as the solvent, to go through it freely, while retaining the solute on the pressurized side. To work effectively, the membrane must have small pores that do not allow large molecules or ions to move through. It is interesting to learn that reverse osmosis membranes have a compressed obstruction film in the polymer matrix which is responsible for most of the separation. This film is only intended to permit water to go through although disallowing the passage of solutes like salt ions. To achieve this, a high pressure of 2–17 bar (30–250 psi) is exerted

Advantages
High efficiency: the system exhibits good removal efficiency even under high loading rates and at low temperatures.
Simplicity: both the building and set-up (operation) of anaerobic reactors are somewhat straightforward.
Flexibility: anaerobic means of treatment can be effortlessly applied on a wide scale, ranging from small to large systems.
Low energy consumption: the reactor's energy consumption is minimal, as it does not require heating up of the influent to attain the operational temperature, and all processes can be driven by gravity.
Energy recovery: the process yields energy in the form of methane, contributing to energy sustainability.
Low sludge production: the system generates a small volume of sludge that is well-stabilized and exhibits favorable dewatering properties.
Low nutrient and chemical requirement: anaerobic treatment, particularly in sewage treatment, often necessitates minimal or no chemical additives. In this context, a sufficient and consistent pH level can be sustained without the need for the introduction of chemicals.
Disadvantages
Low removal of pathogen and nutrient: the removal of pathogens and nutrients is relatively limited in anaerobic treatment, as it is only partial. Consequently, additional post-treatment is necessary to achieve the desired level of removal.
Long start-up: the anaerobic treatment's low startup is attributed to the slow (gradual) rate of growth of the methanogenic organisms, resulting in a prolonged startup period.
Possible bad odor: the production of hydrogen sulfide occurs, necessitating careful management of biogas to prevent unpleasant odors.
Necessity of post-treatment: in order to meet the prescribed standards for the discharge (into the environment) for the organic matter and pathogens, post-treatment of the effluent from the anaerobic process is usually deemed necessary.

Source: [18].

Table 2.
Advantages and disadvantages of using anaerobic method of wastewater treatment.

on the high concentration side of the membrane for fresh and brackish water, and 40–70 bar (600–1000 psi) for seawater, overcoming its natural osmotic pressure of around 24 bar (350 psi) [14].

These practices named above are some of the wastewater treatment practices carried out in Nigeria, the method used vary from place to place and also depend on some factors such as finance available, volume of water treated etc. It all has a single aim, which is getting untreated water treated for reuse or for safe disposal.

4. Emerging technologies for wastewater management in Nigeria

Wastewater management involves collecting, treating, and reusing wastewater to protect water resources, promote environmental sustainability, conserve resources, and safeguard public health. Before the advent of modern technologies, wastewater management relied on simpler and often less efficient techniques. While some of these methods are still used in certain contexts, advancements in science and technology have led to the development of more sophisticated and effective wastewater treatment processes [19]. Some traditional or old wastewater management techniques include; Imhoff tanks (sedimentation tanks) that allow solids to settle at the bottom while liquids move to the upper chamber, septic tanks, open-air lagoons, chemical coagulation, and flocculation. Although old wastewater management techniques have been useful in catering to basic sanitation needs, they often do not meet current environmental standards. Fortunately, the advancements in technology have resulted in more efficient and eco-friendly treatment methods, which ensure better protection of water resources and public health [13].

4.1 Advanced biological treatment

a. Membrane bioreactors: membrane systems have become essential in separation processes, adding new techniques to more traditional processes. Recent developments in membranes have made it possible to achieve design objectives with added flexibility and improved efficiencies [20]. Examples of such innovations can be discovered in membrane bioreactors, seawater desalination, batteries, super-capacitors, solar cells, and microbial fuel cells, which have led to outstanding results.

b. Anammox (anaerobic ammonium oxidation): anaerobic ammonium oxidation is the process of converting ammonium to di-nitrogen gas by means of nitrite equally as an electron acceptor. This biological process removes nitrogen compounds from wastewater more efficiently, reducing the need for energy-intensive nitrification and denitrification processes.

4.2 Resource recovery

a. Phosphorus and nutrient recovery: the reclamation and recycling of nitrogen and phosphorus is extremely important and thus required. With the use of nitrogen- and phosphorus-based synthetic (inorganic) fertilizers, there is a rising trend, and large-scale influx of responsive nitrogen in the atmosphere; this comes with severe health consequences on humans and the environment. Contrary, phosphorus, a non-renewable means of resource, faces a grave threat of depletion. As a result, the retrieval and reuse of both nitrogen and phosphorus is greatly required. There are multiple ways to recover nitrogen, including ion exchange and absorption, bio-electrochemical systems, and air stripping of ammonia. Similarly, there are various methods for phosphorus recovery, such as physical filtration and membrane processes, chemical precipitation, high-temperature acid hydrolysis, biological assimilation, physical-chemical absorption and ion exchange. On Earth, nitrogen and phosphorus play vital roles as macronutrients supporting life. Nevertheless, it is imperative to extract them from waste streams to effectively counteract the potential eutrophication of water-receiving bodies [21].

b. Biogas and energy generation: the reduction in the availability of energy resources and the costs of energy costs have brought about a crucial and critical point of worry for decision-makers and also users about their societal activities, environmental and economic impact. This shift has led wastewater management experts to reconsider wastewater treatment plants (WWTPs) as essential contributors to a bio-based circular economy, rather than viewing them solely as facilities for end-of-life treatment and disposal [5]. Consequently, redesigning and operating modernized wastewater treatment plants has been identified as an important way for resource utilization and energy recovery in wastewater locations. Efficient capture and management of sludge generated at WWTPs have the potential to generate significant energy in the form of biogas, potentially transforming WWTPs into net energy producers rather than consumers [22].

4.3 Decentralized wastewater treatment

This refers to a method of managing wastewater at the location where it is generated, rather than relying on a central sewage treatment plant. These systems such as modular treatment units and containerized plants are gaining more attention due to their potential to provide efficient and sustainable solutions for managing wastewater in various settings [23]. Centralized wastewater collection and treatment systems are costly to build and maintain, particularly in regions with low population densities and scattered households. Developing nations face challenges in securing the necessary funds for constructing centralized facilities and acquiring the technical expertise to effectively manage and operate them. Consequently, decentralized systems emerge not only as a sustainable solution for small communities in the long run but also as a more dependable and cost-effective alternative [24].

4.4 Advanced oxidation processes (AOPs)

In the 1980s, the concept of advanced oxidation processes (AOPs) was first presented for the treatment of potable water. AOPs are defined as oxidation processes that produces a sufficient amount of hydroxyl radicals (OH·) to purify water. Over the years, the concept of advanced oxidation processes (AOP) has evolved to encompass oxidative methods utilizing sulfate radicals (SO_4^-). Unlike conventional oxidants like chlorine and ozone, which serve a dual purpose of decontamination and disinfection, AOPs are primarily employed for the eradication of organic or inorganic contaminants in water and wastewater [25].

It is important to understand that the efficiency of various technologies used to treat industrial wastewater may differ based on the certain characteristic constituent of the wastewater and the regulatory requirements of the industry. Moreover, continuous research and development in this field may result in the discovery of newer and more effective technologies.

5. Anaerobic baffled reactor

The anaerobic baffled reactor (ABR) represents the third generation of anaerobic bioreactors, distinguished by its specific design incorporating a series of vertical baffles within the reactor [26]. Originating from Stanford University in 1983, McCarty and colleagues developed the ABR, conceptualizing it as a sequence of up-flow anaerobic sludge blanket reactors (UASBS), characterized by multiple compartments [8, 27, 28]. Typically, a conventional ABR consists of vertically positioned baffles that guide the wastewater both under and over them as it traverses from the inlet to the outlet. This unique flow pattern minimizes bacteria washout, allowing the ABR to maintain an active biological mass without relying on fixed media.

The gas production in each reactor section leads to the upward movement and settling of bacteria, creating a relatively slow horizontal flow down the reactor.

This results in a Sludge Retention Time (SRT) of 100 days at a Hydraulic Retention Time (HRT) of 20 hours. The deliberate slow horizontal movement ensures intimate contact between wastewater and the active biomass during its passage through the ABR, even with short HRTs ranging from 6 to 20 hours [8, 29]. Additionally, the ABR can be operated with granules or internal media to enhance stability. This separation of Sludge Retention Time (SRT) from Hydraulic Retention Time (HRT) contributes to effective removal of chemical oxygen demand (COD) and solids, low sludge production, and a compact footprint [28].

Thanks to the design of ABR, the reactor exhibits resilience to equally severe hydraulic and organic shock loads yet not compromising performance. As a result, it is gaining widespread use in many developing countries as a cost-effective and effective solution for low-cost sanitation and treatment of industrial wastewater. Furthermore, it often offers the added benefit of energy production [28].

Figure 1 provides a detailed schematic representation of the ABR.

The treatment efficiency of anaerobic baffled reactors (ABRs) typically falls within the range of 65–90% for chemical oxygen demand (COD) removal, equivalent to approximately 70–95% biological oxygen demand (BOD) removal [30, 31]. In the initial sedimentation chamber of the ABR, a significant portion of settleable solids, representing around 50% of the total volume of total suspended solids (TSS), is removed [31]. The unique design allows for enhanced treatment of non-settleable solids, achieving a total suspended solids (TSS) removal of up to 90% [31]. The series arrangement of tanks facilitates the digestion of challenging-to-degrade substances, primarily in the rear part after the digestion of easily degradable materials in the front part [31]. Consequently, recycling of effluent could adversely impact treatment quality. ABRs can be designed to accommodate daily inflows ranging from a few cubic meters per day to several hundred cubic meters per day [31, 32]. The hydraulic retention time (HRT) in ABRs is relatively short, varying from a few hours to two or three days [30–32].

The ABR boasts numerous advantages, including a simple design, absence of moving parts, no need for mechanical mixing, cost-effectiveness in construction, high void volume reducing clogging, minimal sludge bed expansion, low HRT, stability against hydraulic shock loads, prolonged operation without sludge wasting, resilience to organic shocks, low capital and operating costs, low sludge generation, high Solids Retention Times (SRT), and minimal maintenance requirements [32]. However, there

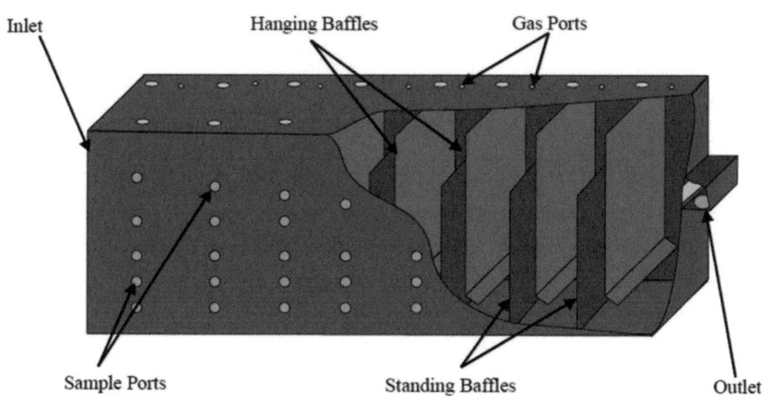

Figure 1.
Diagram of 8-compartment pilot-scale anaerobic baffled reactor with opened section showing internal baffle set up.

are some drawbacks to consider, such as a lengthy start-up phase, the necessity for expert design and construction, limited reduction of pathogens and nutrients, the need for further treatment or appropriate discharge of effluent and sludge, the requirement for a strategy for fecal sludge management (effluent quality deteriorates rapidly without regular sludge removal), and the absence of clear design guidelines [31].

6. Case study of the anaerobic baffled reactor

6.1 Case study I

This case study was carried out to determine the most effective method for removing organics from cassava wastewater in the Ohaji-Egbema Local Government Area of Imo State, Nigeria.

Ten anaerobic baffled reactors (ABRs) of varying sizes were constructed from 0.006 m thick stainless steel, as illustrated in **Figure 2**. All ABRs featured the same number of compartments to ensure proper mixing, each of the compartment is equipped with a port for sampling. Water bath was used for the maintenance of the reactor temperature at 35 ± 0.5°C, and a variable-speed peristaltic pump facilitated the pumping of the influent wastewater. Wastewater sourced from cassava processing units in Ohaji-Egbema Local Government Area, Imo State, underwent a filtration process to reduce total suspended solids to 90% before introduction into the reactor. The bioreactor operated consistently, irrespective of the loading rate, until achieving a steady-state condition. Various combinations of hydraulic retention time (HRT), influent flow rate (Q), and organic loading rate were employed during the operation of the ABRs [33].

The research findings indicated that elevating the number of baffles from 3 to 10 led to a reduction in the inhibitor constant, decreasing from 9.9989 to 1.6101 mg/L.

6.2 Case study II

The research utilized industrial wastewater obtained from a wheat starch factory located in the central province of Iran (Ardineh Starch CO., Isfahan, Iran) as the feed for an anaerobic baffled reactor (ABR).

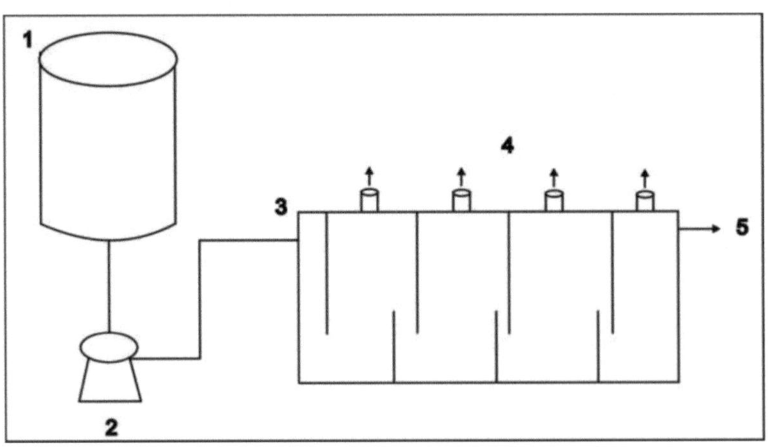

Figure 2.
Scheme of the ABR. Adapted from [33]. 1. Feed tank. 2. Peristaltic pump. 3. Influent. 4. Sampling ports. 5. Effluent.

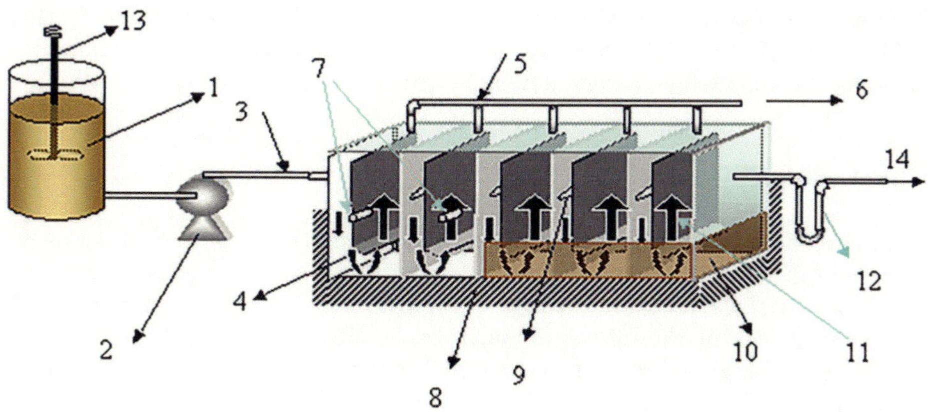

Figure 3.
Schematic diagram of anaerobic baffled reactor adapted from [34]. 1. Feed tank. 2. Peristaltic pump. 3. Influent feed. 4. Flowing fluid. 5. Collection of gas. 6. Generated biogas. 7. Ports for sampling. 8. Bath for water. 9. Passage window. 10. Bed of sludge. 11. Upward flow. 12. Siphoning mechanism. 13. Mixing device. 14. Treated effluent.

The experimental ABR was designed on a laboratory scale, constructed from 0.006 m thick transparent Plexiglas, with external dimensions measuring 0.53 m in length, 0.16 m in width, and a depth of 0.30 m, providing a working volume of 13.5 L. **Figure 3** depicts a schematic diagram of the reactor.

The reactor comprises five equal compartments of 2.7 L each, separated by upright baffles. Each compartment includes downcomer and riser regions formed by additional upright baffles. The upcomer's breadth is approximately 2.6 times that of the downcomer, with the lower portions of the downcomer baffles set at a 45° angle to ensure a uniform flow through the upcomer. This configuration facilitates efficient blending and interaction among the wastewater and anaerobic sludge located at the bottom of each riser. Sampling ports within each compartment enable the retrieval of biological solids and sewage samples.

To maintain a constant operating temperature of 35 ± 0.5°C, the ABR is equipped with a temperature regulator that is placed in a water bath. A variable-speed peristaltic pump (Masterflux L/S) is employed to pump the influent feed, while the outlet is coupled to a glass U-tube for level control and solids trapping. Gas produced during the process is collected through portholes in the upper part of the reactor, and its volume is determined daily using the gas-water displacement technique. The research findings indicate that the ABR effectively treats industrial wastewater from the wheat starch factory, achieving a COD removal efficiency of 61–67% [34].

7. Challenges of using ABR in Nigeria

The use of anaerobic baffled reactors for the treatment of wastewater in Nigeria presents certain potential challenges, and among them is the issue of a delayed initiation period. With the high volume of wastewater being generated, a means of treatment with slow startup time may not be considered ideal enough as faster means of treatment will be preferred. Also, as stated by [31], there are no clear design guidelines yet, this in turn brings about little or no technical know how about the design and installation of the anaerobic baffled reactor in Nigeria. In addition, it has been identified that for removal of nutrients in wastewater, additional treatment means should be added to the ABR which will result into additional cost of treatment and as result could be a challenge to the outright adoption of the ABR in Nigeria.

8. Conclusion

This chapter has reviewed the use of anaerobic baffled reactor in Nigeria for the treatment of industrial wastewater. The use of ABR in Nigeria especially for treatment of industrial wastewater is little or non-existent and as a result more research is needed to be done on the subject of anaerobic baffled reactor in Nigeria; this will in turn lead to scaling up by the industries for commercialization and real plant application.

Author details

Opeyemi K. Olayanju[1*], Kamar T. Oladepo[2], Oluwanishola A. Azeez[1], Adedayo O. Adebusuyi[1] and Oluwademilade Obisakin[1]

1 Department of Civil Engineering, Redeemer's University, Ede, Nigeria

2 Department of Civil Engineering, Obafemi Awolowo University, Ile-Ife, Nigeria

*Address all correspondence to: olayanjuo@run.edu.ng

References

[1] Phair M. Effects of water pollution and control measures. Journal of Pollution Effects and Control. 2022;**10**(1):1-2. DOI: 10.35248/2375-4397.22.10.326

[2] Kulkarni S, Jabade S, Bhatkhande D, Watve H, Khamparia S, Dhaneshwar P. Optimizing design parameters of an anaerobic baffled reactor for better performance: A review. International Journal of Health Sciences. 2022;**6**(S2):14588-14602. DOI: 10.53730/ijhs.v6nS2.8829

[3] Abdullah LG, Idris A, Ahmadun FR, Baharin BS, Emby F, Megat MNMJ, et al. A kinetic study of a membrane anaerobic reactor (MAR) for treatment of sewage sludge. Desalination. 2005;**183**:439-445

[4] Sato N, Okubo T, Onodera T, Ohashi A, Harada H. Prospects for a self-sustainable sewage treatment system: A case study on full-scale UASB system in India's Yamuna River Basin. Journal of Environmental Management. 2006;**80**:198-207

[5] Liu R, Tian Q, Chen J. The developments of anaerobic baffled reactor for wastewater treatment: A review. African Journal of Biotechnology. 2010;**11**:1535-1542

[6] Wang J, Huang Y, Zhao X. Performance and characteritics of an anaerobic baffled reactor. Bioresource Technology. 2004;**93**(2):205-208

[7] Krishna GGVT, Kumar P, Kumar P. Treatment of low-strength soluble wastewater using an anaerobic baffled reactor (ABR). Journal of Environmental Management. 2009;**90**(2):166-176

[8] Barber WP, Stuckey DC. The use of the anaerobic baffled reactor (ABR) for wastewater treatment: A review. Water Research. 1999;**33**(7):1559-1578. DOI: 10.1016/S0043-1354(98)003771-6

[9] Mrafková L, Hutňan M, Drtil M. Behaviour of anaerobic baffled reactor treating non-acidified wastewater. Chemical Papers. 2000;**54**(6b):448-455

[10] Rahul KB, Sajitha M. Design approach and classification of anaerobic baffled reactor for wastewater treatment—A review. In: Paper Presented in the 'Chemical Processing and Process Control' Session of the Kerala State Technological Congress 2019 – KETCON 2019 – Organized at Government Engineering College Thrissur during 15th to 17th February, 2019. India: Kerala Technological University; 2019. pp. 35-41

[11] Adesogan S. Sewage technology in Nigeria: A pragmatic approach. Science Journal of Environmental Engineering Research. 2013:1-9. DOI: 10.7237/sjeer/266

[12] Adeyemi SA. Rethinking wastewater management in Nigeria. The Guardian. 2023. Available from: https://editor.guardian.ng/opinion/rethinking-waste-water-management-in-nigeria/

[13] Armah EK, Chetty M, Adedeji JA, Kukwa DT, Mutsvene B, Shabangu PK, et al. Emerging trends in wastewater treatment technologies: The current perspective. In: Promising Techniques for Wastewater Treatment and Water Quality Assessment. London, UK: InTechOpen; 2020. DOI: 10.5772/intechopen.93898

[14] Okwa JO, Adebanjo A. Sewage and Waste-Water Treatment. Enugu State, Nigeria: National Open University of Nigeria; 2011

[15] Iheukwumere SO, Nwabudike PC, Nkwocha KF, Phil-Eze PO. Domestic

wastewater treatment and reuse in Awka Urban, Anambra State, Nigeria. International Journal of Geography and Environmental Management. 2018;**4**(2):16-24

[16] Metcalf & Eddy Inc. Wastewater Engineering: Treatment and Reuse. 5th ed. McGraw Hill; 2014

[17] van Lier JB, Mahmoud NA, Zeeman G. Anaerobic wastewater treatment. In: Henze M, van Loosdrecht M, Ekama GA, Brdjanovic D, editors. Biological Modelling and Design. London: IWA Publishing; 2008. pp. 415-457

[18] Ahammad SZ, Graham DW, Dolfing J. Wastewater Treatment: Biological. School of Civil Engineering and Geosciences. Newcastle, U.K: Newcastle University; 2013. DOI: 10.1201/9781003045045-61

[19] Adetunji AI, Olaniran AO. Treatment of industrial oily wastewater by advanced technologies: A review. Applied Water Science. 2021;**11**:98. DOI: 10.1007/s13201-021-01430-4

[20] Aslam M, Charfi A, Lesage G, Héran M, Kim J. Membrane bioreactors for wastewater treatment: A review of mechanical cleaning by scouring agents to control membrane fouling. Chemical Engineering Journal. 2017;**307**:897-913

[21] Sengupta S, Nawaz T, Beaudry J. Nitrogen and phosphorus recovery from wastewater. Current Pollution Reports. 2015;**1**(3):155-166. DOI: 10.1007/s40726-015-0013-1

[22] Tetteh EK, Rathilal S. Biogas production from wastewater treatment-evaluating anaerobic and biomagnetic systems. Water-Energy Nexus. 2021;**4**:167-173. DOI: 10.1016/j.wen.2021.11.004

[23] Massoud MA, Tarhini A, Nasr JA. Decentralized approaches to wastewater treatment and management: Applicability in developing countries. Journal of Environmental Management. 2009;**90**:652-659. DOI: 10.1016/j.jenvman.2008.07.001

[24] Chirisa I, Bandauko E, Matamanda AR, Mandisvika G. Decentralized domestic wastewater systems in developing countries: The case study of Harare (Zimbabwe). Applied Water Science. 2017;7:1069-1078

[25] Deng Y, Zhao R. Advanced oxidation processes (AOPs) in wastewater treatment. Current Pollution Reports. 2015;**1**:167-176. DOI: 10.1007/s40726-015-0015-z

[26] Jiang F, Peng Z, Li H, Li J, Wang S. Effect of hydraulic retention time on anaerobic baffled reactor operation: Enhanced biohydrogen production and enrichment of hydrogen-producing acetogens. Process. 2020;**8**(339):1-18. DOI: 10.3390/pr8030339

[27] Manariotis I, Grigoropoulus S. Low strength wastewater using an anaerobic baffled reactor. Water Environmental Research. 2002;**74**(2):170-176

[28] Stuckey DC. Anaerobic Baffled Reactor (ABR) for Wastewater Treatment. 2015. Available from: www.worldscientific.com

[29] Nasr F, Doma H, Nassar H. Treatment of domestic wastewater using an anaerobic baffled reactor followed by a duckweed pond for agricultural purposes. The Environmentalist. 2008;**29**(3):270-279. DOI: 10.1007/s10669-008-9188-y

[30] Morel A, Diener S. Greywater Management in low- and Middle-Income Countries. Review of Different Treatment Systems for Households or Neighbourhoods. Dübendorf, CH: Swiss Federal Institute of Aquatic Science and Technology (Eawag); 2006

[31] Tilley E, Ulrich L, Lüthi C, Reymond P, Zurbrügg C. Compendium of Sanitation Systems and Technologies. 2nd revised ed. Dübendorf, Switzerland: Swiss Federal Institute of Aquatic Science and Technology (Eawag); 2014

[32] Foxon KM, Pillay S, Lalbahadur T, Rodda N, Holder F, Buckley CA. The anaerobic baffled reactor (ABR): An appropriate technology for on-site sanitation. Water SA. 2004;**30**(5):44-50

[33] Ibeje OA, Onukwugha E. A model for optimal treatment of cassava wastewater using anaerobic baffled reactor. Nigerian Journal of Technological Development. 2021;**18**(2):129-132

[34] Movahedyan H, Assadi A, Parvaresh A. Performance evaluation of an anaerobic baffled reactor treating wheat flour starch industry wastewater. Iran Journal of Environmental Health Science & Engineering. 2007;**4**(2):77-84

Chapter 12

Biological Wastewater Treatment

Olawale Theophilus Ogunwumi, Adeyemo Festus Adeniyi, Mezie Chinazor Angus and Oche Sunday Oche

Abstract

Preventing environmental pollution by adequately treating the ever-increasing volume of wastewater generated by the over 8.1 billion (UN 2024 projection) people in the world, meeting governments' often updated effluent quality standards as a result of emerging contaminants in domestic and industrial wastewater, operating wastewater treatment process to generate energy through methane production and capture to save operating costs, and deploying a compact system to fit reducing installation space are some of the daring challenges facing sustainable wastewater treatment technologies today. Hence, there is a need for continued innovation and development of treatment processes. The current chapter discussed advancements in biological wastewater treatment technologies through the years with a focus on reasons for improvements in technologies. Some of the reasons highlighted are capital and operational costs, plant volumetric capacity, effluent quality, efficient nutrient removal, biofouling and membrane clogging, treatment plant installation size, etc. The chapter also discussed biochemical oxygen demand as a measure of water quality for biological treatment systems, the role of genetically engineered microorganisms in biological wastewater treatment, bioremediation as a biological treatment process, treatment plant pilot-scale, and upgrade to full-scale.

Keywords: wastewater, biological treatment process, activated sludge, genetically engineered microorganisms, biochemical oxygen demand, bioremediation, treatment plant pilot-scale, treatment plant full-scale

1. Introduction

Wastewater (**Figure 1**), inevitably produced from domestic, industrial, agricultural, and commercial activities, mixtures from sewer and stormwater runoff [1], has been described as used water generated from the use of freshwater, reclaimed/repossessed water, drinking water, raw and saline water in a variety of applications [2] including dishwashing, laundry, showers [3], food and beverage production, boiler cooling, textile manufacture, oil refining and gas production, mining activity, and petrochemical production [4]. Wastewater usually contains chemical, physical, and biological pollutants [5, 6] and its characteristics such as pH content, heavy metal content, odor, and concentration of suspended solids, depending on its source [7, 8].

Figure 1.
Greywater, a domestic wastewater in a settling tank [1].

1.1 Sources and types of wastewater

Wastewater types can be described in terms of their source which influences the type of pollutants they contain. The common types of wastewater are explained below.

- Domestic Wastewater or Sewage: This is used water from households, commercial, institutional, and public facilities in a municipal [9], transported through a sewer [1]. It is produced from such activities as washing, bathing, cleaning, and flushing. Variants of domestic wastewater include Blackwater (produced from the combination of human waste and flushing water) and greywater (produced from washing, bathing, etc.) [10]. The major pollutants in sewage are pathogens, colloidal matter, grease and oils, nutrients, pharmaceuticals, soaps and detergents, food wastes, suspended solids, minerals [11], microplastics, microfibers, etc. [12].

- Industrial Wastewater: This is the used water generated from the commercial manufacture of a range of industrial products and commodities including food and beverage, textile, shoe and leathery products, electronics and power, metal finishing, petroleum and petrochemicals, mining minerals, and radioactive materials [4]. It is an aqueous discharge and byproduct of industrial processes that is often generated from the cleaning activities associated with these processes [4]. The characteristics, volume, biodegradability, recalcitrance to treatment, and quality of industrial wastewater vary depending on its source industry [4]. Pollutants often found in industrial wastewater include inorganic materials, heavy metals, organic materials, and nutrients including nitrogen- and phosphorus-rich compounds, asbestos, pathogens, oils, and chemicals [13].

- Cooling Water: This wastewater which may cause thermal pollution, is generated from boiler steam condensation and machinery temperature reduction activities [2].
- Leachate: Is the wastewater that is precipitated from water percolation through solid wastes in a landfill or dumpsite [2, 14]. Leachate quality and indeed, its pollutants depend on the type of landfill that generates it. For example, industrial waste landfills generate leachate [15] containing typical industrial pollutants such as heavy metals, radioactive pollutants, and inorganic nutrients [2, 15].
- Surface Runoff: This is the wastewater flow from a mixture of stormwater, rainwater, and snowmelt, which carries pollutants present on the ground along the path of flow [2].
- Urban Runoff: This is the used water generated from urban landscape irrigation and cleaning activities in cities and municipals [2].
- Return Flow: This is the wastewater flow containing pesticides, herbicide residues, animal wastes, dissolved nutrients and minerals, suspended soil, etc., from agricultural activities [2].

1.2 The need for wastewater treatment

Wastewater treatment which involves the removal/reduction of pollutants in wastewater [15], is often considered a form of water use as it is interconnected with other forms of water usage [16] and is necessary to preserve public health, prevent environmental degradation and ensure water security for the ever-growing global population (**Figure 2**) as wastewater is known to cause diseases such as hepatitis, cholera, cancer, reproductive defects, dysentery, and developmental disorder in humans and animals and, can lead to oxygen depletion and biodiversity reduction in natural ecosystems [18].

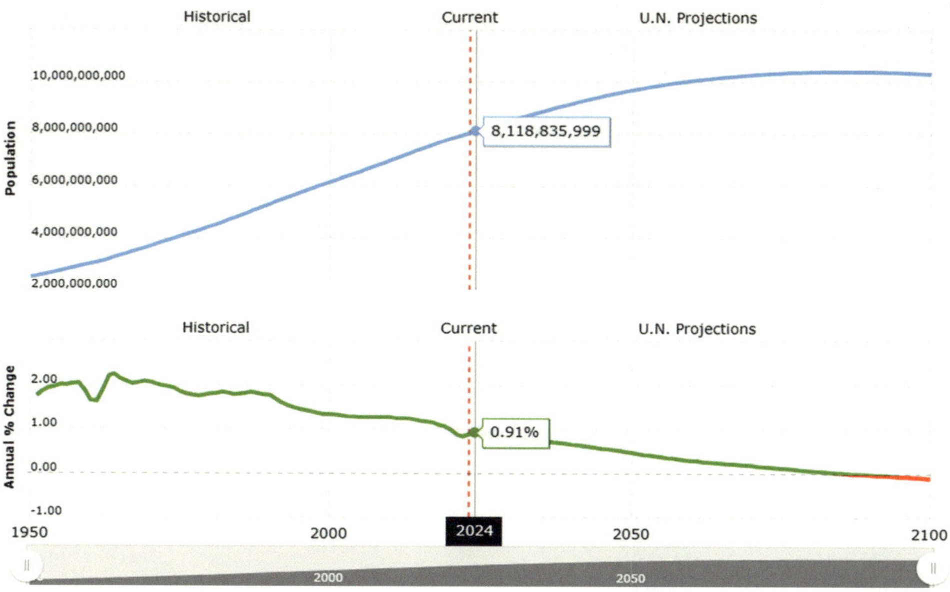

Figure 2.
United Nations projected data for world population from 2024 up to 2100 [17].

Some of the major reasons for treating wastewater are highlighted in the following section.

- Public Health: Untreated wastewater discharged into water bodies in the environment can have devastating health effects on living things from water use due to the presence of pollutants [16].

- Wildlife Ecosystem: Aquatic life and other creatures such as migratory birds are negatively affected when untreated wastewater is discharged directly into their habitat including oceans, rivers, marshes, beaches, and shorelines [16].

- Recreational and Tourism: Clean surface water can be a great recreational playground, sporting arena, and tourist attraction for most people where activities such as fishing, swimming, boating, yachting, boat cruising, and picnicking can be carried out [16]. Pollution by wastewater can prevent sporting activities and recreation [16].

- Economic: Water pollution will affect the economy of a place when it affects public health, prevents recreation and tourism, and affects fishing activities and aquatic life population [19].

2. Basics of biological wastewater treatment

Biological wastewater treatment systems rely on beneficial bacteria and other microbes to decompose organic pollutants in wastewater into simple substances like carbon (IV) oxide (CO_2) and water (H_2O) and generate biological floc or sludge; decomposition may occur aerobically or anaerobically [20]. These microbes might be present in the water naturally or be deliberately introduced during treatment – genetically engineered microbes (GEMs) [20]. Some of the concepts associated with biological wastewater treatment are elaborated in the section below.

2.1 Aerobic vs. anaerobic processes

- Aerobic Processes: In aerobic biological wastewater treatment, bacteria decompose organic pollutants with the aid of oxygen [20]. Reducing organic chemicals to simpler ones defines this process, which ultimately produces carbon dioxide, water, and further microbial biomass [20].

- Anaerobic Processes: Contrarily, oxygen is not utilized during anaerobic biological treatment [20]. Anaerobes hasten the decomposition of organic compounds by taking part in anaerobic digestion and related procedures [20]. Biogas is the final byproduct of decomposing stabilized organic wastes and contains carbon dioxide and methane [20].

2.2 The waste product: sewage sludge

Natural microbes decompose organic pollutants, resulting in the production of sewage sludge as a byproduct [20, 21]. Sludge from sewage treatment plants, often called semi-solid waste, contains a broad range of organic and inorganic materials. The sludge is a reflection of the microbes' leftovers, which include bacteria and undigested organic waste [22].

2.2.1 *Importance of sewage sludge*

- Resource Recovery: Despite being regarded as waste, sewage sludge is a rich supply of materials that can be utilized for resource recovery, which is why it is important [23]. Due to its high organic content, anaerobic digestion offers great promise as a feedstock for producing biogas, which might be used as a thermal storage or energy source [23].

- Provision of Macro-nutrients: Sewage sludge contains numerous nutrients, including nitrogen and phosphorus [23]. One way to boost crop yields while decreasing reliance on synthetic fertilizers is to convert sewage into fertilizer [23].

- Soil Enhancement and Land Use: Biosolids from treated sewage sludge can improve soil quality in many ways [23]. This sustainable farming method improves soil structure, water retention, and plant growth [23].

2.3 Carbon dioxide release: the final stage

Microorganisms digest organic matter and produce sewage sludge, thereby releasing CO_2. The release of this gaseous byproduct marks the end of the biological treatment process, which reduces complex organic molecules to less harmful substances with less impact on the environment [24].

2.4 Impact of biological wastewater treatment

- Greenhouse Gas Emissions: The CO_2 is produced naturally by microbes, but it does contribute to the overall quantity of gases released into the atmosphere [24]. Wastewater treatment plants make great efforts to minimize their environmental impact, but the fact that they nevertheless emit carbon dioxide into the atmosphere highlights the critical need to discover solutions to lessen this effect [24].

- Carbon Footprint Considerations: The carbon footprint must be considered when evaluating the environmental friendliness of treatment plants [22, 24]. The case of biological wastewater treatment is a prime example of this. Common elements of initiatives to decrease carbon emissions include the optimization of treatment processes, the installation of energy-efficient technology, and the investigation of alternative treatment options [23].

- Energy Generation: Energy generation (**Table 1**) will probably be a key focus in the future of biological wastewater treatment. New methods of energy recovery from wastewater treatment and sewage sludge processing, like enhanced anaerobic digestion and biogas capture, are being developed to reduce energy consumption and increase sustainability [23].

- Treatment Process Renewable Energy Integration: Wastewater treatment plants are giving serious consideration to incorporating renewable energy sources like solar and wind power [28]. In line with the larger objectives of renewable energy adoption, these activities seek to lessen the treatment processes' total impact on the environment [29]. It reduces operating expenses by over 9% and increases renewable energy consumption by over 5%, with thermoelectric ratio adjustment improving further [30].

2.5 Biological wastewater treatment and the circular economy: from waste to valuable resource

The concept of circular economy now finds resonance with the future of biological wastewater treatment. A change in perspective from seeing wastewater as a problem to a valuable resource is evident in the growing interest in recovering resources from sewage sludge, such as nutrients, and creating value-added products [31]. One example of the circular economy application is the concept of biorefinery, which involves using wastewater treatment facilities to produce several goods through integrated operations [31]. The goal of biorefinery strategies is to extract as much useful material

		Energy (kWh/m^3)		
Constituent	**Typical concentrations[a] (mg/L)**	**Maximum potential from organic oxidation[b]**	**Required to produce fertilizing elements[c]**	**Thermal heat available for heat-pump extraction[d]**
Organics (COD)				
Total	500			
Refractory	180			
Suspended	80	0.31		
Dissolved	100	0.39		
Biodegradable	320			
Suspended	175	0.67		
Dissolved	145	0.56		
Nitrogen				
Organic	15		0.29	
Ammonia	25		0.48	
Phosphorus	8		0.02	
Water				7.0
Totals		1.93	0.79	7.0

[a]*After Tchobanoglous and Burton [25].*
[b]*Based on a theoretical 3.86 kWh energy production/kg COD oxidized to CO_2 and H_2O [26].*
[c]*Based on production energy of 19.3 kWh/kg N by Haber-Bosch Process and 2.11 kWh/kg P after Gellings and Parmenter [27].*
[d]*Energy associated with a 6 °C change in water temperature through heat extraction.*
Source: Ref. [28].

Table 1.
Domestic wastewater nutrient energy generation.

as possible from wastewater, whether it is bioenergy or bio-based commodities. [31], in their study, removed over 70% of nutrients and provided valuable biomolecules for bioenergy production and nanoparticle synthesis.

3. Genetically engineered microorganisms (GEMs) for wastewater treatment: unveiling opportunities and navigating challenges

An integral part of environmental engineering, wastewater treatment is undergoing a paradigm shift toward greater use of state-of-the-art technology. Along these lines, GEM integration has been a promising new avenue for improving the longevity and efficiency of wastewater treatment systems [32]. Taking a deep dive into GEMs, this section explores their breadth, definition, and the complex relationship between the benefits and drawbacks of using them to treat wastewater. Genetically modified microorganisms are the result of microbiology and genetic engineering; these bacteria, fungi, and algae actively alter their genes to improve performance [32]. With their ability to increase nutrient removal, tolerate harsh environmental conditions, and augment pollutant degradation, GEMs provide a molecular canvas for tackling the complex problems of modern wastewater treatment [32].

3.1 Possible applications of GEMs to wastewater treatment

- Precision in Pollutant Degradation: GEMs enable pinpoint pollution degradation in a new way. Through the intentional insertion of genes that produce specific enzymes or metabolic pathways, these microbes can break down toxins more thoroughly and efficiently [33].

- Adaptability to Diverse Environments: GEMs excel at adapting to demanding environmental situations. Genes that provide resilience allow these microbes to adapt to varied habitats with varying temperatures, pH, and salinity. GEMs are useful wastewater treatment tools because they are versatile [32, 33].

- Enhancing Eutrophication and Nutrient Removal: GEMs can improve nutrient uptake and removal from wastewater, especially nitrogen and phosphorus [33]. An essential element in preserving the ecological balance in receiving water bodies, this focused strategy addresses eutrophication concerns [33].

- Synergistic Bioenergy Production: In addition to removing pollutants, [34] revealed that some GEMs can be engineered to produce biofuels while treating wastewater, resulting in synergistic bioenergy production. Wastewater treatment plants may now serve as centers for resource recovery and environmental responsibility thanks to this dual-purpose functioning, which is in line with worldwide efforts toward sustainable energy generation [33].

- Customized Microbial Consortia: GEMs allow for the creation of customized microbial consortiums that are optimized for certain wastewater compositions [33]. Through mutually beneficial relationships and metabolic capacity complementarity, this microbial community orchestration can improve treatment efficacy as a whole [33, 34].

3.2 Disadvantages of using GEMs to treat wastewater

- Protecting the Environment: Exploring Unknown Regions: Complex problems may arise with the discharge of GEMs into the natural environment. To guarantee environmental safety, an understanding and careful approach to GEM application are required due to concerns about unanticipated ecological repercussions, such as possible changes to native ecosystems and gene transfer to indigenous microbes [34].

- Understanding and Navigating Regulatory Frameworks: GEMs operate in a world where regulations are complicated and frameworks vary worldwide. Science, government agencies, and businesses must work together proactively to protect public and environmental interests while also encouraging innovation [34].

- Long-Term Stability and Viability: GEMs encounter a complex problem when used in real-world wastewater treatment scenarios. The effectiveness of GEM-based therapy systems over the long term might be affected by factors like predation, competition with local microorganisms, and changes in environmental circumstances [34].

- Ethical Considerations – Harmonizing Development with Conscientious Innovation: A rigorous evaluation of the limits of scientific intervention is warranted in light of the ethical concerns raised by the intentional alteration of microbial genomes [34]. If we want to see GEMs used responsibly and with social acceptance in wastewater treatment, we need to set explicit ethical standards.

4. Bioremediation of domestic and industrial wastewater: application, problems, and solutions

As a long-term, eco-friendly option for treating domestic and industrial wastewater, bioremediation – a biological method of wastewater treatment and water recovery – is gaining popularity. This section discusses the bioremediation process, some specific examples, difficulties in its implementation, and possible solutions.

4.1 Bioremediation for wastewater treatment

By harnessing the metabolic power of microbes to decompose and convert contaminants into less dangerous forms, bioremediation (**Figure 3**) offers a comprehensive methodology for treating wastewater. This biological intervention has been utilized in a wide variety of wastewater treatments, from simple domestic to intricate industrial processes [35].

4.1.1 Bioremediation of domestic wastewater: examples

- Systems for Septic Tanks: One example of bioremediation in action is the use of septic tank systems in residential areas [35]. Effective treatment of wastewater, before discharge or additional treatment, is ensured by the microbial communities within septic tanks, which break down organic debris [35].

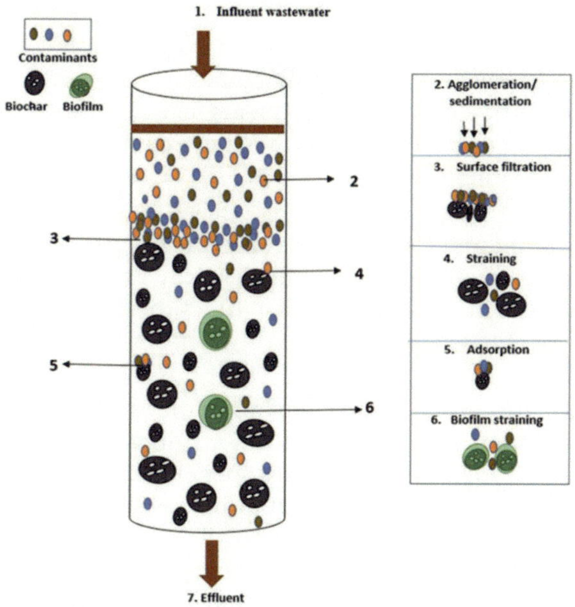

Figure 3.
Bioremediation removes pollutants from onsite municipal wastewater using biochar through several processes [35, 36].

- Realized Wetlands Design: Artificial wetlands purify wastewater with plants and bacteria like natural wetlands. Microbial communities live in wetland system roots and decompose, adsorb, and precipitate pollutants [35].

- Activated Sludge Systems: Activated sludge systems are widely employed in wastewater treatment plants for municipal purposes. The aerobic bacteria in activated sludge break down organic pollutants, cleaning the wastewater before release or treatment [35].

4.1.2 Bioremediation of industrial wastewater: examples

- Textile Industry: Bioremediation has been deployed in the textile industry, which utilizes a great deal of water and produces lots of contaminants. Microbial consortia remove organic solvents, dyes, and other contaminants, a greener treatment than conventional methods [35, 37].

- Petrochemical Industry: This industry struggles with complex effluent streams containing hazardous chemicals and hydrocarbons. Bioaugmentation or the introduction of certain microbes may improve hydrocarbon breakdown and reduce the environmental impact of petrochemical wastewater [37].

- Food Processing Plants: Wastewater from these industries typically contains significant levels of nutrients and organic materials. Bioremediation methods like aerobic treatment and anaerobic digestion reduce environmental impact by promoting organic breakdown [35].

4.2 Bioremediation challenges

- Pollutant Complexity: Wastewater pollutants are complex and variable, making bioremediation difficult. Industrial effluents may require specialized microbial communities and treatment methods to remove complex chemicals, heavy metals, and other contaminants [35].

- Variable Environmental Conditions: Temperature, pH, and nutrient availability affect bioremediation efficiency. Treatment results may not be repeatable due to changes in these factors that affect the efficiency of microbial activity [37].

- Microbial Competition and Inhibition: In situations where microbial populations are diverse, inhibition and competition among microbes within the treatment system are possible [35]. The overall treatment effectiveness can be impacted by imbalances caused by pollutants that promote the growth of particular microbes [35].

- Slow Treatment Rates: Treatment rates in bioremediation techniques can be slower than in chemical or physical methods, especially when dealing with stubborn chemicals [35]. In situations where quick and effective treatment is crucial, this restriction could be a problem.

4.3 Innovative solutions and future directions

- Enhancing Biodegradation using Genetic Engineering: The use of genetic engineering can boost microbes' biodegradation capacities. To address some of the difficulties caused by complicated wastewater compositions, microorganisms that have been engineered (GEMs) with specific enzymes or metabolic pathways can be used to target particular contaminants [37, 38].

- Developing Microbial Consortium for Bioaugmentation: To improve wastewater treatment systems, specialized microbes are purposefully introduced, a process known as bioaugmentation [30]. This process enables the optimization of treatment systems tailoring microbial consortia, which in turn addresses specific contaminants and enhances overall system resilience [30].

- Enhanced Control and Monitoring Technologies: Improvements in bioremediation process management are made possible by new developments in real-time monitoring and control systems. Insights into microbial activities are obtained by sensor technology and data analytics, enabling proactive modifications to maximize treatment efficiency [30].

- Combined Treatment Approaches: Bioremediation, when combined with additional treatment methods like enhanced oxidation techniques or physicochemical processes, provides a holistic and synergistic approach. Overcoming the limits of individual treatments, this integrated strategy can boost overall treatment performance [30].

5. Biochemical oxygen demand (BOD) in wastewater treatment

Water quality evaluation and wastewater management are crucial to aquatic ecosystem sustainability and environmental engineering. Biochemical oxygen demand (BOD) is a key indicator of biological wastewater treatment system performance and organic matter pollution [39]. This section examines biochemical oxygen demand (BOD), its importance to wastewater treatment, and its role as a key indicator of water management strategy and sustainability.

5.1 BOD: an overview

Biochemical Oxygen Demand (BOD) can estimate how much oxygen bacteria need to break down organic matter in water. BOD, which measures oxygen consumption by microbes breaking down complex organic compounds, originated in the bacterial breakdown of organic substances [39]. Organic pollutants including sewage, industrial effluents, and agricultural runoff raise wastewater BOD. Because microorganisms utilize dissolved oxygen, oxygen levels in treated effluent drop as organic pollutants are broken down [39]. If oxygen loss is substantial, aquatic animals and the environment can suffer [39].

5.2 Biological wastewater treatment efficiency by BOD measurement

- Treatment Efficiency by Procedure: The effectiveness of biological wastewater treatment can be gauged by measuring BOD. Aerobic digestion, activated sludge treatment, biofiltration, and other processes rely heavily on microorganisms to decompose organic contaminants in treatment facilities [39]. By keeping tabs on BOD levels throughout treatment, we can learn how well these biological mechanisms are working.

- Efficiency of Activated Sludge (aerobic) System: Microbial colonies in activated sludge systems in municipal wastewater treatment facilities aerobically decompose organic pollutants [40]. We can measure BOD to assess how effectively the system supports microbial growth. Operators can optimize conditions for microbial activity and alter treatment parameters accordingly by tracking BOD reductions through each stage of the process [40].

- Efficiency of Anaerobic Treatment: BOD readings are used as a measure of the overall effectiveness of the anaerobic treatment procedures that purposely reduce oxygen [39]. Anaerobic digestion produces biogas from organic matter. Anaerobic reactors that monitor BOD keep methane-producing bacteria at ease and remove organic contaminants effectively [39].

- Effectiveness of Constructed Wetlands: Using plants and bacteria in an analogous way that natural ecosystems work, artificial wetlands are utilized to treat wastewater [40]. The biological breakdown of organic compounds by measuring the BOD in the inflow and outflow of man-made wetland systems can provide knowledge of the efficiency of these systems. The success of the treatment process and microbial communities in eliminating contaminants from the water is indicated by the decrease in BOD levels [40].

5.3 BOD implications for water sustainability

- Effects on the Environment and the State of the Ecosystem: The environmental impacts of water bodies with high BOD levels is substantial. The degradation of aquatic environments by dissolved oxygen depletion impacts fish populations, biodiversity, and ecological equilibrium. Sustainable water management requires monitoring BOD levels to preserve healthy aquatic habitats [40].

- Regulatory Compliance and Water Quality Standards: As an important criterion for wastewater discharge permits, BOD limitations are included in numerous regulatory frameworks globally [39]. Businesses and cities must follow these rules. To preserve water quality and lessen environmental impacts, sustainable water management goes beyond what is required by regulations [39].

- BOD as a Pollution Source Indicator: One diagnostic tool for locating the origins of water pollution is BOD analysis. The presence of organic contaminants can be identified by analyzing variations in BOD profiles [39, 40]. This allows for the implementation of targeted actions to reduce pollution where it is most prevalent. Proactively identifying and treating pollution causes is essential for sustainable water management [40].

- Using BOD in Representing Water Quality: To mimic the process of organic pollutant breakdown in treatment systems, sophisticated water quality models use BOD data [41]. The goal of these models is to help decision-makers develop sustainable water management plans by projecting future BOD levels under different scenarios [41]. Stakeholders can develop effective policies and procedures by comprehending the possible effects of different elements on BOD.

5.4 Difficulty in measuring and interpreting BOD

- Changes with the Seasons and Other Factors: Temperature, nutrient variations, and seasonal fluctuations are some of the elements that can influence the variability in BOD values [39]. To make educated decisions about water management based on accurate interpretations, it is crucial to understand these variances.

- Non-specificity of BOD: Although BOD gives a general idea of how much organic pollution there is, it does not break it down into kinds. Because of this lack of specificity, it is difficult to identify which contaminants may necessitate tailored treatment approaches [39, 41].

- Time and Resource Intensiveness: The time-consuming process of traditional BOD determination methods requires samples to be incubated for many days. Consistent monitoring is essential, but it might be difficult for smaller treatment centers to allocate the necessary resources [39].

- Emerging Contaminants and BOD Limitations: The proliferation of new pollutants, such as medications and personal care items, poses problems that BOD measurements cannot completely solve [39]. While these compounds might not have a major impact on BOD, they can still harm aquatic ecosystems, which is why we need other ways to measure their impact [39].

5.5 Innovations and future directions in BOD analysis

- Sensor Technologies and Real-time Monitoring: Technological advancements in sensors have made it possible to track BOD levels in real time. More adaptive and responsive water management strategies are made possible by the continuous data provided by online sensors, which allow for rapid responses to BOD changes [40, 41].

- Molecular Methods and Biosensors: Rapid and specific BOD measurements are made possible by biosensors and molecular methods, which target key microbial markers [42]. These advancements in technology allow for more precise and efficient BOD readings, which in turn shed light on the microbiological processes at work in wastewater [42].

- Incorporating AI: Improving BOD prediction models is an area where artificial intelligence (AI) shows enormous potential. Machine learning algorithms are capable of sifting through massive datasets, finding patterns, and making condition-specific predictions on BOD levels. Proactive decision-making in water management is facilitated by this integration, which also improves the precision of forecasts [40, 42].

6. Wastewater treatment plants and sewage sludge systems

If the ever-increasing global wastewater generation in different regions (in billion m^3) (**Figure 4**) and the large yearly per capital production (in m^3/capita/year) (**Figure 5**), are to be reused or discharged safely into the environment without

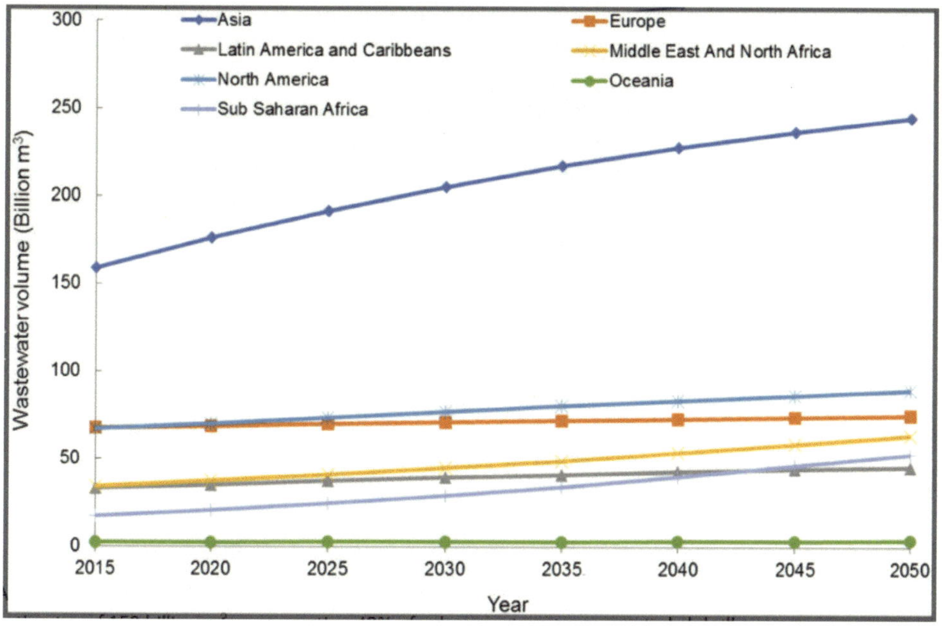

Figure 4.
Global regional wastewater generation projection to 2050 [43].

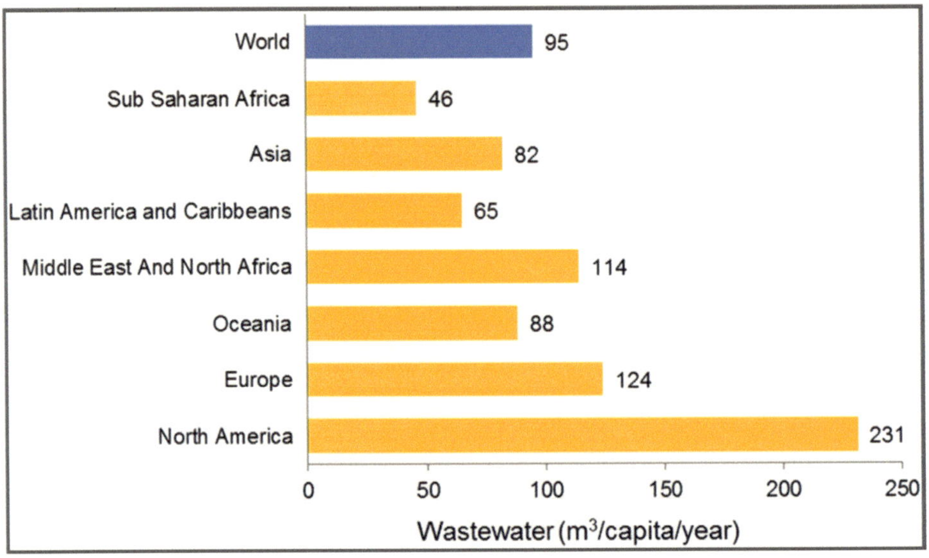

Figure 5.
Yearly global regional wastewater generation per capita [43].

harmful, detrimental health effects, then wastewater must be treated to remove pollutants. This can be achieved in specialized process plants referred to as Wastewater Treatment Plants (WWTPs).

Wastewater treatment plants are facilities that receive and treat domestic and industrial wastewater by removing contaminants, such as coarse debris, grease and oils, sand, organic and inorganic pollutants, and biodegradable materials, in wastewater [44]. When these plants treat wastewater, they produce a mud-like, slurry residue or sewage sludge as the byproduct of the treatment process due to biomass activities of feeding on organic pollutants in the wastewater [44]; hence, treatment plants are also sewage sludge systems.

The sludge produced during the wastewater treatment process contains pollutants like heavy metals, pathogens like viruses, and a large mass of bacteria that can stabilize organic matter in wastewater using aeration, hence the term "activated sludge" [44, 45]. The sludge also contains valuable nutrients like phosphorus and nitrogen that have important agricultural applications in being used as farmland improver and fertilizer [44].

6.1 Wastewater treatment plant operations

The wastewater treatment plant facility undergoes different processes/stages of operation (**Figure 6**) to achieve an acceptable level of wastewater treatment suitable for reuse and discharge to the environment. These stages of operation, which include preliminary treatment, primary, secondary, and tertiary treatments are described below.

- Preliminary Treatment Operation: This is aimed at separating suspended, gross solids such as large plastics, wood logs, fabric, and bottles, excess grease or oil, and grit from raw wastewater from the sewer by channeling it through a trough and moving screen or by cutting the large, foreign objects in small pieces with a comminutor or a macerator and then removing them [44, 46].

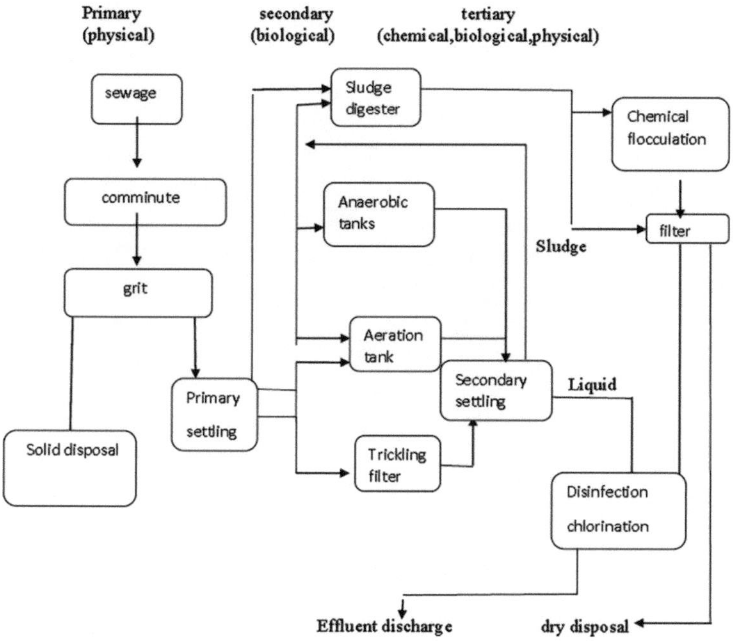

Figure 6.
Wastewater treatment processes [46].

- Primary Treatment Operation: This is a physicochemical wastewater treatment operation that involves allowing relatively smaller suspended and settleable solids to precipitate out of the water by gravity settling and sedimentation, mechanical flocculation, or chemical coagulation [44, 46]. Here, further screening and comminution of smaller solid materials from the preliminary operation can be carried out (**Figure 7**) [47]. This operation is aimed at reducing the strength and biochemical oxygen demand (BOD) of organic matter in wastewater, neutralizing acidic/alkaline wastewater by the addition of appropriate chemicals, removing leftover grease and oil from preliminary treatment, and eliminating volatile pollutants [44]. Also, wastewater equalization, which involves making wastewater effluent homogenous by using mechanical paddles or aeration can be carried out at this stage [46].

- Secondary Treatment Operation: The secondary treatment operation is a biological wastewater treatment that removes dissolved, organic matter or colloids in wastewater when microorganisms consume organic pollutants in an aeration tank (**Figure 7**), producing CO_2 and water H_2O in the process [46, 47]. This operation, which may be an aerobic or anaerobic process, is aimed at removing biodegradable materials and significantly reducing BOD, phenols, and oils in wastewater [46].

- Tertiary Treatment Operation: This wastewater treatment operation is aimed at removing non-biodegradable materials, minerals (such as the hydrated salts of nitrogen and phosphorus known to cause eutrophication in surface water), and heavy metals left in wastewater after the primary and secondary treatment operations [46]. The tertiary treatment is necessary to provide final polished,

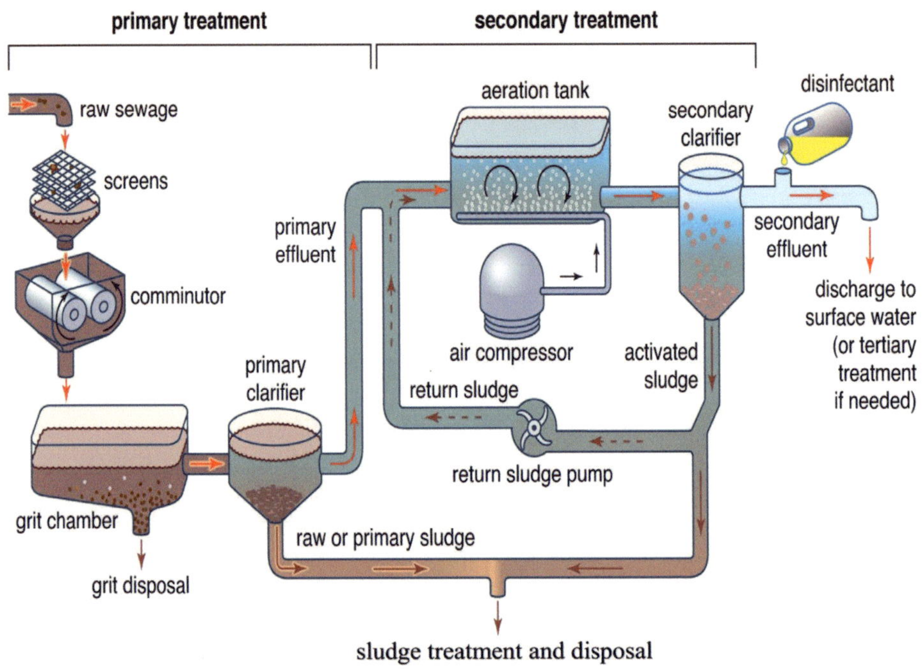

Figure 7.
Primary and secondary operations of wastewater treatment [47].

high-standard effluent before reuse or discharge [48]. This treatment operation can be carried out through ion exchange, reverse osmosis, electrodialysis, adsorption, evaporation, chlorination, and electrolytic recovery processes [46]. The main operations in tertiary treatment are filtering and backwash operations (**Figure 8**) in which secondary effluent is allowed to flow through a filter bed of anthracite, gravel, or sand, and, then passed through a trough containing alum to

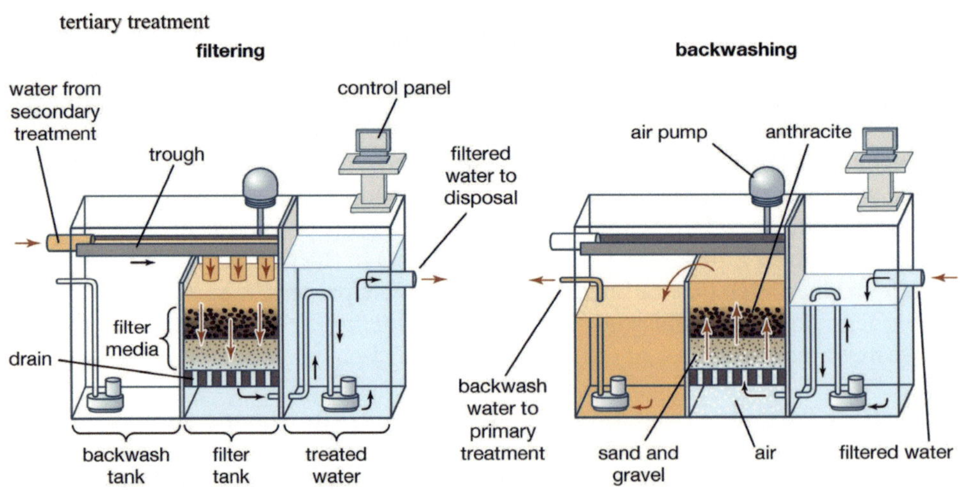

Figure 8.
Tertiary operation of wastewater treatment [47].

filter out finely suspended solids not previously removed [47, 49]. The backwash operation involved removing entrained floc from the filters by allowing filtered water to flow back through the filter media and be recycled [49]. Disinfection of wastewater is achieved by chlorination where chlorine is dozed into the treated effluent to kill any pathogen present and the effluent is dechlorinated to remove excess concentration of injected chlorine in a final step before discharge to the environment [49].

6.2 Biological wastewater treatment technologies

The secondary wastewater treatment process which uses microorganisms and aeration to convert organic pollutants to simple substances, is a biological treatment technology. The four (4) main forms of biological treatment technologies are the Oxidation Pond, Trickling Filter, Rotating Biological Contactor, and Activated Sludge.

- Oxidation Pond: This large, usually shallow pond relies on the interaction between bacteria, algae, and sunlight to treat wastewater. Also called stabilization ponds or lagoons, oxidation ponds, provide an arena for algal growth by using sunlight's energy, CO_2, and inorganic compounds released by biomass in the systems and, the algae in turn, release oxygen (O_2) required by aerobic bacteria to feed on organic matter in wastewater [47]. Oxidation pond designs are made compact by installing external aerators (mechanical air diffusers) in the pond [47]. Dredging, and gravity settling, filtration, or chemical treatment are used to rid the pond of sludge deposits and leftover algae respectively [47].

- Trickling Filter: This biological wastewater treatment technology operates by continuously spraying wastewater over a bed of stone in a tank and collecting the trickling water at the bottom of the tank from where it is sent to a clarifying tank [47]. Microorganisms on the surface of the stone bed consume organic matter in the wastewater as it flows thus, reducing its BOD and treating it [47]. The circulating air passing through the spaces between the stones provides aeration for the microbial activity of feeding on organic matter in the wastewater [47]. The performance of the trickling filter can be optimized by arranging two or more filters in series and allowing treated effluent to recirculate through them [47].

- Rotating Biological Contactor: In this technology, a layer of microbes growing on a sequence of large, flat, plastic disks (in contact with the wastewater to be treated) mounted on a rotating horizontal shaft, consumes the organic matter in the water as each disk in turn, get exposed to atmospheric air due to shaft rotation [47].

- Activated Sludge: Of the biological wastewater treatment technologies discussed so far, the activated sludge technology is the most globally applied and effective [45]. A typical design consists of a primary aeration or mixing tank and a secondary settling tank or clarifier for recycling returned sludge (RAS) for a continuous process and increased efficiency [47]. Wastewater from primary treatment is mixed with fresh sludge and RAS and, aeration from an external air compressor is supplied to the mixture from porous diffusers at the bottom of the mixing tank [47]. Aeration provides oxygen for microorganisms in the sludge which feed on organic materials in wastewater, reducing its BOD and forming a mass of biological floc called Activated Sludge [47]. The treated effluent then flows into the secondary settling tank where clear water is separated from the activated sludge which is returned to the system (RAS) or removed as surplus activated sludge (SAS) [47].

6.3 Developments in biological wastewater treatment process and emerging technologies

Biological wastewater treatment technologies have evolved over the years as industries and households continue to generate large volumes of wastewater and new pollutants are discovered, prompting various national governments to update wastewater effluent discharge quality standards for the protection of human health and the environment [47, 50]. Technologies have also been upgraded for sustainability via energy production and waste resource recovery for a circular economy process [50]. The advancement in treatment technology has also helped to mitigate the effects of future challenges such as increased wastewater generation from an increasing global population, old and aging treatment facilities, limiting plant installation spaces, and climate change, on the quantity of treated wastewater [47].

6.3.1 The stirred tank reactor technology

The activated sludge wastewater treatment technology is an ideal process for treating industrial wastewater in a stirred tank (agitated vessel) seeded with the biological floc (sludge) and aeration at high pressure to supply adequate dissolved oxygen [50]. The stirred tank activated-sludge wastewater treatment technology is one of the oldest biological treatment processes first used at the Davyhulme Wastewater Treatment Facility in the UK between 1912 and 1914 [45, 51, 52]. A typical system design usually referred to as the conventional activated sludge (CAS), consisted of a single mixing tank (agitated vessel), a settling tank (sedimentation tank), and a sludge recycle line (RAS line) (**Figure 9**) [45, 51, 52]. This arrangement allows raw, influent wastewater to be treated and effluent to be separated from the floc (activated sludge) which is recycled or sent to sludge management as waste.

The capacity of the agitated vessel in an activated sludge process is usually large and since oxygen dissolves very slowly in an aqueous solution, a major limitation of the industrial application of the activated sludge process is the high cost of operation arising from operating and maintaining high capacity air compressors [50]. Step aeration technique can be used to improve the performance of the activated sludge

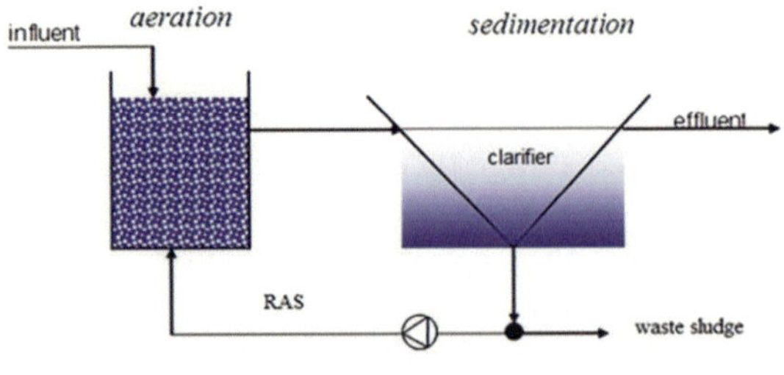

Figure 9.
Conventional activated sludge process [51].

process whereby the process is modified by installing several small-sized, cascading stirred tanks in sequence (The Bardenpho Scheme) instead of a single large tank; the total volume of the small tanks being equal to the volume of the large tank [50]. This enhances higher BOD reduction, denitrification, and phosphorus elimination from wastewater [50]. Higher process performance can also be achieved through a sequential-parallel arrangement of the small tanks.

6.3.1.1 Design and performance characteristics of the activated sludge process

The stirred tank activated-sludge process is designed based on factors such as characteristics of wastewater, government regulation, and treatment objectives [53]. The process design characteristics include:

- Dimension and capacity of the stirred (agitated) tank.
- Transfer system and oxygen requirement of the aeration process.
- Dimension and capacity of the settling tank.
- Design of the sludge return system.
- Design of the waste sludge withdrawal system.
- Mixing pattern of the wastewater and sludge (formation of mixed liquor suspended solid-MLSS).

The process performance of a stirred tank activated-sludge system can be hinged upon various features some of which are:

- Contaminant Reduction: The activated sludge system can reduce up to 99% BOD in secondary treatment and can achieve a high reduction of pathogens in wastewater.
- Loading Rate: The system provides treatment flexibility by operating over a range on hydraulic and organic loading rates.
- Resilience: The system is robust under varying operating conditions as it can withstand different hydraulic and organic shock loads.
- Effluent Quality: Permissible discharge standards of wastewater effluent can be achieved using the activated sludge system.
- Installation Space: The system requires a smaller installation space compared to the system such as a stabilization pond [54].
- Nutrient Elimination: The system is ideal for high removal of nitrogen and phosphorus. Also, the nitrification–denitrification process can be achieved by upgrading with an anoxic chamber inside the aeration vessel [55].

Improvements on the stirred tank activated-sludge system include systems such as the anaerobic-anoxic, oxygen ditch, and sequential batch reactor technologies.

6.3.1.2 Anaerobic-anoxic-oxic (A2O) technology

This activated sludge technique is an improvement on the design of the CAS as it allows easy elimination of phosphorus and nitrogen nutrients in wastewater. Its arrangement (**Figure 10**) consists of an anoxic tank where a Modified Ludzack-Ettinger (MLE) process takes place for denitrification, an anaerobic tank for phosphorus removal by absorption, and an aerobic tank (oxic) for organic pollutant decomposition [51, 56].

6.3.1.3 Oxidation ditch

This is an improvement over CAS where highly efficient rotating biological contactors (RBCs) in the ditch (**Figure 11**) increase surface area and dissolved oxygen available for microbial growth and activities, creating agitation and mixing within the ditch as wastewater flows from a bar screen, leading to effective treatment process [51].

6.3.1.4 Sequential batch reactor (SBR) technology

The SBR technology has the advantages of lower treatment area and cheaper installation cost over the CAS because wastewater equalization, primary/secondary clarification, and microbial treatment take place in a single reactor unit (batch operation) [57]. The SBR operation, commonly referred to as a "fill-and-draw" process, first used a single unit, variable-volume, batch flow system for wastewater

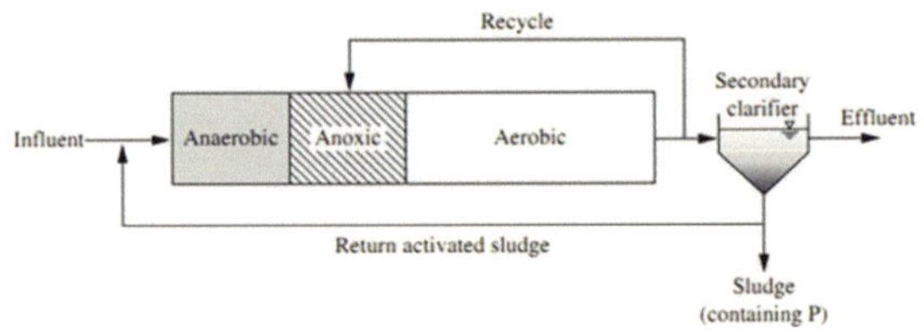

Figure 10.
Typical arrangement of the A2O-activated sludge technology [51].

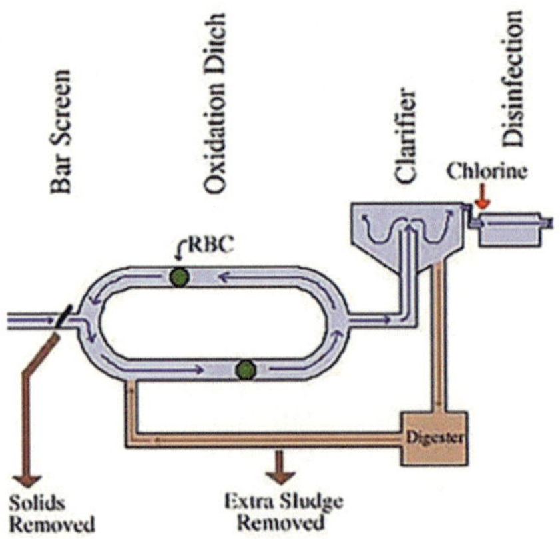

Figure 11.
The oxidation ditch process [51].

treatment, but operational challenges forced an upgrade to a double unit, fixed-volume, continuous flow system with one unit used for aeration and the other for settling [58].

Further upgrades to the SBR technology occurred in the 1950s and 1970s when a continuous-feed batch treatment was incorporated in a variable-volume system (by Pasveer and co-researchers) and in the USA and Australia, with the EPA's grant; the EPA's publication of SBR design manual between 1982 and 1992 led to the global application of the technology [58]. Today, the SBR process consists of a series of tank units, each operated as a batch reactor. A typical treatment cycle goes through five (5) stages such as fill, react, settle, draw, and idle (**Figure 12**).

Modifications of the SBR technology such as the sequencing batch biofilm reactor (SBBR), and the anaerobic sequencing batch biofilm reactor (ASBBR) have recently been developed to mitigate some challenges of the SBR technology including excessive sludge production and high sludge-volume index [57].

6.3.2 Membrane technology

The membrane bioreactor (MBR) technology (**Figure 13**), which is ideal for treating domestic and industrial wastewater, was developed in 1969 under the Dorr-Oliver Scheme by Smith and co-researchers who combined biological treatment (in a bioreactor) with a membrane filtration process – ultrafiltration process (microfiltration is an

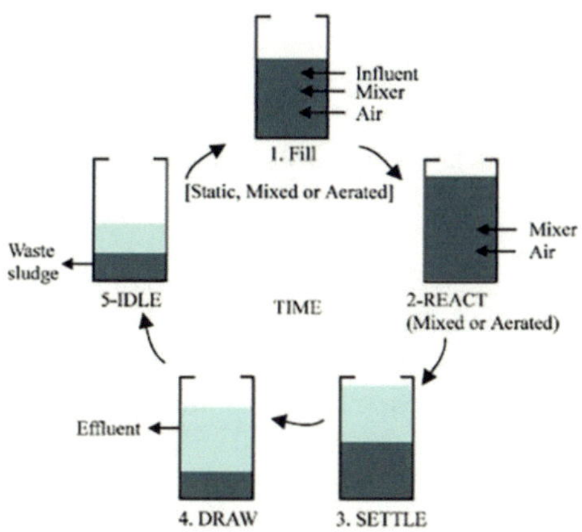

Figure 12.
Stages of the SBR process [58].

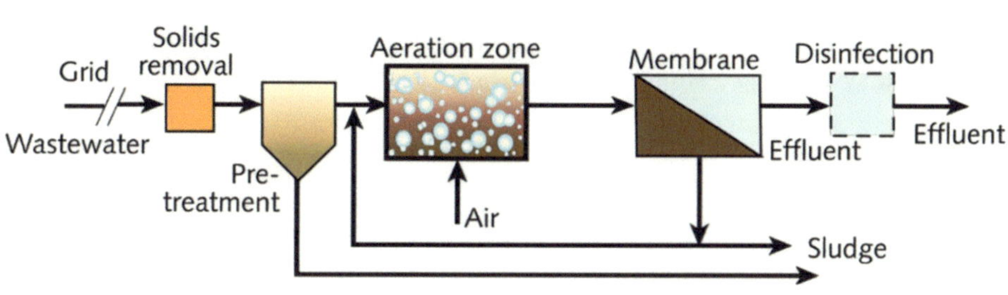

Figure 13.
The MBR process [59].

alternative), instead of the settling tank used in activated sludge system [60]. While the decomposition of organics occurs in the bioreactor, separation of floc from treated water occurs on the membrane (with different pore sizes), this allowed for maximum sludge separation from the treated water than in CAS [60]. Limitations such as membrane fouling that causes clogging, and huge energy costs from membrane scouring, led to the upgrade of the original MBR configuration in 1989 by Yamamoto and other researchers, where a hollow fiber membrane and suction pump (rather than a pressure pump) were placed in the bioreactor where the membrane is immersed [60]. The acceptability of the MBR process increased since the 1990s mainly due to the advancement brought about by the process upgrade (immersion configuration), which provided permeates of high quality and reduced capital costs and clogging [60]. Future MBR process upgrades can be achieved by expanding fiber membrane applications.

6.3.2.1 Design and performance characteristics of the MBR process

Some of the design characteristics of the MBR technology are elaborated in the section below.

- Activated Sludge Process Design: Activated sludge process design is an integral part of the MBR technology in combination with the membrane filtration process. The design of the activated sludge process depends on microbial growth considerations, oxygen supply, and dimension of the bioreactor [50].
- Membrane Material Selection: Polymer or ceramic material is selected for membrane material based on wastewater characteristics.
- Flux Rate of Membrane: Maintaining performance of membrane by balancing the filtration rate.
- Hydraulic Configuration: Hydraulic mechanism design of MBR systems which can be side-streamed or submerged.
- Energy Optimization: Energy optimization during filtration and aeration is a key aspect of MBR operation.
- Cleaning System: The type of cleaning mechanism including air scouring or backwashing should be considered during the design of the MBR system [61].

MBR process performance characteristics include:

- System Footprint: Since a separate settling tank is not required, the operation of the MBR system has a smaller footprint (space) than CAS.
- High Energy Requirement: Membrane scouring process in MBR operation increases the energy requirement for operation and is mitigated by developing membrane with biological and chemical resistance, fouling control mechanisms, and new configuration.
- Quality of Effluent: MBR operation produces non-potable effluent of high quality, which can be treated further for potable water recovery.

- Process Fouling: organic matter from microbial cells is deposited on the membrane during operation. This can be mitigated by selecting new materials and developing new configurations [62].

6.3.3 Biofilm technology

6.3.3.1 Moving-bed biofilm reactor (MBBR) technology

This technology was developed between 1988 and 1990 by Hallvard Odegaard [63] due to the challenges of earlier wastewater treatment reactors including biofouling (MBR), uneven biofilm distribution, and hydraulic instability (biofilm reactors) [64]; malodor pollution and expensive operational costs (trickling filters); mechanical failure, high maintenance, and large area requirement (rotating biological contactor) [65]. In an MBBR process (**Figure 14**), biofilm for wastewater treatment grows on movable carriers (biocarriers) in the aeration tank (oxic zone) and is separated from treated water in the settling tank [65, 67]. These biocarriers come in different shapes, surface areas, and configurations, and while the common configuration is a hollow cylinder, surface area size is very crucial to biofilm formation – a large surface area provides intimate contact for bacteria, wastewater, nutrients, and oxygen in the aeration tank [64, 65]. Operational challenges of the original MBBR system such as inappropriate/slow biofilm growth rate caused by increased solid retention time (SRT), high biofouling, and reduced mass transfer [68], led to the development of a hybrid MBBR system– the moving-bed biofilm membrane reactor (MBBMR), around the mid-1990s [63, 69, 70], which separates biofilm from treated water using a membrane instead of a settling tank [67]. This advanced MBBR system had advantages such as low sludge generation, improved operational efficiency, and reduced biofouling [68]. MBBR system's performance can be evaluated on factors such as biocarrier configuration and design, flexibility of filling fraction (biocarrier size) according to system requirements, biofilm development (whether thick or thin) based on the nature of biocarrier and which affects hydrodynamics of wastewater particles, uniform dissolved oxygen (DO) or aeration supply based on reactor design [65], and hydraulic retention time [71]. Pilot-scale MBBR operations are essential to assess full-scale process cost, scaling factor, feasibility, time, and unforeseen treatment results [72]. Some of the desirable qualities of a good biocarrier are light density,

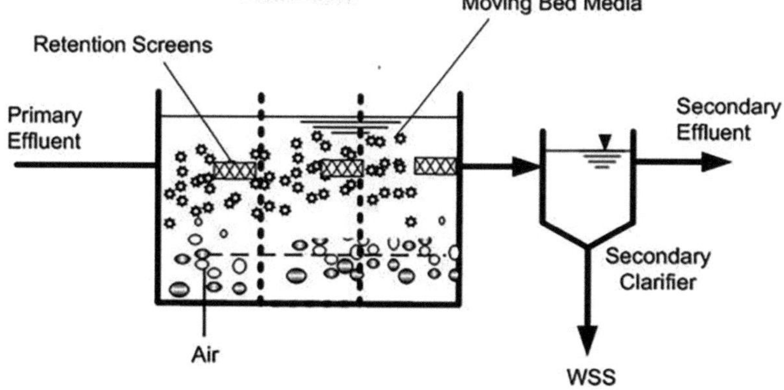

Figure 14.
The MBBR process. WSS = Waste Solid Sludge [66].

non-biodegradable, high resistance to abrasion, insolubility, microbial nontoxicity, and high effective surface area [65, 73]. It is obvious that biocarriers are essential and very critical to the effectiveness and performance of MBBR operations, future improvements should focus on the configuration, geometry, and design of these biofilm carriers.

6.3.3.2 Design and performance characteristics of MBBR process

The MBBR design is an hybrid configuration of biofilter and activated sludge processes for efficient wastewater treatment. The design characteristics are as follows.

- Hybridization Design: A free floating biofilter media for biomass concentration is incorporated into the activated sludge system.
- Scalability: The MBBR technology is designed for easy scale-up and modification.
- Biofilter Design: The configuration and design area of the biocarrier are important aspects of MBBR design.
- Biofilter Mixing: The process is designed for effective biocarrier mixing to distribute biofilm evenly through the CAS system.
- Aeration Supply: The system supports adequate oxygen supply [65].

MBBR process performance can be affected by characteristics such as:

- Biofilter Surface Area: The MBBR process relies on biofilm formation on the biofilter for wastewater treatment. This surface area affects biofilm growth as a larger area enables large bacteria colonies.
- Nutrient Concentration: The process is ideal for high organic (carbonaceous) and ammonia (nitrogenous) wastewater nutrient removal.
- Process Hydraulic Retention Time: The process supports high influent retention time for effective treatment.
- Adequate Process Conditions: The MBBR process can support optimal temperature and pH conditions [65].

6.3.3.3 Integrated fixed-film activated sludge (IFAS) technology

The IFAS technology is a hybridization of the MBBR and CAS technologies, an advancement over MBBR by incorporating suspended and attached growth systems [74] and the most recently invented biological wastewater treatment process having been developed around the early 2000s [75, 76]. This hybrid system involves adding a fixed microbial growth media (MBBR technique) to an activated sludge (AS) tank (CAS technique) to provide a biofilm growth surface and enhance nitrogen elimination by the oxidation process, respectively [75, 76]. This technology (**Figure 15**) allows the suspended-growth microorganisms of CAS and biofilm bacteria to be combined in a single reactor unit, allowing two different kinds of microbes (aerobes and anaerobes)

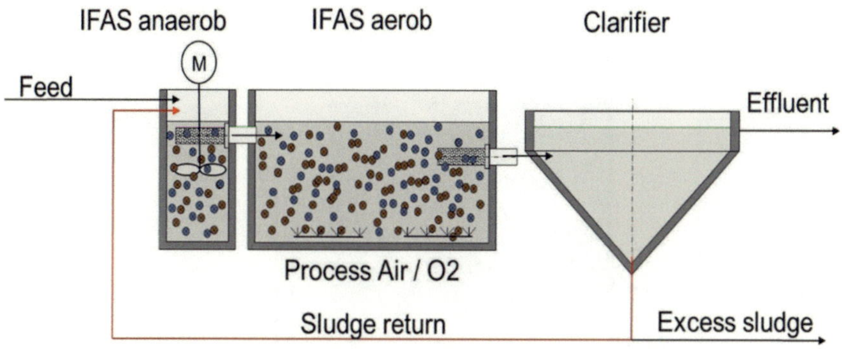

Figure 15.
The IFAS process. M = Stirrer motor [77].

to work in tandem; while the mixed liquor suspended solid (MLSS) biomass (aerobes) decomposes organic wastewater load, the biofilm (anaerobes) sets up a strong nitrifying bacteria to oxidize nitrogenous load [76]. The IFAS technology offers some advantages over traditional MBBR systems including a larger surface area of media for biofilm growth that provides higher efficiency in eliminating phosphorus and nitrogen, higher biomass concentration, organic load, and treatment capacities, more energy efficiency, and less capital cost [77]; although the movable MBBR system's media prevents clogging, promoting diverse bacteria growth, the system is also more compact, and resistant to organic shock load [77]. Over CAS technology, the IFAS system's advantage includes complete nitrification, reduced footprint, longer solids retention time, enhanced anthropogenic composite, and nutrient removal [74]. Pilot- and full-scale IFAS systems' performance can evaluated on factors such as hydraulic retention time (HRT), solid retention time (SRT), biocarrier media, media filling ratio, and dissolve oxygen concentration (DO) [74]. Recent upgrades to the IFAS process incorporated systems for energy generation, methane production, and capturing [74].

6.3.3.4 Design and performance characteristics of the IFAS system

The IFAS system is mainly characterized by the design of a hybrid suspended and attached growth process where a biofilm carrier surface is attached to the activated sludge suspended-growth system to enhance treatment efficiency [78]. The performance of the IFAS system is centered around its operational efficiency, removing above 93% of chemical oxygen demand (COD) in industrial wastewater and, microbial synergy, where suspended- and attached-growth microbes are involved in the treatment process [78].

6.3.3.5 Downflow stationary/fixed-film bioreactor (DSFF) technology

This technology (**Figure 16**), which has been in operation for several decades, is ideal for treating and purifying industrial wastewater by allowing the downward flow of the influent wastewater through a bed of biofilm, causing microorganisms to break down organic matter in the wastewater [78]. This has the advantage of moving wastewater by gravity rather than mechanical agitation, thereby reducing operational costs and ensuring efficient dilution of waste matter in the reactor [78]. Some of the limitations of the DSFF reactor include optimizing biofilter surface area, choosing an appropriate material for film support, the effect of reactor height on its temperature gradient, and wastewater load fluctuations [80].

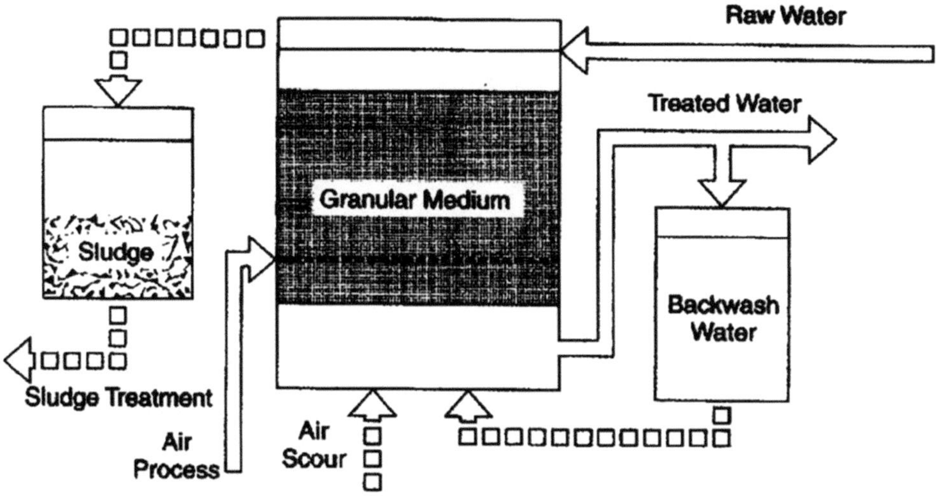

Figure 16.
Schematic diagram of a typical DSFF process [79].

The DSFF technology is an efficient biological treatment technology, and researchers have continued to focus on its development. Future developmental improvements are in the direction of improving biofilm media, developing novel strategies for biofilm attachment, adopting liquid phase oxygen technology for system aeration, and using alternative reactor configuration [81].

6.3.3.6 Design and performance characteristics of the DSFF system

The design characteristics of the DSFF system are as follows.

- Support Media: The system is designed to utilize a support medium including polymer beads, activated carbon particles, or silica granules, for bacteria colonies [50].

- Wastewater Flow Pattern: The downward flow design of the DSFF system allows the dilution of waste as it enters the system [80].

DSFF process performance can be associated with the following characteristics.

- Film Surface-to-Volume Ratio: Optimizing biofilter surface area to biofilm volume is key to the efficient performance of the treatment process.

- Reactor Height and Process Temperature: High bioreactors may have the challenge of distributing substrate uniformly and temperature fluctuations throughout the height of the reactor.

- Biofilm Support Media: The performance of the DSFF system depends on the type of support media selected for the process [50, 80].

6.3.3.7 Packed-bed biofilm reactor (PBBR) technology

This system creates a biofilm for wastewater treatment by filling a packed-bed bioreactor with supporting materials – usually cylindrical glass carriers (**Figure 17**), so that microorganisms attached to the surfaces of these carriers metabolize pollutants in wastewater [83] as they flow through the bed [81]. Recent applications of this technology (a typical example is the trickle-bed biofilm reactor process) have combined biofilm aggregate on carriers with membrane technology processes like ultrafiltration [50, 83]. Although the packed-bed biofilm wastewater treatment technology is an advanced treatment process suitable for industrial wastewater, it has limitations, some of which include restriction of oxygen and wastewater nutrients from the bulk liquid to the biofilm, reduced flowrates due to media clog by organic matter, biofilm detachment, channeling, biofilm susceptibility and slow response to shock loads, and scale-up complexities [83].

6.3.3.8 Design and performance characteristics of the PBBR system

The design attributes of a PBBR process include reactor geometry and configuration, which accounts for the surface property, size, and shape of the bioreactor and can allow maximum nutrient removal without accumulation; media configuration including its surface property, shape, elasticity, size, mass, and absolute size; design of wastewater loading pattern and media remover/cleaning mechanism [84]. PBBR process efficiency is characterized by factors such as optimal biofilm thickness which provides sufficient surface area for microbial activity without clogging, good quality support media, appropriate hydraulic retention time, and adequate aeration through the bulk liquid to the biofilm [84].

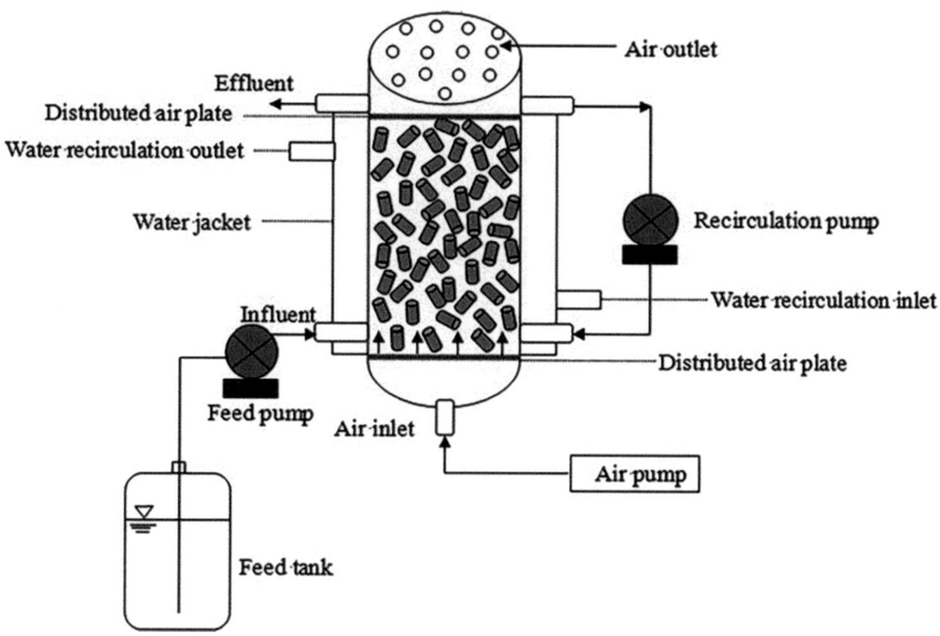

Figure 17.
The PBBR wastewater treatment process [82].

6.3.3.9 Fluidized-bed biofilm reactor (FBBR) technology

This technology is a recent advancement in the biofilm wastewater treatment process and operates by suspending support media (biocarriers) in the reactor with the force of upward-flowing feed wastewater (**Figure 18**) [50]. The FBBR process is applicable for industrial installations and large treatment facilities due to its high fluid flowrates, usually higher than the smallest fluidization velocity but lower than the biofilm aggregates terminal free velocities in the reactor; this causes a suspension (fluidization) of the biofilm aggregates [50]. One advantage of this process is that there is no channeling as biofilm aggregates are surrounded by the wastewater, which improves intimate contact (between bacteria and wastewater organic pollutants) and reactor performance [50]. Also, the process can handle higher wastewater throughput as a result of effective bacteria retention and produces less sludge. Limitations of the FBBR technology include high operating costs from keeping media fluidized by aeration, media fouling, limited nutrient removal, biofilm detachment, scale-up complexities, and other than industrial wastewater with high organic pollutants, it is inadequate for treating other types of wastewater [50].

6.3.3.10 Design and performance characteristics of the FBBR system

A notable design characteristic of the FBBR system is its compact design, which makes it suitable for isolated and localized wastewater treatment applications [86].

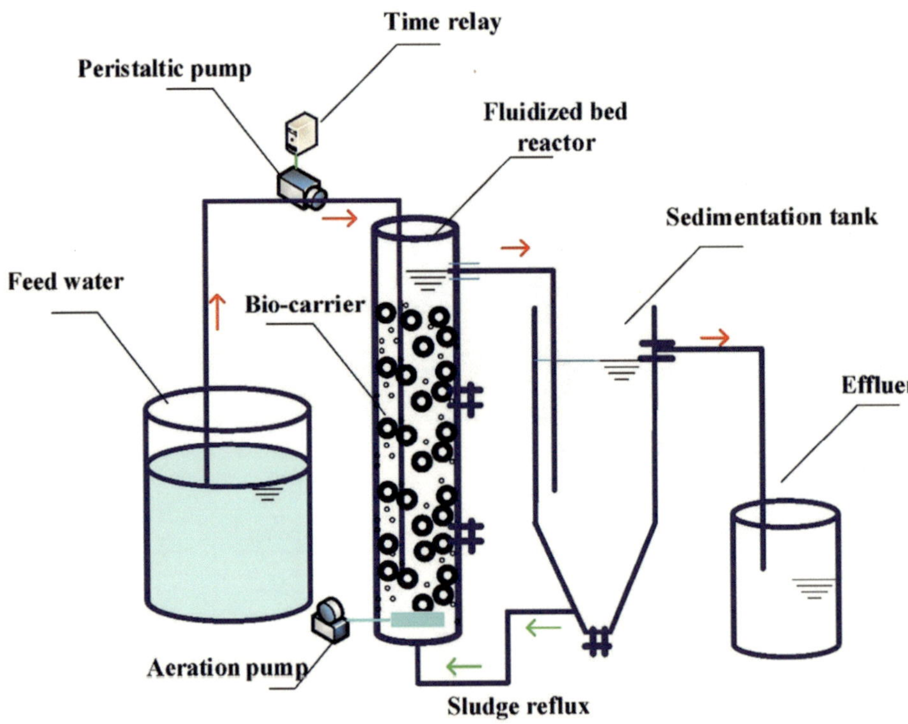

Figure 18.
The FBBR wastewater treatment process [85].

Additionally, the diameter of the media and bioparticle, substrate penetration depth, and internal/external mass transfer coefficients are other important design attributes [86]. The process performance can be attributed to characteristics that include efficient mixing and mass transfer, significantly high wastewater throughput and up-flow velocity, optimal pH and temperature, media porosity, and surface area (a balance between these can achieve efficient treatment), and pollutant loading rate [86, 87].

The semi-fluidized-bed (SFBBR) and inverted fluidized-bed (IFBBR) biofilm wastewater treatment processes are similar to the FBBR process as they both utilize suspended media particles (fluidization) for microbial growth required for wastewater treatment, but while the particles are partially suspended in SFBBR, they are expanded in IFBBR [85–87]. While the FBBR process focuses on biomass holdup (concentration), the SFBBR process emphasizes efficient mass transfer and optimal mixing, and the IFBBR process focuses on faster process startup [88]. The major limitation of the SFBBR process is the limited biomass concentration (retention) in the reactor while optimizing startup time, which is a major problem of the IFBBR process [89].

6.3.3.11 The sludge blanket biofilm reactor (SBBR) technology

This technology utilizes sludge granules (microbial colonies) formed from bacterial cells on support media (biofilm) and the accumulated biomass layer (sludge blanket) at the bottom of the reactor for wastewater treatment [81]. Wastewater passes through the sludge blanket and the biofilm simultaneously for enhanced treatment by suspended growth and biofilm-based microbes, respectively. The SBBR treatment process is similar in design to the up-flow anaerobic sludge blanket process (UASB) (**Figure 19**) however in UASB, treatment is carried by the suspended-growth microorganisms and

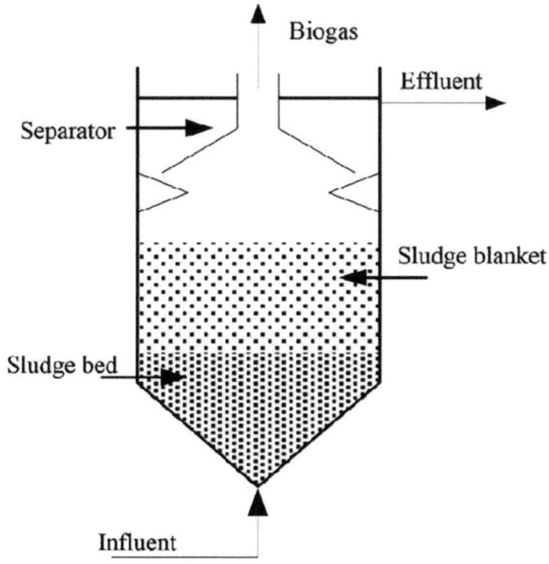

Figure 19.
The UASB process has a similar design to the SBBR [90].

this is done without the supply of oxygen [50, 81]. Some of the limitations of the SBBR process include high operational costs from poor settling of solid pollutants in the sludge blanket and high sludge production, high levels of toxic substance accumulations with the blanket, and engineering design and operation challenges [91].

6.3.3.12 Design and performance characteristics of the SBBR system

The characteristics of SBBR design that contribute to its performance include the reactor configuration, biofilm support media design material, configuration, etc. Attributes such as optimal process pH and temperature, microbial interactions, and sludge settling properties affect SBBR system performance [81, 91].

Bioreactor	Merits	Limitations
Stirred tank	Simple in construction and operation. Uses suspended growth of microbes. Suitable for aerobic and anaerobic processes.	Restricted to low capacities.
Trickle-bed biofilm reactor	Use of attached growth of microbes. Low operating cost due to down flow mode of operation. High cell mass concentration in biofilm promotes rate of bioconversion.	Mainly for aerobic BOD removal. Low capacity due to low feed flow rate maintained.
Moving-bed biofilm reactor (slurry reactor)	Heterogeneous version of stirred tank. High cell concentration in biofilm promotes rate of bioconversion.	Capacity wise inferior to column reactors, Biofilm could get disturbed due to high rate of agitation.
Fluidized-bed biofilm reactor	Operates at high capacities, provides high degree of bioconversion. Once fully fluidized, pressured drop across the bed remains constant and does not increase with increase in feed flow rate. Degree of bioconversion increases with increase in feed flow rate due to bed expansion.	Entrainment loss of particle-biofilm aggregates possible. Operating cost higher than trickle bed (packed bed).
Semifluidized bed biofilm reactor	Higher degree of bioconversion (than fluidized beds) at higher capacities and low reactor volume requirement. Degree of bioconversion increases with increase in feed flow rate, even if reactor volume is kept constant.	Higher operating cost than fluidized beds. Continuous, circulating mode of operation not possible.
Inverse fluidized biofilm reactor	Low operating cost due to down flow mode of operation. Larger-sized particles could be used. Reasonably large degree of bioconversion.	Lower capacity than fluidized/semifluidized bed. Larger reactor volume requirement.
DSFF bioreactor	Simple in construction and operation. No support particles required. Low operating cost due to downflow mode of operation. Multiple tubes/ columns could be used to increase capacity.	Presently restricted to anaerobic operation. Large reactor volume requirement at high capacities.
UASB reactor	Simple in construction. No support particles used. Provides substantially high degree of bioconversion at distinctly high capacities and even with high-strength feedstock.	Restricted to anaerobic processes, employing complex culture of microbes. Enormously large startup time.

Source: Ref. [50].

Table 2.
Comparison of stirred tank and biofilm biological wastewater treatment technologies.

A comparison of the biological wastewater treatment technologies (especially the stirred tank and biofilm technologies) discussed so far in this chapter is provided in **Table 2,** including their merits and limitations. Other emerging biological wastewater treatment technologies for sustainable water treatment and energy generation, some of which do not incorporate the use of bacteria colonies, are phytoremediation, which uses specific plant species to remove wastewater pollutants, and algal-based treatment, which uses algal organisms to metabolize and remove wastewater pollutants; also, the microbial electrochemical system (MES) generates electrical energy from microbial metabolic activities during wastewater treatment [92–94]. These technologies have proven to lessen the impact of wastewater treatment on the environment [92].

7. Conclusion and future perspectives

This chapter discussed the fundamentals of biological wastewater treatment as it relates to the most commonly generated wastewater types – domestic and industrial wastewater. Wastewater is treated to protect public health and wildlife ecosystems, provide recreation and tourism, and increase economic prosperity using biochemical oxygen demand (BOD) analysis as a metric for water quality and treatment effectiveness. Biological wastewater treatment is an eco-friendly process that relies on microorganisms to convert organic pollutants in wastewater to simple substances such as CO_2, H_2O, and energy.

Treatment technologies, which can be grouped as stirred tank reactor technologies (CAS, oxidation ditch, A2O, and SBR), membrane technology (MBR), and biofilm technologies (MBBR, IFAS, DSFF, PBBR, FBBR, SFBBR, IFBBR, and SBBR) have evolved over the years as governments around the world continue to raise the bar of quality standards of wastewater-treated effluents. Advancements in treatment technology have put treatment processes ahead of future challenges such as increased wastewater generation as a result of the increasing global population, old and aging facilities, reduced installation spaces, economic considerations, and climate change. Activated sludge technology (CAS) being the oldest (developed around 1914) biological wastewater treatment technique, has undergone several improvements for enhanced nutrient removal, increased surface area for microbial growth, compact size, and cheaper capital costs, to technologies such as A2O, oxidation ditch, and the SBR. Developments in treatment technologies shifted to better sludge handling between the early 1970s and late 1980s, hence, the development of membrane technologies, that is, the MBR technology, which offered maximum sludge separation from treated water in the membrane media. However, membrane clogging due to biofouling in MBR systems led to the development of MBBR systems around the mid-1990s, with lower biofouling and clogging. The IFAS technology is the most recently developed biofilm technology, and it was deployed around the early 2000s. The technology was developed to mitigate the operational challenge of the slow biofilm growth rate of traditional MBBR systems, is more energy efficient, and has a low capital cost compared to other biofilm technologies including DSFF, PBBR, FBBR, and SBBR; although clogging is better managed in some of these technologies (especially in the MBBR due to the movable biocarrier medium) than in the IFAS system. New developments in IFAS systems are focused on sustainability with the incorporation of mechanisms for methane production and capturing, and energy generation. Overall, biofilm wastewater treatment technology has a better volumetric capacity and biomass retention period, higher biomass concentration, requires smaller installation space, low hydraulic retention time, and sludge production when compared to stirred tank reactor technology, although it has poor mass transfer, which affects the rate of reaction.

Future trends in biological wastewater treatment technologies will focus on process sustainability via energy generation to lower operational costs, compact systems, and large-volume wastewater handling all of which can be provided by the biofilm reactor technologies MBBR, IFA, MES, and even non-bacterial treatment systems like phytoremediation and algal-based treatments. Future designs and optimization of MBBR and IFAS operations should focus on biocarrier configuration, geometry, and design material.

Author details

Olawale Theophilus Ogunwumi[1*], Adeyemo Festus Adeniyi[2], Mezie Chinazor Angus[3] and Oche Sunday Oche[4]

1 Water Quality and Wastewater Treatment Research Laboratory, Chemical and Petroleum Engineering Department, University of Lagos, Nigeria

2 Energy and Sustainability Research Laboratory, Chemical and Petroleum Engineering Department, University of Lagos, Nigeria

3 Department of Corrosion Science and Engineering, Chukwuemeka Odumegwu Ojukwu University (COOU), Anambra State

4 Department of Chemical and Petroleum Engineering, University of Lagos, Nigeria

*Address all correspondence to: ola.ogunwumi@gmail.com; oogunwunmi@unilag.edu.ng

References

[1] Tilley E, Ulrich L, Luthi C, Reymond PH, Zubrugg C. A Compendium of Sanitation Systems and Technologies. 2nd Revised ed. Duebendorf, Switzerland: Swiss Federal Institute of Aquatic Science and Technology (Eawag); 2014. ISBN: 978-3-906484-57-0

[2] Tchobanoglous G, Franklin L, Burton H, Stensel D, Metcalf E. Wastewater Engineering: Treatment and Reuse. 4th ed. Boston: McGraw-Hill; 2003. ISBN: 0-07-041878-0.OCLC 48053912

[3] Andersson K, Rosemarin A, Lamizana B, Kvarnström E, McConville J, Seidu R, et al. Sanitation, Wastewater Management and Sustainability: From Waste Disposal to Resource Recovery. Nairobi and Stockholm: United Nations Environment Program and Stockholm Environment Institute; 2016. p. 56. ISBN: 978-92-807-3488-1

[4] Logitech. 2023. What is industrial wastewater and what are its characteristics. [Internet]. 2023. Available from: https://www.liqtech.com/liquid-filtration-solutions/industrial-water-treatment/what-is- industrial-wastewater-and-what-are-thecharacteristics/#:~:text= Industrial%20wastewater% 20is%20 an%20aqueous,as%20well%20as% 20chemical%20manufacturing [Accessed: December 28, 2023]

[5] Shah A, Arjunan A, Baroutaji A, Zakharova J. A review of physicochemical and biological contaminants in drinking water and their impacts on human health. Water Science and Engineering. 2023;**16**(4):333-344. DOI: 10.1016/j.wse.2023.04.003

[6] Melissa HA, Schleiger R. Water pollutants and their sources. [Internet]. 2023. LibreTexts. Available from: https://www.bio.libretexts.org/Bookshelves/Ecology/Environmental-Science-(Ha-and-Schleiger)/06%3A-Environmental-Impacts/6.02%3A-Pollution/6.2.01%3A-Water-Pollution [Accessed: December 16, 2023]

[7] Scholarly Community Encyclopedia. Wastewater. [Internet]. 2022. Available from: https://www.handwiki.org/wiki/Earth-wastewater [Accessed: December 16, 2023]

[8] Hejabi N, Saghebian SM, Aalami MT, Nourani V. Evaluation of the effluent quality parameters of wastewater treatment plant based on uncertainty analysis and post-processing approaches (a case study). Wastewater Technology. 2021;**83**(7):1633-1648

[9] Von Sperling M. Wastewater characteristics, treatment, and disposal. Water Intelligence Online. 2007;**1**:77-272. ISBN: 978-1-78040-208-6. DOI: 10.2166/9781780402086

[10] Urquhart Leonard Church. Civil Engineering Handbook. 4th ed. New York: McGraw Hill Book Company Inc; 1959

[11] Jain SK, Sing VP. Domestic wastewater. Developments in Water Science Book Series. 2003;**51**:743-786

[12] Popoola LT, Ogunwumi OT, Salami L. A review on the fate and effects of contaminants in biosolid applied on land: Hazards and government regulatory policies. Heliyon. 2023;**9**(10):1-32. ISSN: 2405-3440. DOI: 10.1016/j.heliyon.2023.e.19788

[13] Vysokomornaya OV, Kurilenko YE, Shcherbinina AA. Major contaminants in industrial and domestic wastewater. In: MATEC Web of Conference 23, 00141.

France: EDP Sciences; 2015. DOI: 10.1051/matecconf/20152300141

[14] Ogunwumi OT, Salami L, Adeyemo AF, et al. Treatability Study of Olusosun Dumpsite Leachate Using Activated Sludge Technology. [under peer review]. Lagos: Chemical and Polymer Engineering Department, Lagos State University, Epe; 2023

[15] Ogunwumi OT, Salami L. Perspective chapter: Industrial waste landfills. In: Zailani S, Sarvajayakesavalu S, editors. Solid Waste and Landfills Management - Recent Advances. London: InTechOpen; 2023. pp. 1-32. ISBN: 978-1-80356-327-5. DOI: 10.5772/intechopen.108787

[16] Water Science School. Wastewater Treatment and Water Use. [Internet]. 2018. Available from: https://www.usgs.gov/special-topics/water-science-school/science/wastewater-treatment-water-use#overview [Accessed: December 22, 2023]

[17] Macrotrends. World Population between 1950 and 2100. [Internet]. Seattle, USA: Macro-trends LLC; 2023 Available from: https://www.macrotrends.net/countries/WLD/world/population; 2023 [Accessed: December 28, 2023]

[18] Kumar S. Effects of water pollution: Causes, consequences, and solutions on the environment. [Internet]. 2023. Available from: https://www.scaleclimateaction.org/environment/effects-of-water-pollution-causes-consequences-and-solutions. [Accessed: January 02, 2024]

[19] Amadeo K. Water Pollution Economic Effects, Causes, and Solution. [Internet]. New York, USA: The Balance Dotdash Meredith Publishing; 2021. Available from: https://www.thebalancemoney.com/water-pollution-effects-causes-and-solution-4735830#:~:text=water%20quality%20of%20streams%2C%20and%20the%20environment

[20] Poulsen T, Bester K. Organic micropollutant degradation in sewage sludge during composting under thermophilic conditions. Environmental Science and Technology. 2010;**44**(13):5086-5091. DOI: 10.1021/es9038243

[21] Lindholm-Lehto P, Ahkola H, Knuutinen J. Procedures of determining organic trace compounds in municipal sewage sludge—A review. Environmental Science and Pollution Research. 2017;**24**:4383-4412. DOI: 10.1007/s11356-016-8202-z

[22] Roohollah M, Mostafa A, Reza A, Mohammad G, Zahra K. Simulating the Dispersion of Poisonous Organic Chemical Compounds in Wastewater Treatment Process through the Active Sludge Method Using the TOXChem Model. USA: HSPI; 2021. DOI: 10.29328/journal.abse.1001014

[23] Loosdrecht M, Brdjanovic D. Anticipating the next century of wastewater treatment. Science. 2014;**344**:1452-1453. DOI: 10.1126/science.1255183

[24] Zhu L, Li Y, Liu C, Li G. Intermittent micro-aeration technology to enhance the carbon source release of particulate organic matter in domestic sewage. Water. 2023;**14**(12):1876. DOI: 10.3390/w14121876

[25] Tchobanoglous G, Burton F. Wastewater Engineering: Treatment, Disposal, Reuse. 3rd ed. New York: McGraw Hill; 1991

[26] Owen WF. Energy in Wastewater Treatment, Energy in Wastewater Treatment [Internet]. 1982. Available

from: https://www.osti.gov/biblio/5555462 [Accessed: June 12, 2024]

[27] Gellings CW, Parmenter KE. Energy efficiency in fertilizer production and use. In: Efficient Use and Conservation of Energy. Encyclopedia of Life Support Systems (EOLSS). Developed under the Auspices of UNESCO. Oxford, UK: EOLSS Publishers; 2004

[28] McCarty P, Bae J, Kim J. Domestic wastewater treatment as a net energy producer-can this be achieved? Environmental Science and Technology. 2011;**45**(17):7100-7106. DOI: 10.1021/es2014264

[29] Nakkasunchi S, Hewitt N, Zoppi C, Brandoni C. A review of energy optimization modelling tools for the decarbonisation of wastewater treatment plants. Journal of Cleaner Production. 2020;**279**:123811. DOI: 10.1016/j.jclepro.2020.123811

[30] Wang Y, Jiapu L, Wang S, Yang J, Qi C, Hongzhen G, et al. Operational optimization of wastewater reuse integrated energy system. Energy. 2020;**200**:117309. DOI: 10.1016/j.energy.2020.117309

[31] Goswami R, Mehariya S, Verma P, Lavecchia R, Zuorro A. Microalgae-based biorefineries for sustainable resource recovery from wastewater. Journal of Water Process Engineering. 2020;**40**(101747):1-18. DOI: 10.1016/j.jwpe.2020.101747

[32] Urgun-Demirtas M, Stark B, Pagilla K. Use of genetically engineered microorganisms (GEMs) for the bioremediation of contaminants. Critical Reviews in Biotechnology. 2006;**26**:145-164. DOI: 10.1080/07388550600842794

[33] Zhao F, Wang B, Huang K, Yin J, Ren X, Wang Z, et al. Correlations among antibiotic resistance genes, mobile genetic elements and microbial communities in municipal sewage treatment plants revealed by high-throughput sequencing. International Journal of Environmental Research and Public Health. 2023;**20**:1-20. DOI: 10.3390/ijerph20043593

[34] Rene E, Veiga M, Kennes C. Recent trends in biogenic gas, waste and wastewater fermentation. Fermentation. 2022;**8**(8):347. DOI: 10.3390/fermentation8080347

[35] Quispe JB, Campos LC, Mašek O, Bogush A. Use of biochar-based column filtration systems for greywater treatment: A systematic literature review. Journal of Water Process Engineering. 2022;**48**:102908. DOI: 10.1016/j.jwpe.2022.102908

[36] Muoghalu C, Owusu P, Lebu S, Nakagiri A, Semiyaga S, Iorhemen O, et al. Biochar as a novel technology for treatment of onsite domestic wastewater: A critical review. Frontiers in Environmental Science. 2023;**11**:1-22. DOI: 10.3389/fenvs.2023.1095920

[37] Hoey G, Borja Á, Birchenough S, Buhl-Mortensen L, Gaur N, Narasimhulu K, et al. Recent advances in the bioremediation of persistent organic pollutants and its effect on environment. Journal of Cleaner Production. 2018;**198**:401-416. DOI: 10.1016/J.JCLEPRO.2018.07.076

[38] Kuo E, Yang R, Chin Y, Chien Y, Chen Y, Wei C, et al. Multiomics approaches and genetic engineering of metabolism for improved biorefinery and wastewater treatment in microalgae. Biotechnology Journal. 2022;**17**:1-10. DOI: 10.1002/biot 202100603

[39] Jin-Ping B. Electrical control system of BOD rapid determination instrument

using biosensor. Transducer and Microsystem Technologies. 2011

[40] Li L, Rong S, Wang R, Yu S. Recent advances in artificial intelligence and machine learning for nonlinear relationship analysis and process control in drinking water treatment: A review. Chemical Engineering Journal. 2021;**405**:126673. DOI: 10.1016/j.cej.2020.126673

[41] Shrivastava R, Singh N. Assessment of water quality of textile effluent and its treatment by using coagulants and plant material. Materials Today Proceedings. 2021:3318-3321. DOI: 10.1016/J.MATPR.2021.02.373

[42] Nourani V, Asghari P, Sharghi E. Artificial intelligence-based ensemble modeling of wastewater treatment plant using jittered data. Journal of Cleaner Production. 2021;**291**:125772. DOI: 10.1016/J.JCLEPRO.2020.125772

[43] Quadir M, Drechsel P, Blanca JC, et al. Global and regional potential of wastewater as a water, nutrient, and energy source. Natural Resources Forum. United Kingdom. 2020;**44**(1):40-51. DOI: 10.1111/1477-8947.12187

[44] Novo C. What Is a Wastewater Treatment Plant [Internet]. Smart Water Magazine; 2024. Available from: https://www.smartwatermagazine.com/q-a/what-a-wwtp#:~:text=1%20.andtext=That%20is%2C%20a%20WWTP%20trXDeats,Materials%20dissolved%20in%20the%20water

[45] Yasa ET. Activated Sludge Process and Types for Wastewater Treatment. Biology Reader. USA. [Internet] 2022. Available from: https://www.yasa.ltd/post/activated-sludge-process-and-types-for-waste-water-treatment [Accessed: December 22, 2023]

[46] Rai M, Bhattaarai N. Wastewater Treatment Plant Activated Sludge Method. Maitighar, Kathmandu: Guheshwori Field Report; 2016

[47] Britannica. Wastewater Treatment. [Internet]. 2023. Encyclopedia Britannica. Available from: https://www.britannica.com/technology/wastewater-treatment/oxidation-pond [Accessed: December 22, 2023]

[48] UND EERC. Tertiary Wastewater Treatment. [Internet]. 2024. University of North Dakota Energy and Environmental Research Center. Available from: https://www.undeerc.org/water/Decision-Support/Treatment-Technologies/Tertiary-Wastewater-Treatment.aspx#:~:text=The%20purpose%20of20tertiary%20tratment,hydrogen%20sulfide%20from%20the%20water

[49] AOS. Tertiary Treatment of Wastewater: Methods and Processes. [Internet]. Houston, Texas: AOS Treatment Solutions; 2018. Available from: https://www.aosts.com/tertiary-treatment-of-wastewater-process-methods/#:~:text=Tertiary%20Wastewater%20Treatment%20Methodsandtext=Alum%20also%20Causes%20any%20solids,filters%20to%20continue%20operating%20effectively; 2018 [Accessed: December 18, 2023]

[50] Narayanan CM, Narayan V. Biological wastewater treatment and bioreactor design: A review. Sustainable Environment Research. 2019;**29**(33):1-17. DOI: 10.1186/s42834-019-0036-1

[51] Waterman Engineers. Activated Sludge Process for Wastewater Treatment Technologies. [Internet]. New South Wales, Australia: Waterman Engineers Australia; 2023. Available from: https://www.watermanaustralia.com/

activated-sludge-process-for-wastewater-treatment-technology/ [Accessed: December 10, 2023]

[52] Water N. What is the activated sludge process? – Explained with diagrams. [Internet]. 2022. Available from: https://www.netsolwater.com/understanding-the-workings-of-activated-sludge-process.php?blog=1404 [Accessed: November 28, 2023]

[53] Jamal H. Activated sludge process design. [Internet]. 2017. About civil. Available at: https://www.aboutcivil.org/design-of-activated-sludge-process.html [Accessed: January 02, 2024]

[54] Eawag, Dorothee S. Activated Sludge. [Internet]. Compendium of Sanitation Systems and Technologies. Dubendorf, Switzerland. 2020. Available from: https://www.sswm.info/factsheet/activated-sludge [Accessed: February 02, 2024]

[55] Wiki. Activated Sludge. [Internet]. Maryland, USA: Wikipedia; 2024. Available from: https://www.en.wikipedia.org/wiki/Activated_sludge

[56] Euroteck. Advanced Activated Sludge Process (A2O)/(AO) Process. [Internet]. Hyderabad, India: Euroteck Environmental Ltd; 2021. Available from: https://www.euroteckindia.com/products/treatment-technologies/a20-ao-process/#:~:text=The%20Advanced%20Activated%20Sludge%20Process,(aerobic)%20tanks%20in%20series

[57] Kim K. The Characteristics of Sequencing Batch Reactor (SBR), Anaerobic Sequencing Batch Reactor (ASBR), and Sequencing Batch Biofilm Reactor (SBBR). [Internet]. Wastewater Technology Factsheet: Sequencing Batch Reactors, Environmental Protection Agency (EPA). Washington D.C. 2008. Available from: https://www.engineering.istate.edu [Accessed: January 06, 2024]

[58] IWA. Sequencing Batch Reactor. [Internet]. London, United Kingdom: The International Water Association (IWA) Publications. 2024. Available from: https://www.iwapublishing.com/news/sequencing-batch-reactor/ [Accessed: January 05, 2024]

[59] Pandey A, Singh RK. Industrial wastewater treatment by membrane bioreactor system. Elixir Chemical Engineering. 2014;**70**:23772-23777

[60] Al-Asheh SA, Bagheri M, Aidan A. Membrane bioreactor for wastewater treatment: A review. Case Studies in Chemical and Environmental Engineering. 2021;**4**:100109. DOI: 10.1016/j.cscee.2021.100109

[61] EPA. Wastewater Management Factsheet: Membrane Bioreactors. Washington D.C: United States Environmental Protection Agency; 2007

[62] Asif AM, Zhang Z, Vu M, Mohammed J, et al. Membrane bioreactor for wastewater treatment: Current status, novel configuration and cost analysis. In: Nasr M, Negm AM, editors. Cost-Efficient Wastewater Treatment Technologies. The Handbook of Environmental Chemistry. Vol. 118. Cham: Springer; 2022. pp. 147-167. DOI: 10.1007/698-2022-871

[63] Odegaard H, Rusten B, Westrum T. A new moving-bed biofilm reactor – Application and results. Water Science and Technology. 1994;**29**(10-11):157-165. DOI: 10.2166/wst.1994.0757

[64] Alessandro DB, Maciej K, Tanner D, Jan O. Moving-bed biofilm reactor technology in municipal wastewater:

A review. Journal of Environmental Management. 2019;**247**:849-866

[65] Madan S, Madan R, Hussain A. Advancement in biological wastewater treatment using hybrid moving-bed biofilm reactor (MBBR): A review. Applied Water Science. 2022;**12**:141. DOI: 10.1007/s13201-022-01662-y

[66] AES. Moving-bed Bioreactor (MBBR). [Internet]. AES Arabia Ltd. Available from: https://www.aesarabia.com/moving-bed-bioreactor-mbbr/; 2022 [Accessed: January 02, 2024]

[67] Leyva-Diaz JC. Moving-bed biofilm reactor to treat wastewater. International Journal of Environmental Science and Technology. USA: Springer; 2017;**14**:881-910

[68] Sohail N, Ahmed S, Chung S, Nawaz MS. Performance comparison of three different reactors (MBR, MBBR, MBBMR) for municipal wastewater treatment. Desalination and Water Treatment. 2020;**174**:71-78

[69] Bassin JP, Dezotti M. Moving-bed Biofilm Reactor (MBBR) – An Advanced Biological Process for Wastewater Treatment. Champ: Springer; 2018. pp. 37-74

[70] Odegaard H, Rusten B, Siljudalan. The development of the moving-bed biolfilm process: From idea to commercial product. European Water Management. 1999;**2**(3):36-43

[71] Bakar SNHA, Hasan HA, Mohammed AW, Abdullah SRS, et al. Performance of a laboratory-scale moving-bed biofilm reactor (MBBR) and its microbial diversity in palm oil mill effluent (POME) treatment. Process Safety and Environmental Protection. 2020;**142**:325-335. DOI: 10.1016/j.psep.2020.05.004

[72] Creative Engineers. Pilot Plant and Pilot Scale Research. [Internet]. USA. Available from: https://www.creativeengineers.com/chemical-innovation/pilot-scale/#:~:text=What%20does%20%E2%80%9Cpilot%20scale%E2%80%9D%20mean,a%20few%20kilograms%20of%20material: Creative Engineers International; 2024 [Accessed: December 28, 2023]

[73] Diaz J, Bellingham M, Hassan J, Barret M, Stephenson T, Soares A. Impact of carrier media on oxygen transfer and wastewater hydrodynamics on a moving attached growth system. Chemical Engineering Journal. 2018;**351**:399-408

[74] Waqas S, Bilad MR, Man Z, et al. Recent Progress in integrated fixed-film activated sludge process for wastewater treatment: A review. Journal of Environmental Management. 2020;**268**(110718):1-28. DOI: 10.1016/j.jenvman.2020.110718

[75] Transcend. MBBR: A Revolution in Wastewater Treatment. [Internet]. Taipei, Taiwan: Transcend Software Inc.; 2024. Available from: https://www.transcendinfra.com/mbbr-wastewater/ [Accessed: January 04, 2024]

[76] Frankel T. What Is IFAS and How Does it Work?. [Internet]. New York, USA: SSI Smart Ideas for Water; 2022. Available from: https://www.ssiaeration.com/what-is-ifas-how-does-it-work/#gref

[77] FiltSep. Biological Wastewater Treatment: MBBR and IFAS Technologies. [Internet]. United Kingdom: Filtration Separation. 2024. Available from: https://www.filtsep.com/content/features/biological-water-treatment-mbbr-and-ifas-technology/ [Accessed: January 04, 2024]

[78] Li C et al. Performance and microbial characteristics of integrated fixed-film activated sludge system treating

industrial wastewater. Water Science and Technology. 2012;**66**(12):2785-2792. DOI: 10.2166/wst.2012.421

[79] Christine D, Frank R, et al. Factors affecting the design and operation of biologically active filters. Proceedings of the Water Environment Federation. 2010;**7**:567-596. DOI: 10.2175/193864710798208494

[80] Van Den B, Kennedy KJ, Samson R. Anaerobic downflow stationary fixed film reactors performance under steady-state and non-steady conditions. Water Science and Technology. 1985;**17**(1):89-102. DOI: 10.2166/wst.1985.0007

[81] Lan W, Wei W, Juan X. Denitrifying biofilm processes for wastewater treatment: Developments and perspectives. Royal Society of Chemistry: Environmental Science Water Resources Technology. 2021;**7**:40-67. DOI: 10.1039/D0EW00576B

[82] Lin Y-H. Biodegradation of 2,4-dichlorophenol by mixed culture in an aerobic fixed-biofilm process—Kinetic model and reactor performance. Environmental Engineering Science. 2015;**32**(6):150218130700006. DOI: 10.1089/ees.2014.0482

[83] Díaz O, González E, Vera L, Fernández LJ, Díaz-Marrero AR, Fernández JJ. Recirculating packed-bed biofilm photobioreactor combined with membrane ultrafiltration as advanced wastewater treatment. Environmental Science and Pollution Research. 2023;**30**:69977-69990

[84] Afandizadeh S, Foumeny FA. Design of packed-bed reactors: guides to catalyst shape, size, and loading selection. Applied Thermal Engineering. 2001;**21**(6):669-682. DOI: 10.1016/s1359-4311(00)00072-7

[85] Ding Y, You H, Sun W, et al. Research on biological fluidized-bed system treatment performance and nitrogen removal process for seafood processing wastewater with different operation. Water. 2021;**13**(9):2630. DOI: 10.3390/w13192630

[86] Green Chemical Engineering. Fluidized-bed bioreactor applications for biological wastewater treatment: A review of research and developments. Engineering. 2017;**3**(3):330-342. DOI: 10.1016/j.ENG.2017.03.021

[87] Jaafari J, Mesdaghinia A, Nabizadeh R, Hoseini M, Kamani H, Mahvi AH. Influence of upflow velocity on performance and biofilm characteristics of anaerobic fluidized bed reactor (AFBR) in treating high-strength wastewater. Journal of Environmental Health Science and Engineering. 2014;**12**(139):1-10

[88] Dechsiri C. Particle Transport in Fluidized Beds: Experiments and Stochastic Models [thesis]. Department of Chemical Engineering: University of Groningen; 2004

[89] Maurya A, Kumar R, Raj A. Biofilm-based Technology for Industrial Wastewater Treatment: Current technology, applications and future perspectives. World Journal of Microbiology and Biotechnology. 2023;**39**(112):1-25

[90] Abhishek D, Davies C, Ikumi D. Performance of upflow anaerobic sludge blanket (UASB) reactor and other anaerobic reactor configurations for wastewater treatment: A comparative review and critical updates. Journal of Water Science. London, United Kingdom. 2018. DOI: 10.2166/AQUA.2018.090

[91] Sehar S, Naz I. Role of biofilms in wastewater treatment. In: Dhanasekaran D, Thajuddin N, editors.

Microbial Biofilms – Importance and Applications. London, UK: IntechOpen; 2016. pp. 1-20. DOI: 10.5772/63499

[92] Upadhyay A, Singh R, Singh D. Phytoremediation and sustainable developmental policies and practices. In: Restoration of Wetland Ecosystem: A Trajectory towards Sustainable Environment. Singapore: Springer; 2019. DOI: 10.1007/978-981-13-7665-8_2

[93] Li M, Zamyadi A, Zhang W, Dumee LF, Gao L. Algae-based water treatment: A promising and sustainable approach. Journal of Water Process Engineering. 2022;**46**:102630. DOI: 10.1016/j.jwpe.2022.102630

[94] Hernandez CA, Osma JF. Microbial electrochemical systems: Deriving future trends from historical perspectives and characterization strategies. Frontiers in Environmental Science. 2020;**8**(44):1-19. DOI: 10.3389/fenvs.2020.00044

Chapter 13

Reuse of Wastewater as Non-Conventional Water: A Way to Reduce Water Scarcity Crisis

Mohammadhassan Gholami-Shabani and Katayoon Nematpour

Abstract

Water reuse is no longer a choice, it's an essentiality. Sewage is considered as one of the stable and significant sources of alternative water. Limitation of water resources and fluctuations and climate changes, uneven distribution of water in the world, increasing population, pollution of surface and underground water, and reaching the stage of water stress in many countries made water managers and planners seek to use non-conventional sources of water to achieve sustainable development. Therefore, the use of wastewater as a reliable source of water in terms of quantity in various sectors, including agriculture and industry, has received serious attention. In line with the optimal management of water resources and achieving the desired situation and providing water needs in the future, measures such as demand management and increasing social awareness, supply management, prioritizing consumption, using returned water, increasing water productivity, etc., it is essential. One of the important solutions in this field is the recycling and reuse of wastewater, which leads to the reduction of surface and underground water consumption. The use of treated wastewater has many advantages, including a higher availability of water, sustainable utilization of water resources, reduced energy consumption, decreased nutrient loads, and increased production. It also promotes environmental protection and boosts employment and the local economy. The advantages of reusing wastewater will be discussed in this chapter.

Keywords: reuse, pathogen, salinity, water recycling, alternative water, effluent

1. Introduction

The problem of water scarcity is present due to both natural and man-made factors. Countries or regions around the world are currently experiencing chronic water shortages, which affect millions of people [1]. Severe water scarcity for at least 1 month each year is experienced by almost two thirds of the world's population, comprising almost four billion people [2, 3]. Countries with inadequate water supply are home to over two billion people [4]. By 2025, water shortages could affect half of the world's population [5]. By 2030, intense water shortage could result in the displacement of around 700 million people [6]. By 2040, it is estimated that one out of every four children worldwide will be living in areas with very high water stress

[7]. Countries facing water crises should be concerned, as evidenced by the statistics [2]. When freshwater is used for domestic, sanitary, industrial, and service purposes, the remainder is disposed of as waste (wastewater) after consumption [8]. In numerous countries, wastewater is disposed of in natural watercourses, either as untreated or as treated effluent [9]. The investigators of the water and wastewater treatment industry think that natural resources, including freshwater, will remain under increasing pressure as a result of the growth of population for various uses, such as households and businesses [10]. The world is facing acute shortages of clean water and its population is affected by diseases associated with contaminated water due to the lack of a sustainable source of water [6]. With the intensification of the shortage of fresh and clean water in the world and with the increase in demand for water and due to drought conditions and climate changes, wastewater treatment has become one of the appropriate options to increase water sustainability [10]. Wastewater is the term used to describe water that has been polluted by humans. In other words, the remaining water after use in any of the domestic, industrial, commercial, or agricultural activities is called wastewater [11]. The used water from toilets, showers, baths, kitchen sinks, and laundries is known as home wastewater. Every day, the average amount of wastewater produced by domestic households is 200–300 L per person. The wastewater consists of 90% water and 10% contaminated waste [12]. The chemical contaminants in drinking water pose a threat to millions of people. These contaminants may be associated with naturally occurring inorganic chemicals such as arsenic and fluoride, which cause diseases such as cancer and tooth damage and/or skeletal damage, respectively [13]. Conversely, a wide range of severe health consequences may result from the inappropriate management of water supply in cities and industries or agricultural wastewater as well as prolonged exposure to them [13]. Therefore, this type of wastewater should often be directed to sewage treatment plants through collection pipes [10]. Taking early measures, including wastewater treatment, may help in reversing the catastrophic effects of living without clean and safe water for drinking, agriculture, or industrial use [10]. Industrial facilities also generate considerable amounts of wastewater that have chemicals or contents specific to their manufacturing process [14]. A source of wastewater, which comes from hospitals, offices, hotels, restaurants, airports, and many such areas, is called commercial wastewater that should be cleaned and made available for the next use [10]. The "fit-for-purpose" specifications for a specific next use are met by adequately treating these sources of water [15]. "Fit-for-purpose specifications" are the treatment requirements to bring water from a particular source to the quality needed, to ensure public health, environmental protection, or specific user needs [16]. For instance, the utilization of reclaimed water for agricultural purposes necessitates ensuring its quality is adequate to safeguard plants and soils, uphold food safety standards, and preserve the health of farm laborers [17]. When there is a higher level of human exposure to water, it may require more treatments. Direct potable wastewater reuse is an emerging water recycling technique that is still in its early stages of development. Untreated or contaminated water can transmit diseases, such as diarrhea, cholera, dysentery, typhoid, and polio and expose consumers to various public health challenges [18]. Grit, debris, suspended solids, pathogens that cause disease, decomposing organic waste, nutrients, and roughly 200 different chemicals have been identified as being present in wastewater [19]. The discharge of suspended solids, nutrients, organic chemicals, and metals into bodies of water is largely attributed to wastewater. The discharge of wastewater can result in various esthetic issues, including unpleasant odors and changes in coloration [19]. Disease-causing microorganisms, including

bacteria and viruses, have the potential to render water unsuitable for both human and animal consumption [20]. The overabundance of nutrients can result in the excessive stimulation of aquatic vegetation growth. The decomposition of organic waste can deplete dissolved oxygen levels and pose a threat to the survival of aquatic organisms [21]. Aquatic organisms can be adversely affected by toxic chemicals, posing a threat to their well-being. Furthermore, excessive sedimentation can lead to the suffocation of fish's feeding and spawning grounds [22]. Water recycling from wastewater, also known as reclaimed water, is a suggested approach to address the scarcity of water and combat drought conditions. This method involves the treatment and reuse of various types of wastewater, such as municipal, sanitary, and industrial wastewater. By implementing water recycling, we can effectively tackle the challenges posed by water shortages and ensure a sustainable water supply for various purposes [23]. Water recycling is underscored by several factors, including the scarcity of water supply, the absence of drinkable water, the presence of a substantial population, and its significance in safeguarding natural resources and the environment [23]. Water recycling involves the utilization of treated wastewater for beneficial applications, including agriculture, irrigation, and industrial processes [24]. Water recycling is a crucial component of effective water and wastewater management [23]. The utilization of contemporary techniques in wastewater treatment processes has led to remarkable advancements, thereby facilitating the widespread adoption of recycled water [23]. Nowadays, water treatment procedures have the capability to eliminate

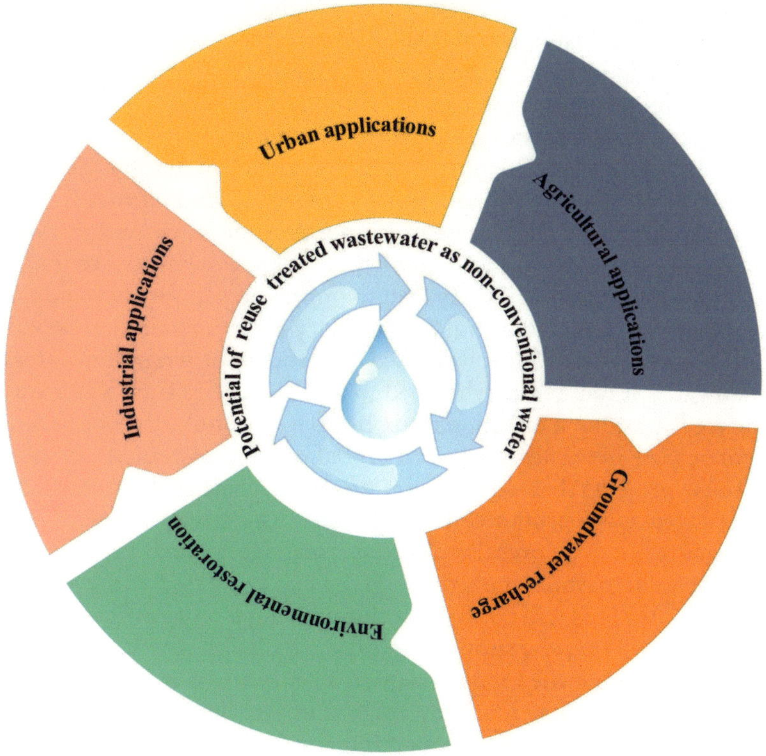

Figure 1.
Important applications of wastewater reuse: wastewater reuse can be classified into several groups, including urban applications, industrial applications, agricultural applications, groundwater recharge, and environmental restoration.

biodegradable compounds, essential nutrients, and harmful microorganisms from wastewater, resulting in purified water that can be utilized in various ways [9]. The primary objective of wastewater treatment is to segregate contaminants, eliminate suspended particles, eradicate toxins, and eradicate pathogens to enable the discharge of the resulting purified water, referred to as effluent, back into the environment for various applications [9, 10, 23]. Water recycling is a sustainable approach and is economical in the long run [25]. Water recycling offers numerous environmental benefits, including the reduction of wastewater discharge into the ecosystem and the mitigation of pollution. Moreover, it provides advantages in terms of water resource management [24]. **Figure 1** shows wastewater reuse can be classified into several groups, including urban applications, industrial applications, agricultural applications, and groundwater recharge [8]. In this chapter, we will discuss the advantages of reusing the wastewater.

2. Benefits of wastewater treatment reuse in agriculture processes

2.1 Sustainable irrigation

For irrigation in agriculture, reuse of wastewater may be a reliable and sustainable source of water [26]. In particular in areas affected by water scarcity or drought, it will help farmers maintain crop yields and reduce reliance on freshwater resources [27]. Farmers are able to maintain agricultural productivity in the event of a scarcity or unavailability of freshwater, if they use untreated wastewater [28]. The need for freshwater resources is reduced when treated wastewater is diverted to agriculture. This makes a difference lighten the strain on conventional freshwater sources, such as streams, lakes, and groundwater, which are frequently overexploited or helpless to climate change-induced inconstancy in precipitation designs. The nutrient content of treated wastewater can benefit agricultural crops by providing essential macro- and micronutrients. In the case of proper treatment, wastewater may be an excellent source of nutrients for plant growth as well as a substitute to that which is used by industrial fertilizers. This may lead to improved soil fertility and crop yields, contributing to increased agricultural productivity [29]. Farmers can benefit economically from access to a reliable and consistent source of irrigation water through reuse of wastewater. It will reduce the need to utilize a costly freshwater water system and compensate for taking a toll on manure buys. In addition, the opportunities for farmers to participate in wastewater treatment and disposal could be created through wastewater reuse projects that would generate new income streams. Sustainable water management practices are promoted through the reuse of wastewater in agriculture [30]. Water resources can be preserved for essential needs, such as drinking water supply and ecosystem maintenance, through the use of treated wastewater instead of freshwater to irrigate. This will contribute to achieving a more efficient and effective use of water resources, thereby supporting the agricultural sector's sustained resilience over time. By closing the loop in resource use, reuse of wastewater is compatible with the principle of a circular economy. Instead of treating wastewater as a waste product to be discharged, it is transformed into a valuable resource for agricultural irrigation. This promotes the recovery of resources, reduces waste, and contributes to a more sustainable and environmentally friendly approach to water and nutrient management [28–30]. Furthermore, reuse of wastewater is a significant solution to maintaining crop yields and the sustainability of agriculture, particularly for water scarcity regions, through provision of reliable and uniform sources of irrigation water, reducing pressure on freshwater resources, increasing farm productivity as well as promotion of responsible water management.

2.2 Nutrient-rich irrigation

Wastewater contains valuable nutrients, such as nitrogen, phosphorus, and potassium (NPK), which can be beneficial for growing plants. These nutrients may be used effectively in agriculture, reducing the need for fertilizers, through reuse of wastewater and encouraging healthy farming practices [8]. The reuse of wastewater allows for the recycling and recovery of nutrients in water, closing a nutrient cycle. Plants can use the nutrients contained in wastewater, such as nitrogen and phosphorus, to grow and develop [31]. This will reduce the need for synthetic fertilizers, which are energy intensive to produce, and may result in negative environmental impacts such as water pollution caused by nutrient discharges. The use of recycled water as an input source for nutrients can significantly reduce reliance on chemical fertilizers. Water pollution, eutrophication of water bodies, and the release of greenhouse gases (GHGs) can be caused by synthetic fertilizers that are not properly managed [10]. Reusing wastewater helps to reduce environmental risks by stopping nutrients from running off and decreasing the amount of harmful substances released into the environment. Treating wastewater properly makes the nutrients in it helpful for making soil more fertile [10, 30]. Plants need nitrogen, phosphorus, and potassium to grow. When these nutrients are present in wastewater, they can help the soil get more nutrients. This helps plants grow better and makes the soil healthier. This can make plants grow better, produce more crops, and create healthier farming areas [32]. By using wastewater that has nutrients in it, farmers can use less expensive fertilizers and save money. Using wastewater to provide nutrients to crops can help farmers save money and support a sustainable economy in farming [30]. Reusing wastewater on farms fits with sustainable farming, like organic and regenerative agriculture. It helps the environment by using wastewater in a smart way, instead of just throwing it away [10, 30]. Reusing wastewater in farming helps manage nutrients, reduces nutrient loss, and makes the environment healthier. Using cleaned wastewater for watering plants saves freshwater for important things. This is particularly relevant in regions that are suffering from water scarcity or where the availability of water resources is reduced [33]. The reuse of wastewater in agriculture promotes sustainable farming practices, reduces reliance on synthetic fertilizers, improves soil fertility, and supports environmental protection efforts by effectively exploiting the valuable nutrients contained in wastewater. It contributes to the sustainability and resilience of agriculture as a whole by providing an important source of nutrients for agricultural management.

2.3 Soil fertility improvement

Wastewater contains organic matter that can enhance soil fertility and structure [34]. The application of treated wastewater to agricultural lands can improve soil quality, leading to healthier and more productive crops. Wastewater contains organic compounds derived from human and industrial activities. When properly treated, these organic matter components can serve as an invaluable source of nutrients and improve the soil's organic matter content [35]. Organic matter enhances soil fertility by increasing nutrient availability, promoting soil microbial activity, improving water-holding capacity, and enhancing soil structure. Treated wastewater contains a range of essential macro- and micronutrients, including nitrogen, phosphorus, potassium, and various micronutrients. These nutrients are critical for plant growth and are readily available for uptake by plants when wastewater is applied to agricultural lands. This nutrient provision can supplement or replace the use of synthetic fertilizers, reducing the reliance on chemical inputs and contributing to sustainable farming practices. The organic matter in wastewater helps improve soil structure by enhancing aggregation, water infiltration, and moisture retention. It contributes to the formation of stable soil aggregates, which improve soil porosity and aeration, root penetration, and nutrient exchange. Improved soil structure enhances root development, leading to healthier plants with increased access to soil nutrients and water. The organic matter present in wastewater acts as a food source for soil microorganisms, stimulating their growth and activity. Soil microorganisms play a crucial role in nutrient cycling, decomposition of organic matter, and the synthesis of compounds that enhance plant growth. The application of treated wastewater can promote a diverse and active soil microbial community, supporting nutrient availability and overall soil health. The presence of organic matter in wastewater can help prevent soil erosion by improving soil stability and increasing the binding capacity of soil particles [36]. Organic matter enhances the formation of soil aggregates that resist erosion caused by wind or water. Additionally, the improved water-holding capacity of soils due to organic matter can help retain moisture for longer periods, reducing irrigation needs and enhancing drought resilience. In rundown, by enriching soil organic matter, providing essential nutrients, improving soil structure, enhancing microbial activity, controlling erosion, and aiding in soil remediation, the application of adequately treated wastewater to agricultural lands promotes soil fertility and agricultural productivity. It supports sustainable farming practices, reduces reliance on synthetic inputs, and contributes to the overall resilience and long-term sustainability of agricultural systems.

2.4 Cost savings

Using treated wastewater for irrigation can help reduce the costs associated with purchasing and transporting chemical fertilizers and freshwater for irrigation [37]. This can be particularly beneficial for small-scale farmers or those operating in water-stressed regions, where accessing freshwater can be expensive. Treated wastewater contains valuable nutrients, including nitrogen, phosphorus, and potassium, which are essential for plant growth [37]. By utilizing treated wastewater for irrigation, farmers can reduce their reliance on chemical fertilizers. This, in turn, lowers the costs associated with purchasing and applying synthetic fertilizers, which can be a significant expense for farmers, especially in areas with high fertilizer prices. The use of treated wastewater as an irrigation source eliminates the need for purchasing freshwater for irrigation purposes. Accessing freshwater can be expensive, particularly in water-scarce regions or areas with limited infrastructure for water supply [37, 38]. By reusing treated wastewater, farmers can significantly reduce or even eliminate the costs associated with obtaining freshwater, directly impacting their operational expenses. Treating and supplying freshwater for irrigation often requires energy-intensive processes, such as pumping, purification, and distribution [39]. By opting for treated wastewater, farmers can save on the energy costs associated with these processes, contributing to overall cost savings. Additionally, treating wastewater can often be done locally, reducing the need for long-distance transportation and associated energy expenditures [40, 41]. Utilizing treated wastewater for irrigation can also help reduce the need for extensive infrastructure development for freshwater supply.

Building irrigation infrastructure, such as pipelines, canals, or reservoirs, can be capital intensive and require ongoing maintenance [42]. By utilizing treated wastewater, farmers can leverage existing wastewater treatment infrastructure or establish decentralized treatment systems, reducing the need for additional infrastructure investments and minimizing maintenance costs. Water-stressed regions often experience higher freshwater costs during drought conditions due to increased demand and reduced supply [43]. By using treated wastewater for irrigation, farmers can reduce their vulnerability to drought and associated costs. Treated wastewater provides a reliable and consistent source of irrigation water, even during periods of water scarcity, ensuring crop productivity and reducing expenses that may arise from surging freshwater prices. In some cases, farmers may have the opportunity to generate revenue through wastewater reuse projects. By participating in the treatment and distribution of wastewater for agriculture, farmers can generate income through service fees or by selling surplus treated wastewater or treated wastewater byproducts. These additional revenue streams can offset agricultural production costs and contribute to farm sustainability [44]. In rundown, by reducing fertilizer expenses, eliminating freshwater costs, saving energy, cutting infrastructure expenses, improving drought resilience, and potentially generating additional revenue, the use of treated wastewater for irrigation provides substantial cost-saving benefits for farmers, particularly for small-scale farmers and those operating in water-stressed regions. It enhances economic viability, promotes sustainable farming practices, and contributes to the long-term sustainability of agricultural systems.

2.5 Reduced groundwater depletion

By reusing wastewater for irrigation, farmers can reduce their reliance on groundwater sources, which are often overexploited. This helps in preserving and managing groundwater resources more sustainably. Wastewater reuse allows farmers to use treated wastewater as an alternative water source for irrigation, reducing their reliance on groundwater [45]. Groundwater sources are often overexploited due to various factors, such as increasing agricultural demands, urbanization, and industrial activities. By using treated wastewater, farmers can significantly decrease their dependence on groundwater, relieving the pressure on these already stressed water sources. In some cases, treated wastewater can be used for groundwater recharge, replenishing depleted aquifers. This process involves intentionally injecting treated wastewater into the ground to replenish underground water reserves. Groundwater recharge helps restore groundwater levels and maintain their long-term sustainability [46]. By recharging groundwater with treated wastewater, farmers can actively contribute to the restoration and management of these vital water resources. Wastewater reuse for irrigation can reduce the risk of contamination and pollution of groundwater sources. When wastewater is not correctly treated or managed, it can contain harmful substances, pathogens, and pollutants that can leach into the soil and contaminate groundwater supplies [30]. By treating wastewater to high standards before reuse and implementing appropriate irrigation practices, the risk of groundwater pollution is minimized, ensuring the protection and long-term health of aquifers. Integrating wastewater reuse into agricultural practices supports a more sustainable water management approach. Instead of solely relying on dwindling groundwater resources, treated wastewater provides an additional and alternative water source for irrigation needs. This diversification of water supply helps balance water demands, reduces overreliance on a single resource, and promotes the efficient and sustainable use of water in agriculture [30, 33]. Climate change has brought about uncertain rainfall patterns, prolonged droughts, and water scarcity in many regions. By reusing wastewater, farmers can adapt to climate change impacts by having a reliable water source for irrigation, even during times of reduced rainfall or prolonged dry spells. This helps farmers maintain agricultural productivity and resilience in the face of climate-related challenges, ensuring the continuous growth of crops and food security [47]. In rundown, by reducing groundwater extraction, facilitating groundwater recharge, protecting aquifers from contamination, promoting sustainable water management, and aiding in climate change adaptation, wastewater reuse in agriculture plays a vital role in preserving and managing groundwater resources more sustainably. It supports long-term water security, reinforces agricultural sustainability, and contributes to the overall resilience of water systems in the face of growing water scarcity challenges.

2.6 Water conservation

Wastewater reuse in agriculture reduces the need for extracting freshwater from rivers, lakes, or underground sources. This conserves water resources and helps maintain environmental flow in water bodies, benefiting aquatic ecosystems and downstream users [30, 48]. By utilizing treated wastewater for irrigation, farmers can minimize the amount of freshwater they need to extract from natural sources. This reduction in water withdrawal helps to conserve water resources, ensuring that these sources remain available for other essential purposes, such as drinking water supply, industrial use, and maintaining healthy aquatic ecosystems. Environmental flow refers to the quantity, timing, and quality of water flows required to sustain aquatic ecosystems [35]. By reducing freshwater extraction for agriculture through wastewater reuse, more water is kept in the natural environment, allowing for the maintenance of adequate environmental flow. This is crucial for the health and biodiversity of rivers, lakes, wetlands, and other aquatic habitats, supporting the survival of various plant and animal species. Wastewater reuse practices in agriculture can alleviate pressure on freshwater ecosystems by decreasing the demand for water from natural sources. This helps in safeguarding the integrity of aquatic ecosystems and their associated habitats, allowing them to thrive in their natural state [30]. Furthermore, by reducing pollution and contaminants that might be present in the wastewater through proper treatment processes, the ecological health of water bodies can be improved, benefiting both aquatic species and downstream users. By reusing wastewater in agriculture, less water is withdrawn from freshwater sources upstream. This ensures that downstream users, such as communities, industries, and agriculture located further along the watercourse, have sufficient water available for their own needs [49]. By maintaining a sustainable water supply downstream, conflicts over water allocation can be minimized, supporting overall water resource management and promoting equitable distribution among different users. During drought conditions, when freshwater sources are scarce, wastewater reuse in agriculture becomes invaluable. By utilizing treated wastewater for irrigation, farmers can offset the reduced availability of freshwater, ensuring that agricultural activities continue even during these challenging periods. This helps maintain food production, farmer livelihoods, and overall regional stability [50, 51]. In rundown, by reducing freshwater extraction, preserving environmental flow, protecting and restoring aquatic ecosystems, sustaining downstream water availability, and mitigating the impact of droughts, wastewater reuse in agriculture plays a crucial role in conserving water resources and maintaining the ecological balance of water bodies. It supports the long-term sustainability of water systems, enhances water security for both human and environmental needs, and fosters resilience in the face of water scarcity and climate change challenges.

2.7 Diversification of water sources

When wastewater is reused in agriculture, it provides an additional source of water that can supplement traditional sources. This diversification helps reduce the vulnerability of agriculture to water scarcity and climate change impacts. As water scarcity becomes more prevalent due to factors like population growth, urbanization, and climate change, securing an adequate water supply for agriculture becomes increasingly challenging. Reusing wastewater provides farmers with an additional source of water that can supplement traditional freshwater sources. This diversification of water sources reduces the vulnerability of agriculture to water scarcity, ensuring a more consistent water supply, even during periods of reduced rainfall or water shortages. Climate change brings about increased uncertainty in rainfall patterns, temperature fluctuations, and extreme weather events. These impacts can disrupt agricultural activities and exacerbate water scarcity challenges. By incorporating wastewater reuse in agriculture, farmers can adapt to these climate change effects [27]. The availability of an alternative water source helps buffer the impacts of irregular rainfall, prolonged droughts, or changing hydrological patterns, enabling farmers to maintain agricultural productivity and resilience in the face of a changing climate. The integration of wastewater reuse in agricultural water management practices promotes a more efficient and sustainable use of water resources [10]. By utilizing treated wastewater for irrigation, farmers can optimize water allocation, ensuring that higher-quality freshwater sources are preserved for drinking water supply or other vital uses. This improved water management contributes to the long-term availability and sustainability of water resources, benefiting agriculture, communities, and ecosystems alike. Wastewater reuse offers farmers the opportunity to diversify their water supply sources [52]. By utilizing treated wastewater in conjunction with traditional freshwater sources, farmers reduce their dependence on a single water source. This diversification helps mitigate the risks associated with relying solely on freshwater, such as water scarcity, contamination, or price volatility. By harnessing multiple water sources, farmers can better manage water-related risks and adapt to changing conditions more effectively. Wastewater reuse supports sustainable agricultural practices and ensures food security. By providing an alternative water source, wastewater reuse enables farmers to sustain agricultural production and maintain crop yields, even when conventional water supplies are limited [28, 30]. This contributes to the overall stability and security of food systems, supporting local economies, farmer livelihoods, and food availability for communities, particularly in water-stressed regions. In rundown, by providing an additional water source, facilitating climate change adaptation, improving water management, promoting resource diversification, and ensuring sustainable agriculture and food security, wastewater reuse reduces the vulnerability of agriculture to water scarcity and climate change impacts. It strengthens the resilience of agricultural systems, enhances the ability to cope with water-related challenges, and supports the long-term sustainability of food production.

2.8 Local economic development

Wastewater reuse projects in agriculture can create employment opportunities and stimulate local economies. For example, the construction and operation of wastewater treatment facilities, as well as the distribution of treated wastewater to farmers, can generate jobs and boost economic growth [30]. Establishing wastewater treatment facilities requires skilled labor and expertise, creating employment opportunities during the construction phase. Engineers, technicians, construction workers, and other professionals are needed to design, build, and operate these facilities. This leads to the creation of jobs in the local economy, boosting employment rates and providing income opportunities for the workforce. Once wastewater treatment facilities are operational, ongoing maintenance and operation activities are necessary to ensure their efficient functioning [53]. This requires a workforce to monitor and maintain the infrastructure, perform regular inspections, conduct quality control checks, and manage the treatment processes. These operational roles offer employment opportunities and contribute to the local economy by providing stable jobs. Treated wastewater needs to be efficiently distributed to farmers for agricultural irrigation. This requires a network of distribution infrastructure, including pipelines, pumps, and storage facilities. The management and maintenance of this distribution system can create employment opportunities, such as technicians, operators, and logistics personnel responsible for ensuring the proper delivery and utilization of treated wastewater. Wastewater reuse projects often lead to the development of supporting industries and businesses. Local suppliers of materials, equipment, and services, such as pipelines, pumps, agricultural machinery, monitoring systems, and irrigation technologies, can benefit from increased demand and business opportunities. Additionally, businesses, such as wastewater testing laboratories, consultants specializing in wastewater management, and agricultural training and advisory services, can thrive, further contributing to the local economy [54]. Reusing wastewater in agriculture can stimulate the growth of agribusiness sectors by supporting increased crop production and diversification. Expanded agricultural activities driven by wastewater reuse can lead to the development of related industries, such as food processing, packaging, storage, transportation, and marketing. These value chains can create employment opportunities and generate economic growth, particularly in rural areas where agriculture is a primary source of livelihood. The economic benefits of wastewater reuse in agriculture extend beyond the immediate employment and business opportunities. The increased agricultural productivity resulting from reliable water supply stimulates the local economy through increased farm income, enhanced trade, and improved food security. The additional economic activity generated by wastewater reuse can have multiplier effects, supporting other sectors of the economy, such as tourism, hospitality, and retail, and contributing to overall regional development [55]. In rundown, by creating jobs in construction, operation, and maintenance activities, stimulating supporting industries and businesses, promoting agribusiness development, and generating multiplier effects, wastewater reuse projects in agriculture can have significant economic impacts. They contribute to local employment, income generation, and economic growth, fostering sustainable development and improving the livelihoods of communities involved in agriculture and wastewater management.

2.9 Climate change resilience

Wastewater reuse in agriculture can contribute to climate change adaptation by providing a more resilient water supply for irrigation. It helps farmers cope with variable and unpredictable weather patterns, reducing their vulnerability to droughts and other climate-related risks [51, 56]. By incorporating wastewater reuse in agriculture, farmers can diversify their water supply sources. This reduces their dependence on traditional freshwater sources, such as rivers or groundwater, which may become increasingly unreliable due to changes in precipitation patterns or reduced availability. The availability of an alternative water source through wastewater reuse helps farmers maintain a more consistent water supply, even during periods of drought or water scarcity caused by climate change. Climate change is leading to more frequent and intense droughts in many regions. Wastewater reuse provides an additional water source that can supplement conventional sources during periods of reduced rainfall or water scarcity [51]. By utilizing treated wastewater for irrigation, farmers can ensure a continuous water supply for their crops, mitigating the impacts of droughts on agricultural productivity and food security. Wastewater reuse often involves using treated wastewater for localized irrigation, targeting specific crops or areas with high water demand. This precision irrigation approach improves water use efficiency by delivering water directly to plant roots, minimizing losses due to evaporation, runoff, or overspray. With water scarcity becoming more prevalent under climate change, maximizing water use efficiency is vital for sustainable agriculture. Wastewater reuse helps farmers make the most of their available water resources, ensuring optimal use even under challenging climate conditions [57]. Wastewater contains valuable nutrients, including nitrogen, phosphorus, and potassium. When reused in agriculture, these nutrients can reduce the dependence on chemical fertilizers, contributing to sustainable and climate-smart farming practices. By using treated wastewater for irrigation, farmers can optimize nutrient management, promoting soil fertility and crop productivity. This nutrient recycling approach not only adapts to climate change but also reduces the environmental impact of fertilizers on water bodies and contributes to mitigating greenhouse gas emissions associated with fertilizer production. In areas prone to heat waves or rising temperatures due to climate change, wastewater reuse for irrigation can help regulate soil temperature and create a microclimate that is more favorable for plant growth. The application of treated wastewater can provide cooling effects, reducing heat stress on crops, and improving their resilience to extreme temperature events. Maintaining optimal soil moisture levels through wastewater reuse can also help buffer against temperature fluctuations, providing a more stable environment for plant growth. Implementing wastewater reuse in agriculture requires proper planning and infrastructure development. This process encourages farmers and communities to assess potential climate change risks, incorporate adaptation measures, and build resilience. By integrating wastewater reuse into their farm management practices, farmers adopt a proactive approach to climate change adaptation, reducing their vulnerability to climate-related risks and enhancing their long-term sustainability [38, 58]. In rundown, by diversifying water sources, ensuring water availability during droughts, improving water use efficiency, promoting nutrient recycling, providing temperature regulation benefits, and facilitating adaptation planning, wastewater reuse in agriculture contributes to climate change adaptation. It helps farmers cope with the impacts of climate change, reduces vulnerability to water-related risks, and promotes resilient agricultural systems that can withstand variable and unpredictable weather patterns.

3. Benefits of wastewater reuse in industrial processes

3.1 Water conservation

Reusing wastewater in industries helps save freshwater by using less of it [24, 59]. This is very important for industries that need a lot of water for their work, like manufacturing, producing energy, and mining. Companies can use wastewater again in their factories to use less freshwater. Instead of always getting lots of new water, they can clean and use the water they already have from their activities. This reduces the strain on local water resources, especially in areas which are stressed by drought or where there is a shortage [60]. By conserving freshwater resources, industries help sustainable management of water and prioritize its availability for other essential needs. Instead of putting wastewater into water bodies and ecosystem or sewage systems, industries can clean it and use it again in their work. This reduces the amount of liquid waste released, helping wastewater plants and lessening the harm that untreated or poorly treated wastewater can do to rivers and lakes [61]. By reducing the amount of wastewater they release, industries help keep water and aquatic life safe and healthy. Using recycled wastewater in industrial processes usually helps to use water more efficiently. Industrial companies can use new technology and methods to save water, using less water overall. This can mean making changes to how things are done, improving equipment, and finding ways to use less water [62]. By using water efficiently, industries can use less freshwater and help protect our water supply. Wastewater from factories can have useful stuff like energy, nutrients for plants, and metals, which can take it out and use it again [63]. The industry can obtain and recover these resources through wastewater treatment and reuse, thereby minimizing the need to extract or process new sources of materials. The circular economy principles, waste reduction, and the conservation of natural resources such as freshwater are promoted through this resource recovery approach. Reusing wastewater in factories can help the environment and save money [64]. Industries help the environment by saving water, keeping water clean, and cutting back on pollution related to using water. Regulatory requirements and sustainability objectives are often aligned with reuse of wastewater in industrial processes. In a lot of places, businesses have to use less water, put less water back into the environment, and find ways to use water again. By using wastewater again, industries can follow the rules and show they care about using water in a good way [65]. Finally, reuse of wastewater from industry plays a crucial role to preserve freshwater resources by reducing the consumption of freshwater, minimizing wastewater discharges, increasing water efficiency, promoting resource recovery, and achieving compliance with sustainability targets. It contributes to the sustainable management of water, supports ecosystems' overall health, and ensures their availability for other essential purpose.

3.2 Cost savings

Reusing cleaned wastewater can save money on buying and treating freshwater for industrial use. The cost of water treatment and disposal is reduced, thus reducing reliance on expensive freshwater sources. Industries that rely on freshwater sources for their operations often incur considerable expenses associated with purchasing water [66]. The use of treated wastewater can decrease or remove the need for freshwater purchases, thus leading to savings in costs. It can be costly to treat freshwater in order to comply with the quality requirements of industrial processes. In order to be used for industrial applications, the water must go through a series of treatment processes including filtration, disinfection, and sometimes desalination [67]. Industries can save money by using treated wastewater, instead of treating a lot of freshwater. It may cost and be an environmental burden to dispose of wastewater produced by industrial processes. Getting rid of wastewater by putting it into rivers or sewage systems can cost money in charges, permits, and making sure you follow the rules. By using treated wastewater again, industries can decrease the amount of wastewater they have to get rid of, and save money on disposal. This can reduce the financial burden on wastewater treatment and disposal, thereby reducing costs. Energy savings for industry can also be achieved by the reuse of treated wastewater. Treated wastewater is already partly cleaned or filtered, so it takes less energy to treat it more compared to freshwater. Furthermore, water transport and distribution costs can be reduced due to the fact that untreated wastewater is able to originate from local sources, which eliminates the need for transporting freshwater in distant locations. In the case of industrial wastewater reuse, these energy savings contribute to overall cost reductions. As has been pointed out, there may be valuable resources like energy, nutrients, and metals in the wastewater. Industries have the possibility to recover and reuse those resources through wastewater treatment and recycling. This not only decreases the need for purchasing additional resources but also makes potential revenue streams from selling or utilizing the recovered resources. The revenue created from resource recovery can offset the costs related with wastewater treatment and fund to cost savings. In industrial plants, the reuse of wastewater can result in efficiency gains. The industrial sector can streamline its water management systems, optimize water use, and improve process efficiency through direct use of wastewater in a number of processes. This could lead to improvements in productivity, reduced delays, and increased operational efficiency, which would ultimately contribute to a reduction in costs over time [68]. In rundown, by decreasing freshwater purchase, minimizing water treatment costs, reducing disposal expenses, saving energy, generating revenue through resource recovery, and enhancing operational efficiency, reusing treated wastewater can significantly reduce costs for industries. These cost savings can enhance the economic viability of industrial operations and promote sustainable water management practices.

3.3 Enhanced efficiency

The overall efficiency of water use can be improved by using processed wastewater in industrial processes. Uninterrupted operation, minimizing interruptions and growing productivity, is ensured through the availability of a stable and reliable water supply [60]. Industries can ensure a stable and continuous supply of water to their operations with the use of recycled wastewater as an aquatic source. Treated wastewater may provide a stable water supply compared to the use of only changing freshwater resources. This stability will decrease the risk of interruptions in manufacture due to water scarcity or disruption, thus minimizing interruption and enhancing total operational efficiency. Industry can optimize their water management systems by implementing the reuse of treated wastewater. They are capable of designing and implementing processes to maximize the effectiveness of water use, reduce waste, and minimize overall water consumption. The adoption of water efficiency technologies, the implementation of closed-loop systems, and better management of water supply and storage practices are also part of this. In order to increase productivity and reduce water consumption, industry can make the most of available water resources by optimizing water reuse. In order to provide a more precise water application system, treated wastewater can be adapted for particular manufacturing processes. This precision enables the industry to supply water directly where it is required, minimizing losses by evaporation, leakage, or spraying. Industries can optimize their water use and reduction in the overall water consumption through the prevention of unwanted water loss. The efficient use of water resources will be ensured through this targeted approach in the application of water, leading to increased efficiency and productivity gains. Through closed-loop systems, treated wastewater may be recycled in industrial processes. In order to minimize the need for additional freshwater intake, water used in one part of the process may be treated and reused in another part of the process. This closed-loop method decreases the, in general, water request and improves water utilize effectiveness. The volume of water required for industrial operations can be substantially reduced through the implementation of closed-loop methods, increasing sustainability and minimizing water waste. Inspection, repair, and maintenance of water infrastructure and distribution systems are often involved in recycled wastewater reuse. Industries can detect and remedy water leakages via leaks and ineffective processes if they focus on the efficiency and integrity of their water supply systems. Loss and wastage of water are reduced through such an active approach to managing the water, resulting in improved efficiency as well as cost savings. Appropriate water management planning and monitoring is essential for the implementation of recycled wastewater treatment. In this way, industry can progress with their understanding of water consumption designs and identify areas for enhancement and implement water saving measures. Industry can improve water efficiency and ensure sustainable consumption through the optimization of water management practices, such as leak detection, metering, and real-time monitoring systems [69]. Finally, the use of untreated wastewater in industrial processes improves the overall water efficiency by improving the reliability and consistency of water supply, optimizing water reuse, enabling precise application of water, introducing close loop systems, limiting leakages, and managing water more effectively. This improved efficiency contributes to the sustainability and profitability of industry by ensuring uninterrupted operation, minimizing disruption, and increasing productivity.

3.4 Reduced environmental impact

Reusing wastewater mitigates the environmental impact associated with the discharge of untreated or partially treated wastewater into water bodies [70]. Treating and reusing wastewater helps in preserving water quality, preventing pollution, and protecting aquatic ecosystems. Untreated or inadequately treated wastewater discharge into water bodies can introduce pollutants, contaminants, and pathogens, negatively impacting water quality and aquatic ecosystems. By treating wastewater before reuse, industries remove or reduce harmful substances, such as organic matter, nutrients, heavy metals, and pathogens. This reduces the pollution load associated with wastewater discharge, preventing potential harm to water bodies and the organisms that depend on them. Discharging untreated wastewater can disturb the natural balance of aquatic ecosystems. High levels of nutrients, such as nitrogen and phosphorus from untreated wastewater, can lead to eutrophication, causing harmful algal blooms, oxygen depletion, and the degradation of aquatic habitats. By treating and reusing wastewater, industries minimize the release of excessive nutrients into water bodies, protecting the delicate balance of aquatic ecosystems and safeguarding the biodiversity they support. Wastewater reuse reduces the demand for freshwater sources, preserving these valuable resources for other essential needs. This is particularly important in areas facing water scarcity or regions where freshwater availability is limited. By reusing treated wastewater, industries decrease their reliance on freshwater extraction, preventing overexploitation of local water sources. Preserving water resources ensures long-term sustainability and supports the availability of water for diverse ecosystems, human consumption, and agriculture. Untreated or poorly treated wastewater can contain harmful pathogens, posing risks to public health through waterborne diseases. By treating wastewater before reuse, industries remove or inactivate pathogens, minimizing the potential for disease transmission. This protects public health, reduces the burden on healthcare systems, and ensures the safety of water sources used for various purposes, including drinking water supplies. Untreated wastewater may contain various chemicals, including industrial pollutants, pharmaceuticals, and personal care products. Discharging such wastewater into water bodies can result in chemical contamination, posing risks to aquatic organisms and potentially entering the food chain. Through proper treatment before reuse, industries can remove or reduce these chemical contaminants, preventing their negative effects on the environment and minimizing potential ecological and human health risks. Wastewater reuse aligns with regulatory requirements and sustainability goals related to water management and environmental protection. Many jurisdictions have strict regulations on the discharge of untreated or partially treated wastewater, encouraging industries to invest in wastewater treatment and reuse. By complying with these regulations and embracing wastewater reuse, industries demonstrate their commitment to sustainable practices and contribute to the overall protection of the environment [70]. In rundown, by treating and reusing wastewater, industries mitigate the environmental impact associated with untreated or partially treated wastewater discharge. They prevent water pollution, protect aquatic ecosystems, preserve water resources, prevent waterborne diseases, prevent chemical contamination, and comply with regulations and sustainability goals. Wastewater reuse is an essential component of sustainable water management, promoting environmental stewardship and contributing to a healthier and more sustainable planet.

3.5 Resource recovery

Wastewater contains valuable resources like energy, nutrients, and chemicals that can be recovered and reused in industrial processes. For example, the extraction of biogas from wastewater can generate renewable energy, and the recovery of nutrients can be used as inputs in fertilizer production or other industrial applications. Wastewater contains organic matter that can be converted into biogas through anaerobic digestion. Biogas, primarily composed of methane, can be captured and used as a renewable energy source for heating, electricity generation, or even as a vehicle fuel [64, 71]. The extraction of biogas from wastewater not only reduces the environmental impact of wastewater treatment but also contributes to energy independence and cost savings for industries. Wastewater is rich in nutrients like nitrogen and phosphorus, which are essential for plant growth. Instead of being discharged into water bodies, these nutrients can be extracted from wastewater and reused in agricultural fertilizer production. Recovered nutrients can also be utilized as inputs for hydroponic systems, aquaculture, or in the production of bioplastics, contributing to resource conservation and reducing the reliance on synthetic fertilizers. Wastewater can contain various chemicals and compounds that can be recovered and reused in industrial applications. For example, some industries can recover and reuse chemicals present in wastewater as inputs for their processes, reducing the need to purchase new chemicals. This not only results in cost savings but also reduces the reliance on virgin materials and minimizes the environmental impact associated with chemical production. Although water itself is not a resource recovered from wastewater, reusing treated wastewater can help preserve freshwater resources. By utilizing treated wastewater for non-potable purposes like irrigation, industrial processes, or toilet flushing, industries reduce their reliance on freshwater sources. This conservation of freshwater resources ensures their availability for other essential needs and minimizes the strain on water-stressed regions. Wastewater treatment processes often generate sludge or biosolids that can be further treated and reused. These biosolids can be converted into fertilizers or soil amendments, used in land reclamation projects, or employed in the production of biofuels. By recovering and reusing biosolids, industries can divert waste from landfills, reduce greenhouse gas emissions, and contribute to a circular economy approach [72]. In rundown, the recovery and reuse of resources from wastewater contribute to the transition toward a more sustainable and circular economy. It reduces the reliance on virgin resources, reduces waste generation and disposal, minimizes environmental impacts, and provides economic opportunities. By tapping into the valuable resources present in wastewater, industries can enhance resource efficiency, reduce costs, and support sustainable development.

3.6 Compliance with regulations

Wastewater reuse can help industries comply with stringent environmental regulations related to water usage and discharge. By reusing treated wastewater, industries can demonstrate their commitment to sustainability and environmental stewardship. Many regions face water scarcity or have limited access to freshwater sources. By reusing treated wastewater, industries reduce their reliance on freshwater extraction, thereby conserving water resources [73]. This aligns with environmental regulations that aim to promote responsible water management and ensure the sustainable use of water resources. Environmental regulations, such as effluent discharge limits, aim to protect water bodies from pollution. By treating wastewater before reuse, industries remove or significantly reduce pollutants, including organic matter, nutrients, heavy metals, and pathogens. This ensures that the discharged wastewater meets regulatory standards, preventing harmful impacts on receiving water bodies and reducing the risk of associated penalties or legal actions. Discharging untreated or partially treated wastewater can degrade water quality, affecting aquatic ecosystems and public health. By utilizing treated wastewater, industries contribute to maintaining high water quality standards. The advanced treatment processes applied to wastewater ensure that it is properly treated to remove contaminants and meet stringent regulatory requirements for various uses. By reusing treated wastewater, industries minimize the environmental impact associated with wastewater discharge. This includes reducing the introduction of pollutants and contaminants into water bodies, mitigating the risk of eutrophication, protecting aquatic ecosystems, and preserving biodiversity. By actively implementing wastewater reuse strategies, industries demonstrate their commitment to minimizing their environmental footprint and complying with regulatory standards. Environmental regulations often emphasize sustainable resource management and waste reduction. By recovering resources from wastewater, such as energy, nutrients, and chemicals, industries contribute to circular economy principles and sustainable practices. These resource recovery efforts align with regulatory goals related to waste reduction, resource conservation, and the promotion of a more sustainable industrial sector [74]. Wastewater reuse reflects a proactive approach to environmental management and corporate social responsibility. By implementing measures to reuse treated wastewater, industries demonstrate their commitment to sustainable practices, including reducing water consumption, conserving natural resources, and minimizing their impact on the environment. This proactive stance can enhance their reputation, build trust with stakeholders and the public, and attract environmentally conscious customers [73]. In rundown, wastewater reuse helps industries comply with stringent environmental regulations related to water usage and discharge. By conserving water, limiting pollutant discharge, protecting water quality, reducing environmental impacts, promoting sustainable resource management, and showcasing corporate responsibility, industries can meet regulatory requirements while contributing to a more sustainable future.

3.7 Community relations

Implementing wastewater reuse projects in industries can improve their public image and strengthen relationships with local communities and stakeholders [10]. It demonstrates a commitment to sustainable practices and responsible water management. By incorporating treated wastewater into their operations, industries showcase their dedication to conserving water resources and minimizing their impact on the environment [75]. This commitment resonates with stakeholders who value companies that prioritize sustainability and actively work toward reducing their ecological footprint. Reusing treated wastewater showcases industry's role as an environmental steward. It highlights their proactive approach to minimizing the negative impacts of their operations on natural resources and ecosystems. This fosters goodwill among local communities, environmental groups, and other stakeholders who prioritize or advocate for environmental protection and sustainability. Conserving freshwater resources through wastewater reuse resonates with stakeholders concerned about water scarcity and the responsible use of water. By demonstrating a commitment to responsible water management, industries contribute to the preservation and sustainable use of water resources. This can help build trust and positive relationships with communities facing water scarcity or those who prioritize water conservation efforts. Implementing wastewater reuse projects provides an opportunity for industries to engage with local communities and stakeholders. Through open communication channels, industries can educate communities about their water management practices, the benefits of treated wastewater reuse, and the positive impact it has on local water resources. Engaging in community outreach initiatives and seeking input from stakeholders can foster positive relationships and enhance industry reputation as a responsible and accountable member of the community. Wastewater reuse aligns with regulatory requirements for responsible water management, thereby showcasing industry's commitment to compliance. It demonstrates a proactive approach to meeting environmental standards and goes beyond mere compliance by actively contributing to sustainable water practices. This commitment to regulatory compliance and social responsibility enhances industry's reputation and may result in increased support from regulators, government agencies, and community leaders. Embracing wastewater reuse can be appealing to environmentally conscious customers who prefer to support businesses that prioritize sustainability. By implementing sustainable water management practices, industries can position themselves as leaders in environmental responsibility and attract customers who prioritize environmentally friendly products and services. This can lead to increased market share and enhanced business opportunities [10]. In rundown, implementing wastewater reuse projects in industries improves their public image and strengthens relationships with local communities and stakeholders. It highlights industry's commitment to sustainability, responsible water management, environmental stewardship, and community engagement. By actively demonstrating these values, industries can enhance their reputation, gain public support, and attract customers who prioritize sustainability and responsible business practices.

3.8 Innovation and technological advancements

The development and adoption of innovative technologies and solutions is encouraged by the reuse of wastewater in industrial processes. This will encourage research and development (R&D), enabling industries to explore new ways of optimizing water use, applying effective treatment processes, and improving efficiency in all aspects of operation. Industries invest in investigation and development to improve treatment technologies and production methods to make them more efficient, cost-effective, and capable of eliminating contaminants to meet stringent quality standards [10]. Innovative treatment processes are developing and being adopted to enhance operational sustainability as the industry focuses on optimizing water use. Advanced monitoring and control systems are being used by industry in order to guarantee the suitability and safety of reused wastewater. These systems use real-time sensors, data analysis, and automation to keep a constant monitoring of the water quality parameters as well as detecting any changes in treatment or control processes. Industry is able to maintain the required water quality and optimize its reuse process with the development and deployment of these technologies. Resources, such as nutrients, energy, and chemicals, are valuable and profitable in the wastewater. Technological progress has resulted in the development of processes allowing industries to recover this resource from wastewater. For example, nutrient extraction methods can detect and trap nutrients from wastewater for use in the manufacture of fertilizers or energy recovery systems can generate renewable energy from organic substance in wastewater [76]. The circular economy and the manufacture of additional value from wastewater are supported by investigation and deployment of resource recovery technology. The reuse of wastewater encourages industries to adopt water efficient processes and technologies during their operations. Industries innovate and optimize their manufacturing processes in order to minimize water requirements, with a view to decreasing the use of freshwater. In order to maximize water efficiency, it may include the introduction of closed-loop systems, water conservation equipment, and recycling technology. In order to optimize water resources and reduce their environmental footprint, these technologies are being developed and adopted by industry. Smart water management systems are frequently incorporated into the implementation of wastewater reuse projects. More accurate water management, reduction of water waste, and support for overall sustainability efforts can be achieved through the integration of smart technologies in industrial processes. Collaboration between different stakeholders, such as investigators, engineers, and industry experts, is encouraged when implementing wastewater reuse plans. By making it easier to exchange ideas, experiences, and best practices, knowledge sharing and collaboration is promoting innovation. This collective approach is helping in the growth and progress of technologies and solutions for wastewater reuse, which will improve environmental sustainability in industrial processes [77]. Overall, the reuse of wastewater in industrial processes is stimulating innovation and technological progress in water treatment, resource recovery, water efficiency, and smart water use. These advances are improving the sustainability of operations, increasing water quality monitoring, and contributing to the development of a circular economy. The industries benefit from these innovations, ensuring the appropriate and effective use of water resources while mitigating their impacts on the environment.

4. Benefits of wastewater reuse in groundwater replenishment processes

4.1 Groundwater recharge

The reuse of treated wastewater in groundwater replenishment processes involves injecting it into the aquifers and replenishing water reserves [46, 78]. It also aids to restore depleted aquifers and maintain sustainable groundwater levels. High-quality treated water is pumped into the subterranean aquifers to replenish groundwater [79]. Before injecting it into the ground, the wastewater goes through a thorough cleaning process to take out any harmful stuff like germs and chemicals [80]. It shall ensure that the injected water complies with suitable water quality standards. The depletion of groundwater reserves is restored through the injection of treated wastewater to underground aquifers. Over time, the injection water will mix with groundwater already in place and increase its total amount of water within the aquifer. It contributes to the restoration of groundwater levels in aquifers that were previously drained, thereby making them more resilient and able to cope with water needs for years to come. Using treated wastewater to refill the groundwater helps to keep a steady supply of water balanced with the demand for it. In places where people use a lot of groundwater, taking too much water can cause the groundwater to run out. The water supply is enhanced and the reliance on unsustainably high pumping rates from subterranean sources has been reduced by using treated wastewater to fill aquifers. This will render assistance to decrease the risk of overextraction and promote sustainable groundwater supplies for a long time. Injecting treated wastewater into aquifers for groundwater replenishment can help develop water quality. The injected water undergoes additional treatment by biological, physical, and chemical processes, as it is naturally filtered through the surrounding soil and rock layers [80]. This may result in further elimination of trace contaminants and show progress with the overall water quality when it is used to recharge an aquifer [81]. In coastal areas, the excessive extraction of groundwater can result in seawater intrusion where saline water infiltrates freshwater aquifers [81]. By refilling underground water sources with cleaned wastewater, we can reduce the strain on freshwater supplies and lower the chances of saltwater getting into them. This will help maintain the integrity of groundwater aquifers and ensure a sustainable supply of water which is naturally filtered. The groundwater resources face important challenges due to climate change, which includes long droughts and water scarcity. Using treated wastewater to refill groundwater can help us deal with changing conditions in a way that lasts a long time. During periods of high rainfall or surplus water, regions can effectively store excess water by injecting treated wastewater into aquifers, creating a buffer against future droughts, and ensuring long-term water security [82]. Overall, using treated wastewater to recharge groundwater is a good way to fill up aquifers and manage groundwater in a sustainable way. It's helping to ensure water supply demand is balanced, preventing saltwater intrusion, improving water quality, and adapting to the challenges of climate change. Communities can ensure the long-term availability and sustainability of their groundwater sources through this approach.

4.2 Water resource sustainability

Groundwater is a vital source of freshwater for many communities around the world [83]. By recharging groundwater with treated wastewater, the long-term sustainability of this valuable water resource can be ensured, reducing reliance on surface water or unsustainable extraction practices [83]. By recharging groundwater with treated wastewater, communities can diversify their water sources [83]. This reduces their reliance on surface water bodies such as rivers and lakes, which are often subject to seasonal variations in availability and quality. Groundwater replenishment through wastewater reuse provides an additional, reliable source of freshwater, helping to secure water supply, even during periods of drought or surface water scarcity [82]. Recharging groundwater with treated wastewater reduces the demand for surface water resources. This is particularly important in regions where surface water bodies are already overexploited. By relying on treated wastewater, communities can reduce the pressure on surface water sources, ensuring their long-term sustainability and minimizing the risk of their depletion. Groundwater replenishment through wastewater reuse enhances water security by reducing dependence on external water supplies. It provides a locally available and controlled water source, reducing vulnerability to disruptions in surface water availability (e.g., droughts and pollution) or limitations in water transfer infrastructure. Communities can rely on the recharged groundwater as a consistent and secure resource for their water needs, reducing the risk of water shortages and ensuring stability and resilience in the face of changing conditions. Recharging groundwater with treated wastewater helps to sustainably manage and protect aquifers [84]. Instead of relying solely on groundwater extraction, which can lead to depletion and related problems like land subsidence, communities can utilize treated wastewater to replenish aquifers and maintain sustainable groundwater levels. This approach promotes responsible water management, preserving the long-term viability of groundwater resources for future generations. Treated wastewater used for groundwater replenishment goes through rigorous treatment processes to meet quality standards before injection. This improves the overall quality of the recharged groundwater. By replenishing aquifers with high-quality water, the treated wastewater helps to improve groundwater quality and reduces the need for extensive treatment when the water is later extracted for human consumption or industrial use [85]. Groundwater replenishment through treated wastewater reuse can be a cost-effective and energy-efficient approach to meet water demands. Compared to constructing and maintaining extensive infrastructure for water transfer or desalination, using treated wastewater for groundwater recharge can be economically advantageous. It also reduces energy requirements, as it avoids the need for energy-intensive desalination processes or long-distance water transportation [85]. In rundown, recharging groundwater with treated wastewater provides a sustainable solution to promote the long-term sustainability of this valuable freshwater resource. It reduces reliance on surface water or unsustainable extraction practices, enhances water security, and conserves surface water resources. By utilizing this approach, communities can ensure a reliable and locally controlled water source while preserving and maintaining the integrity of their groundwater reserves.

4.3 Water quality improvement

Treated wastewater used for groundwater replenishment goes through an advanced treatment process, removing contaminants and pollutants. This improves the overall quality of the injected water, ensuring that the replenished groundwater remains clean and safe for use. The treatment process begins with primary treatment, where physical processes such as screening and sedimentation are used to remove large solids, debris, and settleable particles from the wastewater [86]. This initial step helps prevent clogging in subsequent treatment processes and protects downstream equipment. After primary treatment, the wastewater undergoes secondary treatment, which focuses on the biological removal of organic matter and the reduction of nutrients [87]. This typically involves biological processes like activated sludge, trickling filters, or rotating biological contactors. Microorganisms break down organic pollutants, converting them into solids that settle or form flocs. Tertiary treatment is the advanced stage of the treatment process, specifically designed to remove remaining contaminants and pollutants [88]. Various techniques are employed, depending on the specific requirements and the level of water quality desired. Common tertiary treatment processes include filtration, advanced oxidation, membrane filtration (e.g., microfiltration, ultrafiltration, and reverse osmosis), disinfection (e.g., chlorination and ultraviolet (UV) disinfection), and nutrient removal (e.g., denitrification and phosphorus removal). Advanced filtration processes like microfiltration, ultrafiltration, and reverse osmosis play a crucial role in removing even smaller particles, suspended solids, dissolved organic matter, and trace contaminants from the treated wastewater [88]. These processes utilize membranes with fine pores that selectively filter out contaminants based on their size and molecular properties, producing high-quality water. To ensure the safety of the injected water, disinfection is typically performed as the final step of the treatment process. Common disinfection methods include chlorination, UV disinfection, and advanced oxidation processes. These methods target and destroy or deactivate any remaining pathogens, bacteria, viruses, or other harmful microorganisms that may be present in the treated wastewater [89]. It's important to note that the specific treatment processes and technologies may vary depending on local regulations, water quality goals, and the intended use of the replenished groundwater. However, the overarching aim of the treatment process is to ensure that the treated wastewater meets stringent quality standards, guaranteeing the cleanliness and safety of the injected water for the long-term sustainability of groundwater resources [86]. In rundown, by undergoing these advanced treatment processes, the treated wastewater for groundwater replenishment is effectively purified, with contaminants, pollutants, and pathogens removed or greatly reduced. This ensures that the injected water is of high quality and meets appropriate water quality standards. The improvement in water quality safeguards the replenished groundwater, providing clean and safe water for various uses such as irrigation, industrial processes, or even drinking water supply (with additional treatment, if necessary).

4.4 Drought resilience

In times of drought or water scarcity, groundwater replenishment through reuse of wastewater helps to make a buffer [90]. By renewing aquifers, communities can depend on put-away groundwater amid dry periods, diminishing the defenselessness to drought-induced water deficiencies. Groundwater replenishment provides a means of storing excess water during periods of high rainfall or surplus water availability. By infusing treated wastewater into aquifers amid these times, communities can successfully store water underground. In the event of limited or reduced surface water supply, such storage water shall act as a buffer and provide an adequate source of water for subsequent dry periods. Ensuring a sustainable supply of water is also helping communities become more resilient to droughts, in spite of the diminished availability of surface waters. The groundwater storage capacity is increased when the aquifers are replenished with treatment water [85]. Amid times of tall water accessibility, overabundance of water can be infused into the aquifer, successfully reviving it. This helps in the growth of the volume of water stored underground, making a reserve that can be tapped into during droughts. The better the replenishment efforts, the larger the amount of water available in the aquifer, providing a longer-term buffer against water scarcity. Groundwater, when legitimately overseen and revived, offers a consistent and true and reliable water supply amid dry seasons. Not at all like surface water sources that can be intensely affected by varieties in precipitation or drying up amid drawn-out dry spells, renewed groundwater encompasses a slower exhaustion rate. This enables more sustainable and reliable water sources to be relied upon in communities. Groundwater replenishment through wastewater reuse helps decrease vulnerability to water use restrictions. Amid dry spells, water shortage can lead to strict controls and restrictions on water utilization, especially from surface water sources. Communities can reduce their reliance on surface water and, at the same time, avoid or minimize the impact of these restrictions through the use of recharged groundwater. In addition, groundwater replenishment supports local water management and reduces reliance on external water sources. Communities are empowered to manage and control their water resources through the use of treated wastewater within the same catchment area or region. This will reduce the vulnerability of regions to outside factors, such as water transfers, or reliance on distant sources that may be more vulnerable to drought caused shortages. Groundwater replenishment is a long-term solution to ensure water security, in particular in areas with a history of drought and water scarcity [91, 92]. By infusing treated wastewater into aquifers amid damp periods, communities can construct up-saves that can be withdrawn amid times of low precipitation or constrained surface water availability. In the face of climate change, such sustainable practices support to preserve a stable and secure water supply [93]. Overall, when times of drought or water scarcity occur, groundwater replenishment by the reuse of wastewater helps to create a buffer against this. This will increase the storage capacity of aquifers, provide a reliable water supply, reduce vulnerability to water use restrictions, support regional water management, and ensure sustainable water security in the future. The use of this approach allows communities to better manage and cope with drought conditions, mitigate their impact on water scarcity, and ensure a sustainable water supply in the future.

4.5 Reduced wastewater discharge

By reusing wastewater for groundwater replenishment, the discharge of treated wastewater into surface water bodies is reduced [24]. This leads to a decrease in the pollution and ecological impacts associated with the direct release of treated wastewater into rivers or oceans. Reusing treated wastewater for groundwater replenishment eliminates or greatly reduces the direct discharge of treated wastewater into rivers, lakes, or oceans. This prevents the pollution of surface water bodies with potentially harmful substances, pollutants, and nutrients that may still be present in treated wastewater. By diverting treated wastewater to recharge aquifers, the risk of contaminating surface water, harming aquatic ecosystems, and compromising water quality downstream is mitigated [94]. Discharging treated wastewater directly into water bodies can have detrimental impacts on aquatic ecosystems. High nutrient levels, such as nitrogen and phosphorus, can cause eutrophication, leading to oxygen depletion and the formation of harmful algal blooms. These episodes can harm fish populations, degrade water quality, and disrupt the balance of aquatic ecosystems. By diverting treated wastewater for groundwater replenishment, these negative impacts on surface water ecosystems are minimized, preserving the health and biodiversity of these environments. When wastewater treatment plants release treated wastewater into rivers or oceans, there is still the potential for downstream pollution as the water flows through different regions. By reusing wastewater for groundwater replenishment, the risk of downstream pollution is significantly reduced. The treated wastewater is injected into the ground, effectively removing the potential for contamination and pollution downstream, where it could impact other communities, water users, and sensitive ecosystems. In coastal areas, the discharge of treated wastewater into the ocean can harm delicate marine environments, including coral reefs, seagrass beds, and coastal habitats. Nutrient enrichment, increased turbidity, and the introduction of contaminants in treated wastewater can disrupt these ecosystems, affecting marine life and biodiversity. By recharging groundwater with treated wastewater, the discharge into coastal waters is reduced, minimizing the ecological impacts on these valuable ecosystems [95]. Surface water bodies often serve as sources of drinking water for communities. The direct discharge of treated wastewater into these water bodies can introduce contaminants and pollutants that may pose risks to public health and require extensive additional treatment for their removal. By reusing wastewater for groundwater replenishment, the potential for contamination of drinking water sources is reduced [24]. Treated wastewater undergoes advanced treatment processes to meet stringent quality standards before being injected into aquifers, ensuring the protection of groundwater resources that may be tapped for drinking water supply. Reusing wastewater for groundwater replenishment is an integral part of the circular water management approach, where water is treated, reused, and recycled in a sustainable manner. By closing the loop and reusing treated wastewater, the valuable water resource is conserved, significantly reducing the reliance on freshwater extraction and the need for additional water supplies, thus contributing to a more sustainable and resilient water management system [96]. In rundown, reusing wastewater for groundwater replenishment helps decrease the discharge of treated wastewater into surface water bodies, reducing pollution and ecological impacts. It protects surface water quality, preserves aquatic ecosystems, prevents downstream pollution, safeguards coastal environments, protects drinking water sources, and promotes circular water management. This sustainable practice contributes to the overall health and sustainability of water resources, benefiting both human and ecological communities.

4.6 Potable water supply augmentation

Groundwater replenishment processes, when coupled with advanced treatment technologies, can produce high-quality water suitable for potable water supply [97]. This increases the overall freshwater supply and diversifies sources, reducing the reliance on traditional water sources and minimizing the need for costly infrastructure projects. Groundwater replenishment typically involves the use of advanced treatment technologies, such as membrane filtration, reverse osmosis, UV disinfection, and advanced oxidation processes (AOPs) [98–100]. These processes are capable of removing a wide range of contaminants, such as suspended solids, pathogens, nutrients, salts, and trace organic compounds. The combination of multiple treatment barriers ensures that the treated water meets stringent water quality standards and can be safely used for various purposes, including potable water supply. The incorporation of advanced treatment technologies enables the production of high-quality water that meets or surpasses regulatory standards for drinking water. Through robust treatment processes and multiple safeguards, including rigorous monitoring and quality control measures, the final product from groundwater replenishment can be consistently safe and reliable for human consumption. This assurance of water quality allows for its integration into the potable water supply system. Groundwater replenishment, when used for potable water supply, expands the available freshwater supply [101]. By utilizing advanced treatment technologies to convert treated wastewater into a reliable source of potable water, communities can augment their water resources and overcome water scarcity challenges [102]. This is particularly important in areas facing limited freshwater availability, growing population demands, or environmental constraints on traditional water sources [103]. Groundwater replenishment offers the advantage of diversifying water sources for potable water supply [91]. By adding treated wastewater into the water supply portfolio, communities can reduce their reliance on traditional sources such as surface water or groundwater extraction from diminishing aquifers [104]. This diversification enhances water supply resilience, ensuring a more secure and sustainable water system. It also helps mitigate the risks associated with single-source dependence or potential vulnerabilities to drought, contamination, or climate change impacts. Groundwater replenishment can be a cost-effective alternative compared to developing traditional water supply infrastructures that involve long-distance water transfers, desalination plants, or large-scale dam projects. By utilizing existing or upgraded wastewater treatment facilities and utilizing the natural process of aquifer storage and recovery, the need for extensive and expensive infrastructure projects can be minimized. This can result in significant cost savings for communities while still ensuring a safe and adequate potable water supply. Groundwater replenishment for potable water supply offers environmental advantages [105]. It encourages the reuse and recycling of treated wastewater, reducing the demands on freshwater sources and minimizing the discharge of treated wastewater into surface water bodies. This contributes to the conservation of freshwater resources, protects ecosystems from pollution, and promotes sustainable water management practices. In rundown, coupling groundwater replenishment processes with advanced treatment technologies enables the production of high-quality water suitable for potable water supply. This increases the overall freshwater supply, diversifies water sources, reduces reliance on traditional sources, and minimizes the need for costly infrastructure projects. Through this innovative approach, communities can address water scarcity, enhance water supply resilience, and contribute to sustainable water resource management.

5. Conclusion

In water-scarce regions such as the Middle East and North Africa (MENA), water has a valuable value and must be accounted for every drop. It is therefore necessary to reclassify wastewater as a renewable source of water rather than waste, in order to make it more effective at increasing water availability and preventing environmental pollution. Utilization of this asset requires collection, treatment, and distribution of all produced wastewater. In spite of the fact that reuse of wastewater is recognized in most water-scarce nations, the reuse of wastewater is still exceptionally very low. It is generally discarded as waste once freshwater has been used for an economic or beneficial purpose. In many countries, these wastewaters are discharged, either as untreated waste or as treated effluent, into natural ecosystem like watercourses, from which they are abstracted for further use after undergoing "self-purification" within the ecosystem. Wastewater may be reused more than a dozen times before it is released as part of the system of indirect reuse. However, it is also possible to use it more directly. Wastewater treatment contributes to economic and environmental sustainability, such as converting waste into clean water, converting into energy, thereby reducing the impact on the environment caused by wastewaters or building sustainable industrial cities and towns. For the purpose of providing clean drinking water and other uses, wastewater treatment removes pollutants, neutralizes particulate matter, removes toxicants, and kills pathogens. Wastewater treatment eliminates waste by ensuring that valuable products, such as energy, clean water, and fertilizers, are converted into dirty water that would otherwise be disposed of. The treatment of wastewater also provides sufficient nutrients to increase crop yields when used for agricultural purposes. The treatment of wastewater prevents exposure to potentially toxic chemicals, which can lead to severe diseases. One of the most sustainable ways to achieve a sustainable energy supply is to recover energy from wastewater. Constructing wastewater treatment plants is the first step to reducing waste and making sure that things that would have been wasted are reused. Wastewater treatment plants can make enough clean water for cities and suburbs that do not have enough water. Moreover, cleaning wastewater helps to prevent harm to the environment when it's released into rivers, lakes, and oceans. Making more clean water available would mean less scarcity and lessening the strain on natural resources. Managing resources sustainably means making sure we always have enough for the future, so we do not use them all up. Therefore, institutions and policy capacity need to be enhanced, public awareness of related issues needs to be strengthened, and appropriate funding mechanisms set up in order for the successful implementation of wastewater reuse strategies.

Conflict of interest

Authors declare no conflict of interest.

Author details

Mohammadhassan Gholami-Shabani[1,2]* and Katayoon Nematpour[2]

1 Department of Mycology, Pasteur Institute of Iran, Tehran, Iran

2 Department of Labour and Environment, Ministry of Industry, Mine and Trade, Tehran, Iran

*Address all correspondence to: shabani@kimo.com

References

[1] Kumari U, Swamy K, Gupta A, Karri R, Meikap B. Global water challenge and future perspective. In: Dehghani M, Karri R, Lima E, editors. Green Technologies for the Defluoridation of Water. Amsterdam, The Netherlands: Elsevier; 2021. pp. 197-212. DOI: B978-0-323-85768-0.00002-6

[2] Mekonnen MM, Hoekstra AY. Four billion people facing severe water scarcity. Science Advances. 2016;**2**:e1500323. DOI: 10.1126/sciadv.1500323

[3] Boretti A, Rosa L. Reassessing the projections of the world water development report. npj Clean Water. 2019;**2**:15. DOI: 10.1038/s41545-019-0039-9

[4] Martínez-Santos P. Does 91% of the world's population really have "sustainable access to safe drinking water"? International Journal of Water Resources Development. 2017;**33**:514-533. DOI: 10.1080/07900627.2017.1298517

[5] Makarigakis AK, Jimenez-Cisneros BE. UNESCO's contribution to face global water challenges. Water. 2019;**11**:388. DOI: 10.3390/w11020388

[6] Lal BS. Water for life: Issues and challenges. International Journal of Science and Research. 2019;**8**:1949-1957. DOI: 10.21275/ART20199011

[7] du Plessis, A. Water as a source of conflict and global risk. In: du Plessis A, editor. Water as an Inescapable Risk. Springer Water. Cham: Springer; 2019. pp. 115-143. DOI: 10.1007/978-3-030-03186-2_6

[8] Kesari KK, Soni R, Jamal QM, Tripathi P, Lal JA, Jha NK, et al. Wastewater treatment and reuse: A review of its applications and health implications. Water, Air, & Soil Pollution. 2021;**232**:1-28. DOI: 10.1007/s11270-021-05154-8

[9] Wilas J, Draszawka-Bolzan B, Cyraniak E. Wastewater reuse. World News of Natural Sciences. 2016;**5**:33-41

[10] Silva JA. Wastewater treatment and reuse for sustainable water resources management: A systematic literature review. Sustainability. 2023;**15**:10940. DOI: 10.3390/su151410940

[11] Jiménez B, Drechsel P, Koné D, Bahri A, Rashid-Sally L, Qadir M. Wastewater sludge and excreta use in developing countries: An overview. In: Drechsel P, Scott A, Rashid-Sally L, Redwood M, Bahri A, editors. Wastewater Irrigation and Health: Assessing and Mitigating Risk in Low-Income Countries. London, UK: Earthscan; 2009. pp. 3-27. DOI: 10.4324/9781849774666-2

[12] Wärff C. Household wastewater generation model. In: Report LUTEDX/(TEIE7279)/1-29/(2020), Division of Industrial Electrical Engineering and Automation. Lund, Sweden: Lund University; 2020

[13] Toolabi A, Bonyadi Z, Paydar M, Najafpoor AA, Ramavandi B. Spatial distribution, occurrence, and health risk assessment of nitrate, fluoride, and arsenic in bam groundwater resource, Iran. Groundwater for Sustainable Development. 2021;**12**:100543. DOI: 10.1016/j.gsd.2020.100543

[14] Thomas O, Thomas MF. Industrial wastewater. In: Thomas O, Christopher Burgess C, editors. UV-Visible Spectrophotometry of Waters and Soils. Amsterdam, The Netherlands: Elsevier; 2022. pp. 385-416. DOI: 10.1016/B978-0-323-90994-5.00013-7

[15] World Health Organization. Safety and Quality of Water Use and Reuse in the Production and Processing of Dairy Products: Meeting Report. Food and Agriculture Organization of the United Nations. Geneva, Switzerland: FAO eBooks; 2023. DOI: 10.4060/cc4081en

[16] Martin B, Messner E. Potable reuse: A pathway for success and sustainability. In: Younos T, Lee J, Parece TE, editors. Alternative Water Sources for Producing Potable Water. The Handbook of Environmental Chemistry. Cham: Springer; 2023. pp. 97-115. DOI: 10.1007/698_2023_1021

[17] Bouwer H, Idelovitch E. Quality requirements for irrigation with sewage water. Journal of Irrigation and Drainage Engineering. 1987;**113**:516-535. DOI: 10.1061/(ASCE)0733-9437(1987)113:4(516)

[18] Unz RF. Disease transmission by contaminated water. In: Nemerow NL, Agardy FJ, Sullivan P, Salvato JA, editors. Environmental Engineering: Prevention and Response to Water-, Food-, Soil-, and Air-Borne Disease and Illness. Vol. 7. New Jersey, United States: Wiley; 2009. pp. 1-98. DOI: 10.1002/9780470432815.ch1

[19] Reena DA. A review on advancement in waste water treatment. Journal of Positive School Psychology. 2022;**6**:3895-3910

[20] Haseena M, Malik MF, Javed A, Arshad S, Asif N, Zulfiqar S, et al. Water pollution and human health. Environmental Risk Assessment and Remediation. 2017;**1**:16-19

[21] Chapra SC. Water quality. In: Kutz M, editor. Handbook of Environmental Engineering. New Jersey, United States: Wiley; 2018. pp. 333-349. DOI: 10.1002/9781119304418.ch11

[22] Karri RR, Ravindran G, Dehghani MH. Wastewater—Sources, toxicity, and their consequences to human health. In: Rao R, Ravindran G, Dehghani MH, editors. Soft Computing Techniques in Solid Waste and Wastewater Management. Amsterdam, The Netherlands: Elsevier; 2021. pp. 3-33. DOI: 10.1016/B978-0-12-824463-0.00001-X

[23] Jodar-Abellan A, López-Ortiz MI, Melgarejo-Moreno J. Wastewater treatment and water reuse in Spain. Current situation and perspectives. Water. 2019;**11**:1551. DOI: 10.3390/w11081551

[24] Jhansi SC, Mishra SK. Wastewater treatment and reuse: Sustainability options. Consilience. 2013;**10**:1-15

[25] Voulvoulis N. Water reuse from a circular economy perspective and potential risks from an unregulated approach. Current Opinion in Environmental Science & Health. 2018;**2**:32-45. DOI: 10.1016/j.coesh.2018.01.005

[26] Helmecke M, Fries E, Schulte C. Regulating water reuse for agricultural irrigation: Risks related to organic micro-contaminants. Environmental Sciences Europe. 2020;**32**:1-10. DOI: 10.1186/s12302-019-0283-0

[27] Ingrao C, Strippoli R, Lagioia G, Huisingh D. Water scarcity in agriculture: An overview of causes, impacts and approaches for reducing the risks. Heliyon. 2023;**9**:e18507. DOI: 10.1016/j.heliyon.2023.e18507

[28] Khan MM, Siddiqi SA, Farooque AA, Iqbal Q, Shahid SA, Akram MT, et al. Towards sustainable application of wastewater in agriculture: A review on reusability and risk assessment. Agronomy. 2022;**12**:1397. DOI: 10.3390/agronomy12061397

[29] Drechsel P, Evans AE. Wastewater use in irrigated agriculture. Irrigation and Drainage Systems. 2010;**24**:1-3. DOI: 10.1007/s10795-010-9095-5

[30] Jaramillo MF, Restrepo I. Wastewater reuse in agriculture: A review about its limitations and benefits. Sustainability. 2017;**9**:1734. DOI: 10.3390/su9101734

[31] Ye Y, Ngo HH, Guo W, Chang SW, Nguyen DD, Zhang X, et al. Nutrient recovery from wastewater: From technology to economy. Bioresource Technology Reports. 2020;**11**:100425. DOI: 10.1016/j.biteb.2020.100425

[32] Tymchuk I, Shkvirko O, Sakalova H, Malovanyy M, Dabizhuk T, Shevchuk O, et al. Wastewater a source of nutrients for crops growth and development. Journal of Ecological Engineering. 2020;**21**:88-96. DOI: 10.12911/22998993/122188

[33] Al Hamedi FH, Kandhan K, Liu Y, Ren M, Jaleel A, Alyafei MA. Wastewater irrigation: A promising way for future sustainable agriculture and food security in the United Arab Emirates. Water. 2023;**15**:2284. DOI: 10.3390/w15122284

[34] Hamilton D. Organic Matter Content of Wastewater and Manure. USA: Oklahoma Cooperative Extension Service; 2012

[35] Yahaya SM, Mahmud AA, Abdu N. The use of wastewater for irrigation: Pros and cons for human health in developing countries. Total Environment Research Themes. 2023;**6**:100044. DOI: 10.1016/j.totert.2023.100044

[36] Balaganesh P, Vasudevan M, Natarajan N, Suneeth Kumar SM. Improving soil fertility and nutrient dynamics with leachate attributes from sewage sludge by impoundment and co-composting. Clean–Soil, Air, Water. 2020;**48**:2000125. DOI: 10.1002/clen.202000125

[37] Mishra S, Kumar R, Kumar M. Use of treated sewage or wastewater as an irrigation water for agricultural purposes-environmental, health, and economic impacts. Total Environment Research Themes. 2023;**6**:100051. DOI: 10.1016/j.totert.2023.100051

[38] Hashem MS, Qi X. Treated wastewater irrigation—A review. Water. 2021;**13**:1527. DOI: 10.3390/w13111527

[39] Garcia X, Pargament D. Reusing wastewater to cope with water scarcity: Economic, social and environmental considerations for decision-making. Resources, Conservation and Recycling. 2015;**101**:154-166. DOI: 10.1016/j.resconrec.2015.05.015

[40] Van de Walle A, Kim M, Alam MK, Wang X, Wu D, Dash SR, et al. Greywater reuse as a key enabler for improving urban wastewater management. Environmental Science and Ecotechnology. 2023;**16**:100277. DOI: 10.1016/j.ese.2023.100277

[41] Zarei M. Wastewater resources management for energy recovery from circular economy perspective. Water-Energy Nexus. 2020;**3**:170-185. DOI: 10.1016/j.wen.2020.11.001

[42] Plappally AK, Lienhard JH. Energy requirements for water production, treatment, end use, reclamation, and disposal. Renewable and Sustainable Energy Reviews. 2012;**16**:4818-4848. DOI: 10.1016/j.rser.2012.05.022

[43] Qadir M, Boers TM, Schubert S, Ghafoor A, Murtaza G. Agricultural water management in water-starved countries: Challenges and opportunities. Agricultural Water Management. 2003;**62**:165-185. DOI: 10.1016/S0378-3774(03)00146-X

[44] Hagenvoort J, Ortega-Reig M, Botella S, García C, de Luis A, Palau-Salvador G. Reusing treated wastewater from a circular economy perspective—The case of the real Acequia de Moncada in Valencia (Spain). Water. 2019;**11**:1830. DOI: 10.3390/w11091830

[45] Ramirez C, Almulla Y, Nerini FF. Reusing wastewater for agricultural irrigation: A water-energy-food nexus assessment in the North Western Sahara aquifer system. Environmental Research Letters. 2021;**16**:044052. DOI: 10.1088/1748-9326/abe780

[46] Shawaqfah M, Almomani F, Al-Rousan T. Potential use of treated wastewater as groundwater recharge using GIS techniques and modeling tools in dhuleil-halabat well-field/Jordan. Water. 2021;**13**:1581. DOI: 10.3390/w13111581

[47] Scott CA, Faruqui NI, Raschid-Sally L. Wastewater use in irrigated agriculture: Management challenges in developing countries. In: Scott CA, Faruqui NI, Raschid-Sally L, editors. Wastewater Use in Irrigated Agriculture: Confronting the Livelihood and Environmental Realities. Wallingford UK: Cabi Publishing; 2004. pp. 1-10. DOI: 10.1079/9780851998237.0001

[48] Jiménez B. Irrigation in developing countries using wastewater. International Review for Environmental Strategies. 2006;**6**:229-250

[49] Hettiarachchi H, Caucci S, Ardakanian R. Safe use of wastewater in agriculture: The golden example of nexus approach. In: Hettiarachchi H, Ardakanian R, editors. Safe Use of Wastewater in Agriculture. Cham: Springer; 2018. pp. 1-11. DOI: 10.1007/978-3-319-74268-7_1

[50] Ofori S, Puškáčová A, Růžičková I, Wanner J. Treated wastewater reuse for irrigation: Pros and cons. Science of the Total Environment. 2021;**760**:144026. DOI: 10.1016/j.scitotenv.2020.144026

[51] Ungureanu N, Vlăduț V, Voicu G. Water scarcity and wastewater reuse in crop irrigation. Sustainability. 2020;**12**:9055. DOI: 10.3390/su12219055

[52] Alzahrani F, Elsebaei M, Tawfik R. Public acceptance of treated wastewater reuse in the agricultural sector in Saudi Arabia. Sustainability. 2023;**15**:15434. DOI: 10.3390/su152115434

[53] Hernández-Chover V, Castellet-Viciano L, Hernández-Sancho F. Preventive maintenance versus cost of repairs in asset management: An efficiency analysis in wastewater treatment plants. Process Safety and Environmental Protection. 2020;**141**:215-221. DOI: 10.1016/j.psep.2020.04.035

[54] Renner M. Wastewater and Jobs the Decent Work Approach to Reducing Untreated Wastewater. Geneva, Switzerland: International Labour Organization; 2017

[55] del Villar A, García-López M. The potential of wastewater reuse and the role of economic valuation in the pursuit of sustainability: The case of the canal de Isabel II. Sustainability. 2023;**15**:843. DOI: 10.3390/su15010843

[56] Vo PT, Ngo HH, Guo W, Zhou JL, Nguyen PD, Listowski A, et al. A mini-review on the impacts of climate change on wastewater reclamation and reuse. Science of the Total Environment. 2014;**494**:9-17. DOI: 10.1016/j.scitotenv.2014.06.090

[57] Hettiarachchi H, Ardakanian R. Safe Use of Wastewater in Agriculture: Good Practice Examples; Institute for Integrated Management of Material.

Shibuya, Tokyo: United Nations University; 2016

[58] Freitas TR, Santos JA, Silva AP, Fraga H. Reviewing the adverse climate change impacts and adaptation measures on almond trees (*Prunus dulcis*). Agriculture. 2023;**13**:1423. DOI: 10.3390/agriculture13071423

[59] Klemeš JJ. Industrial water recycle/reuse. Current Opinion in Chemical Engineering. 2012;**1**:238-245. DOI: 10.1016/j.coche.2012.03.010

[60] Gutterres M, de Aquim PM. Wastewater reuse focused on industrial applications. In: Sharma S, Sanghi R, editors. Wastewater Reuse and Management. Dordrecht: Springer Netherlands; 2012. pp. 127-164. DOI: 10.1007/978-94-007-4942-9_5

[61] Pintilie L, Torres CM, Teodosiu C, Castells F. Urban wastewater reclamation for industrial reuse: An LCA case study. Journal of Cleaner Production. 2016;**139**:1-4. DOI: 10.1016/j.jclepro.2016.07.209

[62] Kushwaha JP. A review on sugar industry wastewater: Sources, treatment technologies, and reuse. Desalination and Water Treatment. 2015;**53**:309-318. DOI: 10.1080/19443994.2013.838526

[63] Dutta D, Arya S, Kumar S. Industrial wastewater treatment: Current trends, bottlenecks, and best practices. Chemosphere. 2021;**285**:131245. DOI: 10.1016/j.chemosphere.2021.131245

[64] Estahbanati MK, Kumar S, Khajvand M, Drogui P, Tyagi RD. Environmental impacts of recovery of resources from industrial wastewater. In: Pandey A, Tyagi RD, Varjani Biomass S, editors. Biomass, Biofuels, Biochemicals. Amsterdam, The Netherlands: Elsevier; 2021. pp. 121-162. DOI: 10.1016/B978-0-12-821878-5.00003-9

[65] Lyu S, Chen W, Zhang W, Fan Y, Jiao W. Wastewater reclamation and reuse in China: Opportunities and challenges. Journal of Environmental Sciences. 2016;**39**:86-96. DOI: 10.1016/j.jes.2015.11.012

[66] Mohsen MS, Jaber JO. Potential of industrial wastewater reuse. Desalination. 2003;**152**:281-289. DOI: 10.1016/S0011-9164(02)01075-5

[67] Udaiyappan AF, Hasan HA, Takriff MS, Abdullah SR. A review of the potentials, challenges and current status of microalgae biomass applications in industrial wastewater treatment. Journal of Water Process Engineering. 2017;**20**:8-21. DOI: 10.1016/j.jwpe.2017.09.006

[68] Walsh BP, Cusack DO, O'Sullivan DT. An industrial water management value system framework development. Sustainable Production and Consumption. 2016;**5**:82-93. DOI: 10.1016/j.spc.2015.11.004

[69] Reznik A, Dinar A, Hernandez-Sancho F. Treated wastewater reuse: An efficient and sustainable solution for water resource scarcity. Environmental and Resource Economics. 2019;**74**:1647-1685. DOI: 10.1007/s10640-019-00383-2

[70] De Gisi S, Casella P, Cellamare CM, Ferraris M, Petta L, Notarnicola M. Wastewater reuse. In: Abraham MA, editor. Encyclopedia of Sustainable Technologies. Amsterdam, The Netherlands: Elsevier; 2017. pp. 52-68. DOI: 10.1016/B978-0-12-409548-9.10528-7

[71] Makisha N, Semenova D. Production of biogas at wastewater treatment plants and its further application. In: MATEC Web of Conferences. Vol. 144. France: EDP Sciences: Ulis; 2018. p. 4016. DOI: 10.1051/matecconf/201814404016

[72] Ali S, Peter AP, Chew KW, Munawaroh HS, Show PL. Resource

recovery from industrial effluents through the cultivation of microalgae: A review. Bioresource Technology. 2021;**337**:125461. DOI: 10.1016/j.biortech.2021.125461

[73] Choukr-Allah R. Wastewater treatment and reuse. In: El-Ashry M, Saab N, Zeitoon B, editors. Arab Environment: Water: Sustainable Management of a Scarce Resource. Report of the Arab Forum for Environment and Development (AFED). Lebanon: Beirut; 2010. pp. 107-124

[74] Preisner M, Smol M, Horttanainen M, Deviatkin I, Havukainen J, Klavins M, et al. Indicators for resource recovery monitoring within the circular economy model implementation in the wastewater sector. Journal of Environmental Management. 2022;**304**:114261. DOI: 10.1016/j.jenvman.2021.114261

[75] Strade E, Kalnina D, Kulczycka J. Water efficiency and safe re-use of different grades of water-topical issues for the pharmaceutical industry. Water Resources and Industry. 2020;**24**:100132. DOI: 10.1016/j.wri.2020.100132

[76] Perera MK, Englehardt JD, Dvorak AC. Technologies for recovering nutrients from wastewater: A critical review. Environmental Engineering Science. 2019;**36**:511-529. DOI: 10.1089/ees.2018.0436

[77] Wang X, Daigger G, Lee DJ, Liu J, Ren NQ, Qu J, et al. Evolving wastewater infrastructure paradigm to enhance harmony with nature. Science Advances. 2018;**4**:eaaq0210. DOI: 10.1126/sciadv.aaq0210

[78] El Arabi N. Environmental management of groundwater in Egypt via artificial recharge extending the practice to soil aquifer treatment (SAT). International Journal of Environment and Sustainability. 2012;**1**:66-82

[79] Dillon P. Ground water replenishment with recycled water—An Australian perspective. Groundwater. 2009;**47**:492-495. DOI: 10.1111/j.1745-6584.2009.00587.x

[80] Drewes J. Ground water replenishment with recycled water-water quality improvements during managed aquifer recharge. Ground Water. 2009;**47**:502. DOI: 10.1111/j.1745-6584.2009.00587_5.x

[81] Hussain MS, Javadi A, Sherif M. Artificial recharge of coastal aquifers using treated wastewater to control saltwater intrusion. Quality and Quantity. 2016;**1**:1-4

[82] Scanlon BR, Fakhreddine S, Rateb A, de Graaf I, Famiglietti J, Gleeson T, et al. Global water resources and the role of groundwater in a resilient water future. Nature Reviews Earth & Environment. 2023;**4**:87-101. DOI: 10.1038/s43017-022-00378-6

[83] Foster S, Chilton J, Nijsten GJ, Richts A. Groundwater—A global focus on the 'local resource'. Current Opinion in Environmental Sustainability. 2013;**5**:685-695. DOI: 10.1016/j.cosust.2013.10.010

[84] Masciopinto C, Vurro M, Palmisano VN, Liso IS. A suitable tool for sustainable groundwater management. Water Resources Management. 2017;**31**:4133-4147. DOI: 10.1007/s11269-017-1736-0

[85] El Arabi NE, Dawoud MA. Groundwater aquifer recharge with treated wastewater in Egypt: Technical, environmental, economical and regulatory considerations. Desalination and Water Treatment. 2012;**47**:266-278. DOI: 10.1080/19443994.2012.696405

[86] Smarzewska S, Morawska K. Wastewater treatment technologies. In: Abdel Rahman RO, Hussain CM, editors. Handbook of Advanced Approaches Towards Pollution Prevention and Control. Amsterdam, The Netherlands: Elsevier; 2021. pp. 3-32. DOI: 10.1016/B978-0-12-822121-1.00001-1

[87] Asthana M, Kumar A, Sharma BS. Wastewater treatment. In: Singh R, editor. Principles and Applications of Environmental Biotechnology for a Sustainable Future. Applied Environmental Science and Engineering for a Sustainable Future. Singapore: Springer; 2017. pp. 173-232. DOI: 10.1007/978-981-10-1866-4_6

[88] Rout PR, Zhang TC, Bhunia P, Surampalli RY. Treatment technologies for emerging contaminants in wastewater treatment plants: A review. Science of the Total Environment. 2021;**753**:141990. DOI: 10.1016/j.scitotenv.2020.141990

[89] Salgot M, Folch M. Wastewater treatment and water reuse. Current Opinion in Environmental Science & Health. 2018;**2**:64-74. DOI: 10.1016/j.coesh.2018.03.005

[90] Shevah Y. Water resources, water scarcity challenges, and perspectives. In: Ahuja S, de Andrade JB, Dionysiou DD, Hristovski KD, editors. Water Challenges and Solutions on a Global Scale. ACS Symposium Series. United States: American Chemical Society; 2015. pp. 185-219. DOI: 10.1021/bk-2015-1206.ch010

[91] Hund SV, Allen DM, Morillas L, Johnson MS. Groundwater recharge indicator as tool for decision makers to increase socio-hydrological resilience to seasonal drought. Journal of Hydrology. 2018;**563**:1119-1134. DOI: 10.1016/j.jhydrol.2018.05.069

[92] Sun J, Donn MJ, Gerber P, Higginson S, Siade AJ, Schafer D, et al. Assessing and managing large-scale geochemical impacts from groundwater replenishment with highly treated reclaimed wastewater. Water Resources Research. 2020;**56**:e2020WR028066. DOI: 10.1029/2020WR028066

[93] Pyne RD. Groundwater Recharge and Wells: A Guide to Aquifer Storage Recovery. Boca Raton, Florida, United States: CRC Press; 2017. pp. 1-22. DOI: 10.1201/9780203719718

[94] Dillon P, Kumar A, Kookana R, Leijs R, Reed D, Parsons S. In: Gleton G, editor. Managed Aquifer Recharge-risks to Groundwater Dependent Ecosystems-A Review. Canberra, Australia: Water for a Healthy Country Flagship Report to Land & Water Australia; Commonwealth Scientific and Industrial Research Organisation; 2009. pp. 1-10

[95] Mitra A, Zaman S, Mitra A, Zaman S. Threats to marine and estuarine ecosystems. In: Mitra A, Zaman S, editors. Basics of Marine and Estuarine Ecology. New Delhi: Springer; 2016. pp. 365-417. DOI: 10.1007/978-81-322-2707-6_10

[96] Foster SS, Chilton PJ. Downstream of downtown: Urban wastewater as groundwater recharge. Hydrogeology Journal. 2004;**12**:115-120. DOI: 10.1007/s10040-003-0296-y

[97] Dhal R. Advanced thinking: Potable reuse gaining traction. Environmental Health Perspectives. 2014;**122**:A332-A335. DOI: 10.1289/ehp.122-A332

[98] Patel M. UV/AOP a key part of the groundwater replenishment system. Water Environment Federation. 2011;**2011**:3516-3525. DOI: 10.2175/193864711802721686

[99] Gao J, Qin T, Wacławek S, Duan X, Huang Y, Liu H, et al. The application of

advanced oxidation processes (AOPs) to treat unconventional water for fit-for-purpose reuse. Current Opinion in Chemical Engineering. 2023;**42**:100974. DOI: 10.1016/j.coche.2023.100974

[100] Miklos DB, Remy C, Jekel M, Linden KG, Drewes JE, Hübner U. Evaluation of advanced oxidation processes for water and wastewater treatment–a critical review. Water Research. 2018;**139**:118-131. DOI: 10.1016/j.watres.2018.03.042

[101] Gude VG. Desalination of deep groundwater aquifers for freshwater supplies–challenges and strategies. Groundwater for Sustainable Development. 2018;**6**:87-92. DOI: 10.1016/j.gsd.2017.11.002

[102] Ghernaout D. Increasing trends towards drinking water reclamation from treated wastewater. World Journal of Applied Chemistry. 2018;**3**:1-9. DOI: 10.11648/j.wjac.20180301.11

[103] Okello C, Tomasello B, Greggio N, Wambiji N, Antonellini M. Impact of population growth and climate change on the freshwater resources of Lamu Island, Kenya. Water. 2015;**7**:1264-1290. DOI: 10.3390/w7031264

[104] Lee GF, Jones-Lee A. Water/Aquifer Quality Issues That Need to be Considered in Enhanced Groundwater Recharge Projects. Rancho Cordova, CA: Submitted to the Central Valley Regional Water Quality Control Board Basin Plan Triennal Rewiew; 2005. pp. 1-20

[105] Gao L, Connor JD, Dillon P. The economics of groundwater replenishment for reliable urban water supply. Water. 2014;**6**:1662-1670. DOI: 10.3390/w6061662

Chapter 14

Modeling of a Septic Tank Waste Treatment System for the Production of Biogas

Ibrahima Niane, Amadou Diao, Fakoro Souleymane Dia and Senghane Mbodji

Abstract

In developing countries, in general, especially in universities, the management of wastewater and excreta remains one of the major concerns. This uncontrolled practice can have negative im-pacts on the environment such as deforestation and global warming. The recovery of domestic wastewater through the production of biogas (methanization) is beneficial both on the environmental level and on the energy level (electricity, heat) and social, especially in universities where the increase of the students is accompanied by growth in energy consumption. This article delves into the physico-chemical and biological processes involved in wastewater treatment, while also making predictions about the behavior of various variables such as methane, nitrogen, acetogenic microorganisms, and methanogenic biomasses. This approach allows knowing the variation of the biomass according to the type of bacteria. Indeed, it shows the relation, which exists between the biomass and the organic load. It is firstly noted in the results that the mean value of the biomass of acidogenesis and that of the biomass of methanogenesis are approximately equal to 0.67 d^{-1}. Secondly, the nitric acid concentration is around 47.5 mgN/l/h and that of the viscosity is estimated to 0.046 Pa.s. Afterward, the temperature and the pH favor the activity of anaerobic bacteria. We finally conclude that an experimental device of 71 m^3 intended for the accumulation of sludge will be designed and built.

Keywords: biogas, biomasses acetogenic, methanogenic, excreta, septic tank

1. Introduction

Generally, in Senegal and particularly, in the site of Ngoundiane, septic tanks are built without any rule of design and sizing.

The aim of this study is to propose a mathematical model that helps researchers to better understand the physicochemical and biological processes of wastewater. This model can probably predict the behavior of variables such as methane, nitrogen, acidogenic microorganisms, and methanogenic biomasses of a septic tank digester. This study also aims to determine the growth of anaerobic microorganisms thanks to

IntechOpen

the kinetic constants, which correspond the maximum specific growth rate (μmax), the half-saturation constant (Kds), and the biomass mortality rate) to constitute an essential support for the development of a global model of septic tank operation, which takes into account not only setting but also anaerobic digestion.

Symbols	Values	Units	Senses
S_s	—	m^2	Side surface
k	0. 8	W/m/K	Thermal conductivity
φ	—	Watt	Thermal flux density
V_{stok}	—	m^3/h	Volume gasometer
Q_{pmh}	—	m^3/h	Average hourly quantity of biogas supplied
Q_{cmh}	—	m^3/h	Average hourly quantity consumed
Dcg	—	m	Diameter of gas pipe
J	10	mm	Pressure drop
L_c	—	m	Production-use distance
λ	—	—	Pressure drop coefficient
ρ	1.2	Kg/m^3	Density of biogas
V_{eg}	—	m^3/s	Flow velocity
D_{cg}		m	Gas pipe diameter
g	9.81	m/s^2	Gravity acceleration
μ	$1.297.10^{-5}$	Kg/m/s	Dynamic viscosity
R_e	—	—	Reynolds number
Q_r	97.5	Kg	Daily quantity of feces obtained with us
V_u	—	m^3	Useful volume of the digester
C_{s1}	—	g/l	Initial substrate concentration
K_{ds1}	7.1	g/l	Half-saturation constant
μ_{cmax1}	1.20	d^{-1}	Maximum growth rate of acidogenic
K_{1ac}	0.2	$KgBOD/m^3$	Acetic acid inhibition constant [1]
K_{ds2}	0.55	g/l	Monod half-saturation constant
C_{s2}	—	g/l	Substrate of concentration
μ_{cmax2}	0.74	d^{-1}	Maximum growth rate of methanogenic
K_{i-2}	0.96	g/l	Inhibition constant of the bacteria methanogenic [2]
K_2	$0.01 < K_2 < 19.803$	—	Empirical constant with 20°C < T < 37°C [3]
D_{dig}	4.25	m	Digester diameter obtained with us
ρ_{gaz}	1.2	Kg/m^3	Density of the gas
$C_{c/Kg}$	0.8$ = 462.80FCFA		Mass price [4]
C_{VF/m^3}	3000	FCFA	Volume cost [4]

Table 1.
Table of some symbols.

This research aims to model the digester to understand its physical, chemical, and biological process, as well as to evaluate its financial viability (**Table 1**).

2. Material and method

2.1 Experimental apparatus

2.1.1 The septic tank digester

The idea is always to use the available materials as much as possible in order to reduce manufacturing costs as possible. Lansing et al. [5] stated that most anaerobic digesters encountered in developing countries lack a heating and agitation system for the digestate.

We were, therefore, inspired by the Puxin digester model developed by the company from Shenzhen (China), which is represented in **Figure 1**.

The cylindrical body of the digester is made of reinforced concrete to ensure its tightness and to maintain the temperature of organic matter. In fact, to strengthen thermal insulation, the digester is buried in the ground about 4/5 of its height [6]. The concrete construction not only ensures a long service life of the installation but also avoids investments in the purchase of insulation material (glass wool, etc.)

The useful capacity of a septic tank is calculated based on the volume of sludge accumulation and the volume occupied by the organic matter. This volume also depends on the period between two emptying, the sludge accumulation rate, and the sludge occupation rate, which is not only 50% of the useful volume of the pit at the time of emptying but also on [6]. According to Coulibaly et al. [7], the specific production that corresponds to the quantity of sludge produced per inhabitant and per day is equal to 0.4 liters per day and per inhabitant. The retention time of sludge inside the pits is estimated at 2.5 to 3 years for a more efficient biological activity [8].

The useful capacity of a septic tank is given by the following mathematical relationship [6]:

$$V_u = 0.0004 \times N_a \times 365 \times t_{ret} \times 2 \tag{1}$$

$$V_u = 71 m^3$$

with:

V_u means volume of sludge accumulation in cubic meter,

N_a is the number of inhabitants,

t_{ret} corresponds the sludge retention time in year.

2.1.1.1 Modeling of heat transfers between the inside and the outside of the digester: Steady state

In this study, the aim is to determine the heat flux (W) that passes through a considered cylindrical digester made up of a monolayer wall, a side surface, a wall thickness, and a thermal conductivity of the materials. Here, the digester is concrete, the thermal conductivity is 0.8, and the wall thickness is 15 cm [9].

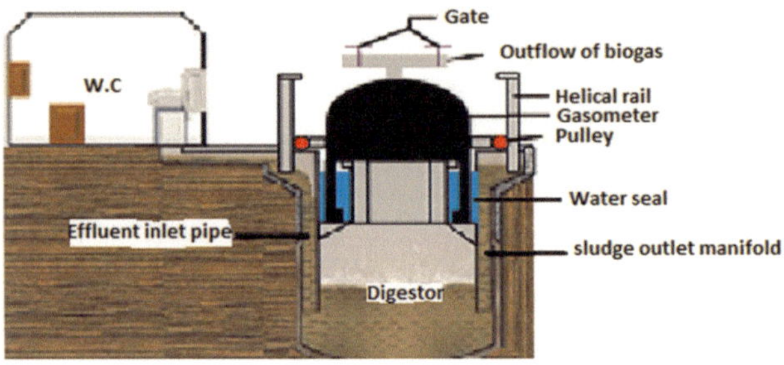

Figure 1.
Model digester septic tank.

The thermal flux density crossing this hollow cylinder is given by Fourier's law presented in Eq. (2) [10]:

$$\varphi = \frac{2\pi \times k \times H}{\ln \frac{r_2}{r_1}} (T_2 - T_1) \tag{2}$$

This becomes:

$$\varphi = \frac{\pi \times k \times H \times D_{dig}}{e_p} (T_2 - T_1) \tag{3}$$

where
φ is the thermal flux density in watt (W),
e_p is the wall thickness in meter (m),
r_1, r_2 are the internal and external radii of the cylinder, respectively,
T_1 and T_2, the temperatures at the interfaces of the walls as shown in **Figure 2**,
H corresponds to the height of the cylinder.

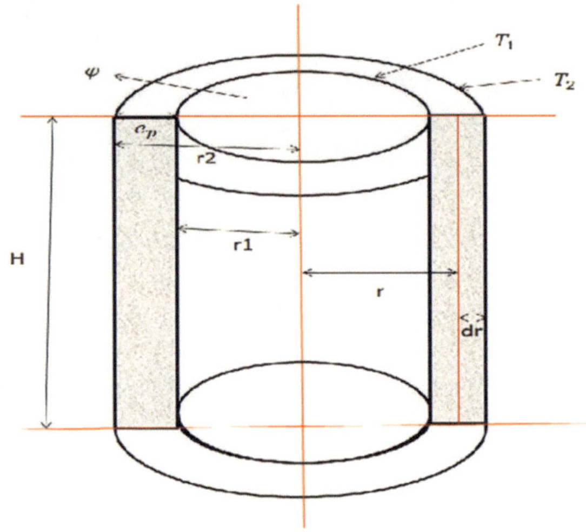

Figure 2.
Cylindrical digester.

2.1.1.2 Conductive thermal resistance of cylindrical digester

Per definition, the thermal resistance is the inverse of the thermal flux [10]. From the Eq. (2), we deduce the expression of the thermal resistance of a digester.

$$R_{Th} = \frac{\ln\left(\frac{r_2}{r_1}\right)}{2\pi \times k \times H} \tag{4}$$

2.1.2 The bell gasometer

The gasometric bell can be manufactured from steel sheets having a thickness of $15/10^{th}$ of mm and dimension of 2 m in length by 1 m in width. The welding will be done by a mechanical welding workshop. To fight against corrosion, we will cover the inside of the bell with liquid rubber (Flint-Kote), and its exterior will be covered with black paint preferably to fight against rust. In addition, the black color participates in the absorption of solar energy to heat the walls of the digester [6]. The gas outlet pipe having a diameter of approximately 10 mm is positioned on a minimum slope of 1% from plastic pipes and galvanized or copper metal pipes compatible with other equipment such as valves, taps, etc.

2.1.2.1 Sizing of a gasometer

A gasometer is a storage tank for biogas obtained from organic waste. Its volume can be calculated according to the average hourly quantity of biogas supplied and the average hourly quantity consumed, which depend on the consumption period daily and their duration [11].

$$V_{stok} = \left(\left(Q_{pmh} - Q_{cmh}\right) \times t_i\right) \times 25\% \tag{5}$$

with t_i is the longest duration of use of the tank (in hours), which is represented in **Figure 3**.

$$Q_{pmh} = \frac{V_{bio}}{24} \tag{6}$$

$$Q_{cmh} = \frac{V_{bio}}{\sum ti} \tag{7}$$

$$Q_{pmh} = 6.824/24 = 0.284375\ m^3/h$$

$$Q_{cmh} = 0.284/8 = 0.03554688\ m^3/h$$

$$V_{stok} = ((0.284\text{–}0.035) * 8)0.25$$

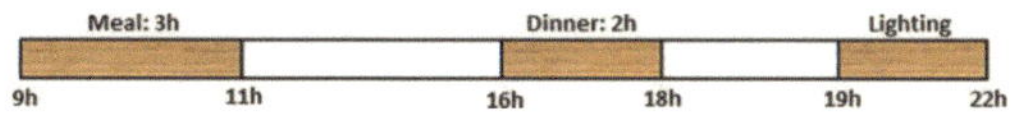

Figure 3.
Timeline of biogas use.

2.1.2.2 Diameter and length of gas pipe

This research is important to test the diameters suitable for our installation compared to those available on the market. The dimensioning of the diameters is very important for the safety of biogas appliances. It is obtained using the pressure drop equation (depending on the variables shown in the table below used in fluid mechanics [12].

$$J = \Delta H = \frac{\lambda \times Lc \times Ve^2}{D_{cg} \times 2g} \tag{8}$$

$$\geq D_{cg} = \frac{\lambda \times L_c \times {V_{eg}}^2}{J \times 2g} \tag{9}$$

$$\geq L_c = \frac{D_{cg} \times J \times 2g}{\lambda \times {V_{eg}}^2} \tag{10}$$

It is important to note that the pressure loss coefficient (λ) is a function of the Reynolds number, which depends on the dynamic viscosity (μ) of the fluid, its density, the diameter of the pipe, and the flow velocity [4].

NB: In cylindrical-shaped reactors, the acceptable pressure drop is small and is estimated at 10 mm water column.

$$R_e = \frac{\rho \times V_{eg} \times D_{cg}}{\mu} \tag{11}$$

$$V_{eg} = \frac{4 \times q_{mgc}}{3.14 \times {D_{cg}}^2} \tag{12}$$

$$q_{mgc} = q_{br} + q_l \tag{13}$$

We are going to use two maximum flow rate burners (q_{br}) of 0.9 m^3/h each and eight flow lamps (q_l) 0.14m^3/h each.

$$q_{mgc} = 2 \times q_{br} + 8 \times q_l \tag{14}$$

$$AN: q_{brm} = 0.9 \times 2 = 1.8\ m^3/h$$

$$q_{lm} = 0.14 \times 8 = 1.12\ m^3/h$$

q_{lm} is the maximum electrical flow
q_{brm} means the maximum kitchen flow

$$q_{mgc} = 1.8\ m^3/h + 1.12\ m^3/h$$

$$q_{mgc} = 2.92\ m^3/h$$

Indeed, the coefficient of friction is determined according to the flow regime of the fluid inside the pipe [4].

If Re < 2400 then one is in laminar mode corresponding to the line of Poiseuille:

$$\lambda = \frac{64}{R_e} \tag{15}$$

If $2000 < R_e < 3000$, we have a transition flow.
If $4000 < R_e < 105$ then the regime is turbulent:

$$\lambda = 0.316/R_e^{(1/4)} \tag{16}$$

2.2 Anaerobic digestion stage

Anaerobic digestion can be defined as the natural transformation of organic waste in the absence of air by anaerobic bacteria to produce the biogas, the carbon dioxide, and the digestate.

In this study, we simplified the anaerobic digestion steps in two and which are namely the acid fermentation (from hydrogenated to acetate) and the methane fermentation (methanogenated phase) to facilitate the digestion by quantifying only the volatile fatty acids and the biogas.

2.3 Modeling of cell growth

In anaerobic digestion, the growth rate of microorganisms can be evaluated by several models, the most used of the Monod model created in [11] and combines the net (specific) growth rate and the substrate concentration (Cs) by considering only two bacterial populations namely acidogenesis and methanogens. It represents the reaction rate of anaerobic bacteria for cell growth that depends on the concentration and is written as:

$$\mu_s = \mu_{cmax} \times \frac{C_s}{K_{ds} + C_s} \tag{17}$$

with

μ_s means the growth rate

K_{ds} is the half-saturation constant (in gram per liter (g/l)). It is the concentration where $\mu_s = \mu_{cmax}/2$.

μ_{cmax} (d^{-1}) corresponds to maximum growth rate.

Therefore, thanks to a model that can generalize the relationship between the growth of bacteria and their concentration, and the modeling will be carried out according to the type of bacteria using the kinetic properties developed by Amoco [2].

To better facilitate the separate modeling of the growth of microorganisms, Amoco model [2], in which two reactions based on acidogenesis reaction and methanogenesis reaction are considered, is used.

2.3.1 Change in acidogenic biomass

The degradation of the acetogenic biomass leads to the production of volatile fatty acids, which become the substrate of concentration (acetate (C_{s2})) and new biomass cells for a pH of 6 [2, 6, 13]. It is calculated using the Monod equation expressed below:

$$\mu_{s1} = \mu_{cmax1} \times \frac{C_{s1}}{K_{ds1} + C_{s1}} \tag{18}$$

At the start of the experiment, the acidogenic biomass is:

$$C_{s1} = \frac{Q_r\,(g)}{V_u(l)} \tag{19}$$

This equation is used to model the growth of a population of bacteria that feed on a substrate and whose maximum growth is limited.

2.3.2 Growth of methanogenic bacteria

To obtain the variation of the methanogenic biomass, it is assumed that the reactor operates at constant temperature and pH (anaerobic conditions). For the methanogenic simulation is shown by Gaston Nsavyimana [6] that the digester will be fed at the flow time of the experiment with varying amounts of acetate. Here, we will work with a half-reaction constant [2] where the concentration of volatile fatty acid (acetate) corresponds to 50% of the concentration of the initial substrate [13].

$$\mu_{s2} = \mu_{cmax2} \times \frac{C_{s2}}{K_{ds2} + C_{s2}} \tag{20}$$

2.3.3 Effect of organic acid inhibitor

Several models have been developed to study the effect of the inhibitor on anaerobic digestion. The first of them is mainly interested in methanogenesis under the hypothesis that it would be the limiting step in the overall anaerobic digestion process [14].

2.3.3.1 Effect of organic acid inhibitor on the performance of the acidogenesis biomass

This effect is modeled as a function of the acetic acid inhibition constant. The reaction rate is given by the following more synthetic model, called AM2.

$$\mu_{s1}{}' = \mu_{cmax1} \times \frac{C_{s1}}{K_{ds1} + C_{s1}}(1 - K_{1ac} \times C_{s1}) \tag{21}$$

with:
$\mu_{s1}{}'$ is the reaction rate of the acidogenesis biomass.

2.3.3.2 Effect of organic acid inhibitor on the performance of methanogenic bacteria

The inhibitory effect of organic acids on methane bacteria is modeled by including an inhibition term. It is in this sense that Haldane [13] proposed an equation called the Haldane equation to represent the inhibition of the growth of methanogens at high substrate concentrations.

The Haldane equation below is used to model the growth of bacteria [1, 13].

$$\mu'_{s2} = \mu_{cmax2} \times \frac{C_{s2}}{K_{ds2} + C_{s2} + \frac{(C_{s2})^2}{K_{i-2}}} \tag{22}$$

with

$$C_{s2} = \frac{C_{s1}}{100} \times 50 \tag{23}$$

The specific growth rate of methanogenic bacteria is chosen according to the temperature. Here, we have taken this value because the internal temperature of the pits is between 20°C and 35°C.

2.4 PH modeling

PH is a parameter that can have a negative impact on anaerobic digestion. It is regulated by volatile fatty acids (VFAs), the bicarbonates, and the ammonia. The inhibition caused by pH is modeled by the AMD1 model and expressed as [15].

$$I_{nib} = \frac{\left(1 + 2 \times 10^{0,5(pH^{+} - pH^{-})}\right)}{\left(1 + 10^{(pH^{+} - pH^{-})} + 10^{(pH^{-} - pH^{+})}\right)} \tag{24}$$

Where pH^+ and pH^- are obtained graphically by the **Figure 2** for an inhibition coefficient of 0.5. More, the values of pH^+ and pH^- are according to the values measured in the center of Ngoundiane represented in the **Table 2.**

So, we have pH^- = 6.2 and pH^+ = 7.8

And I_{nib} is the inhibition,

Infir is the infirmary,

P.A is the Pavillon A,

P.B means the Pavillon B,

B.S.P is the pedagogic service block,

L.R corresponds to the lecture room.

The graphic where the inhibition coefficient is the follower.

2.5 Influence of the temperature

The temperatures have an important place in the methanization. Here, we work in mesophyll with the temperature measured in the center of Ngoundiane. These temperatures are represented in the **Table 3.**

S.T	Infir	P.A	P.B	B.S.P	Tower	Resto	L.R
pH_{av}	7.30	7.01	7.02	7.40	7.42	6.64	7.45

Table 2.
PH measured in the center of Ngoundiane.

Hours (h)	8 h	10 h	12 h	14 h	16 h	18 h
T (°C)	29.5	28.2	28	27.5	30	34.5

Table 3.
Temperature of the septic tank.

2.5.1 Nitrification rate

The performance of a nitrification or denitrification reactor is due to the nitrate formation and nitration rates given, respectively by the following formulas [14].

$$V_{Nti}\ (T) = V_{Nti}\ \left(20^{\circ}C\right) \times 1.07^{(T-20)} \tag{25}$$

and

$$V_{Nta}\ (T) = V_{Nta}\ \left(20^{\circ}C\right) \times 1.04^{(T-20)}) \tag{26}$$

Where the nitrification and nitration formation rates are depended of the temperature and shown in the **Table 4**.

The rate of nitrification is between 7.5 mgN/l/h and 17.6mgN/l/h for temperatures ranging from 25°C to 29°C and for a pH that varies from 8.5 to 8.7 [16].

The kinetics of nitrification can explain an interruption (shunt) of nitrate in the reactor. It also provides an idea of their biodegradability because a decrease in this rate causes a variation in the nitrogen load at the inlet and incomplete nitrification. If we observe an equality of the speeds at the inlet and outlet, we can say that there is no degradation of nitrogen [16].

2.5.2 Kinetic coefficient of growth of bacteria by day

Temperature plays a primary role in anaerobic digestion. It promotes the activity and growth of bacteria. The kinetic coefficient of growth of microorganisms is modeled by the following relation [13]:

$$C_{cinet} = 0.013 \times T\text{–}0.129 \tag{27}$$

2.6 Modeling of actual biogas production

The methanization is not only the conversion of acetate in the presence of water but also the reduction of CO_2 by hydrogen into methane (CH_4) thanks to hydraulic bacteria (hemicellulolytic, anylolytic, and proteolytic bacteria).

$$CH_3COOH + H_2O \rightarrow CH_4 + H_2CO_3\ (70\%CH_4)$$
$$CO_2 + 4H_2 \rightarrow CH_4 + 2H_2O\ (30\%CH_4)$$

This biogas production can be modeled by the time-independent "steady-state" methods based on the mass of biodegradable organic matter and the potential for biogas production.

Phases	Acetogenesis	Methanogenesis
V_{Nti} (mgN/l/h) 20°C et pH = 7.2	22.2±5.1	10.9±1.2
V_{Nta} (mgN/l/h) 20°C et pH = 7.2	8.9±1.9	0

Table 4.
Nitrification rate in 20°C and pH =7.2.

2.6.1 Volume of produced methane

The volume production (P_v) also called technological yield is linked to the daily production of biogas (V_{bio}) and to the useful volume of the digester (V_u) by the following equation:

$$P_v = \frac{V_{bio}}{V_u} = \left(Nm_g{}^3/m_d{}^3\right) \tag{28}$$

2.6.2 Evaluating the produced methane

The methane flow is modeled by the following equation [3].

$$Q_{CH4} = V_{Reac} \times \mu_{s2} \times K_2 \times C_{s2} \tag{29}$$

The characteristics of the digester, such as temperature and organic load, accompany the operation in biogas production. It is in this sense that we made our studies with human wastes and at mesophilic temperatures (T).

2.7 Viscosity

Viscosity (V_{isc}) plays an important role in anaerobic digestion because it allows us to have an idea about the resistance to the movement of the fluid in its reactors.

According to Heléne Caillet, viscosity increases with increases with solid and volatile solid concentration but decreases with increased temperature due to enhanced mobility of polymer chains. Indeed, the viscosity can also be predicted from the pH by applying the equation, which is presented as follows [17]:

$$V_{isc}\ (Pa.s) = -0.0135pH^2 + 0.1965pH–0.667 \tag{30}$$

2.8 Investment cost of the digester

The investment cost of the methanizer is the sum between the acquisition and installation costs of the digester.

First, the construction cost is defined as the costs for labor written as [3].

$$C_{cost} = C_{VF/m^3} \times V_u \tag{31}$$

V_u is the capacity of the reactor.

Secondly, the acquisition costs of the digester (C_{ac}) represents the expenses related to the construction materials of the digester [6].

$$C_{ac} = 3.14 \times D_{aq} \times \rho_{gaz} \times H_{dig} \times C_{c/Kg} \tag{32}$$

and

$$D_{aq} = D_{dig} \times \left(2.5 \times D_{dig} + 1.2\right) \tag{33}$$

where

D_{aq} is the acquired diameter,

H_{dig} corresponds to digester height.

After, we make a research on the digester excavation cost (C_{exa}) in FCFA. The digester cost of digging the pit, $C_{exa,}$ is the expressed as:

$$C_{exa} = \frac{\pi}{4} \times \left(D_{digest} + 0.33 \times D_{digest}\right)^2 \times H_{digest} \times C_{VF/m^3} \quad (34)$$

Also, the coating cost represents the product of the necessary volume of reinforced concrete times its volume cost [4].

$$C_{coat} = 3.14 \times D_{coat} \times H_{dig} \times C_{VF/m^3} \quad (35)$$

$$D_{coat} = \left(D_{dig} + 0.165 \times Ddig\right) \times 0.165 \times D_{dig} \quad (36)$$

D_{coat} means the coating surface.

3. Results and discussions

The volume of the reactor is sufficient to fight against any overflow of organic matter. The capacity of the gasometer is

$$V_{stok} = 0.5\ m^3/h = 12m^3/d.$$

The values for gas pipe diameter and length can be found in **Table 5** for clarity.

Concerning the value of the resistance found in our study, it is more important than those found by Aline Lebranchu [18]. We will increase the thickness of the digester by applying coating to it in order to adjust its resistance. The resistance is expressed in kelvin per watt (K/W).

The variations in acetogenesis and methanogenesis biomasses in the anaerobic digestion of septic tank are represented in the **Table 6**.

This table is obtained using Excel. It helped us to better see the evolution of microorganisms within methanization represented in the **Figures 4** and **5** (obtained by the software: origin-pro).

Therefore, the results obtained on the modeling of methanogenic biomass are shown in the **Figure 6**.

For D_{cg} = 12.5 mm		For D_{cg} = 16 mm	
V_{eg} (m/s)	5.97868577	V_{eg}(m/s)	3.649150743
R_e	5813.972142	R_e	4542.165736
	R_e˃2400; turbulent regime		R_e˃2400; turbulent flow
λ	0.036188355	λ	0.038492076
L_c(m)	1.895905487	L_c(m)	6.124406929
Interpretation	L_c = 1.89 m is very small; 12.5 mm diameter is excluded	L_c = 6.12 m is sufficient; We take D_{cg} = 16 mm	

Table 5.
Determination of the diameter of gas pipes and their length.

							average
C_{s1} (g/l)	3	6	9	12	15	18	10.5
C_{s2} (g/l)	1.5	3	4.5	6	7.5	9	5.25
μ_{s1} (d^{-1})	0.356	0.55	0.671	0.754	0.814	0.860	0.668
μ_{s2} (d^{-1})	0.541	0.625	0.659	0.678	0.689	0.697	0.67
μ'_{s2} (d^{-1})	0.253	0.172	0.127	0.101	0.83	0.071	0.134
μ'_{s1} (d^{-1})	0.148	0.114	0.093	0.078	0.068	0.06	0.094

Table 6.
Variation in acetogenesis and methanogenesis biomasses in the anaerobic digestion of septic tank residues at the Ngoundiane site.

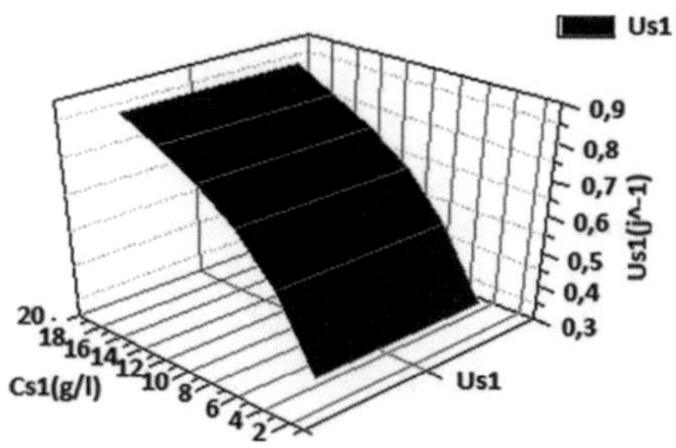

Figure 4.
The growth function of acidogenesis bacteria as a function of substrate concentration.

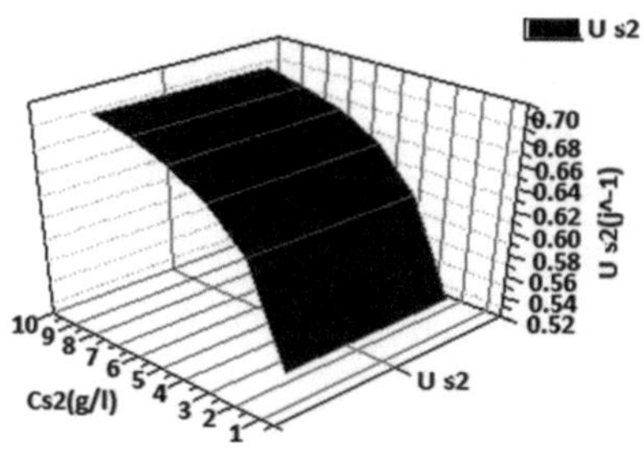

Figure 5.
The growth function of methanogenesis biomass as a function of substrate concentration.

In the absence of factors inhibiting the microbial mortality, the microbial growth rate $\mu_s (d^{-1})$ increases at the proliferation profile of the concentration of the substrate introduced into the anaerobic reactor in mesophyll, as shown in the **Figures 4** and **5**. The maintenance of the growth of microorganisms can be ensured by the addition of

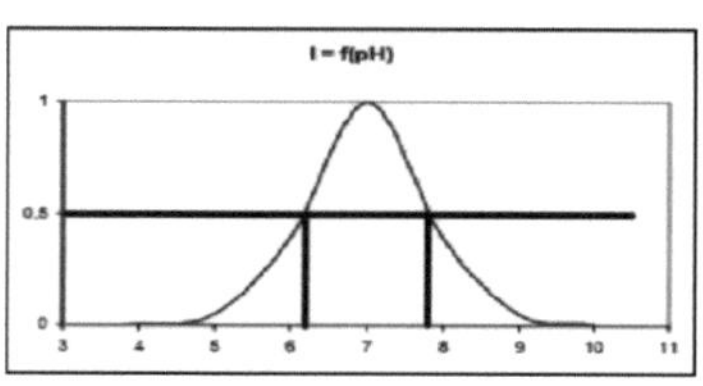

Figure 6.
Values of the inhibition coefficient of methanogenic bacteria as a function of Ph.

additives, such as substrates rich in sugar (carrots) and cooked foods rich in starch (rice) [19], the solution of sodium hydroxide (N_aOH) with a concentration of 0.5 mol/L, the sodium carbonate (Na_2CO_3) towards neutrality (pH = 7). These parameters increase the C/N ratio and stabilize the pH for the good progress of anaerobic fermentation [19, 20]. Our values are identical to those calculated by Laurent Lardon [2] which are between 0.2 d^{-1} and 0.8 d^{-1}.

About the function of acidogenesis and methanogen biomass taking into account inhibition, the **Figures** 7 and **8** show the operating diagrams of the system with inhibition for increasing values of organic matter entering the device.

Here, we find that the growth of microorganisms (μ_S) is a function of substrate. It increases to a certain concentration (3 g/liter for the substrate or organic matter, 2 g/liter for acetate) and then decreases. This decrease in the level of microbial biomass $\mu_s'(d^{-1})$ becomes more important by increasing the value of the concentration of the substrate (acetate). A high concentration of hydrogen can also inhibit the growth of acetogenic bacteria. This hydrogen comes from volatile fatty acids (VFAs) and from the simple substrate (C_S). Indeed, the removal of bacteria can be explained by the acceleration of their rate of division. On the other hand, a high concentration of acetate can inhibit the growth of hydrogenotrophic methanogenic bacteria. In fact, anaerobic digestion (anaerobic degradation) is accompanied by the production of intermediate substrates called volatile fatty acids. However, in the literature, a quantity of volatile fatty acids (VFAs) greater than 3000 mg/l leads to an inhibition of the methanization process of this substrate in accordance to our study as it is seen **Figure** 7. Our values are identical to those calculated by Laurent Lardon and Jonathan Hess [2, 14] which are between 0.05 d^{-1} and 0.35 d^{-1}.

Moreover, the inhibition linked to pH obtained on the calculation of this inhibition is 10^{-7}. It is very small and cannot affect the growth of anaerobic microorganisms.

Microorganisms need nitrogen for their cell structure.

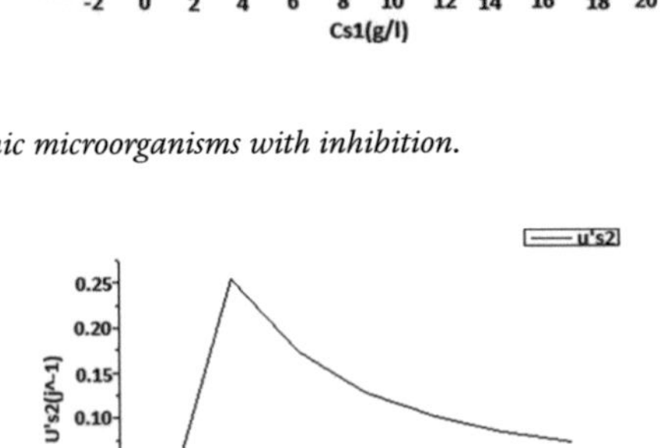

Figure 7.
The growth function of acidogenic microorganisms with inhibition.

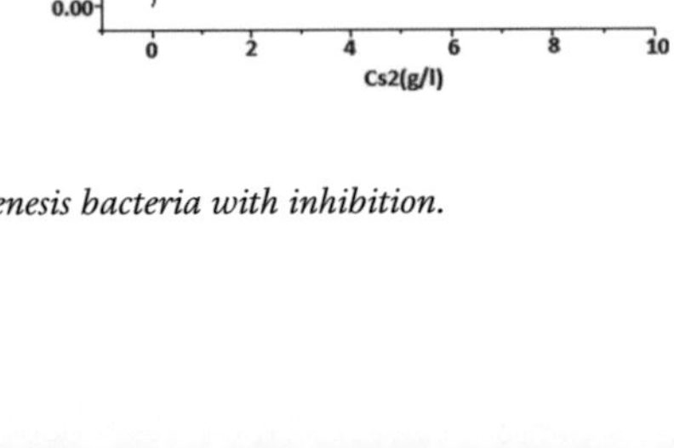

Figure 8.
The growth function of methanogenesis bacteria with inhibition.

In a nutshell, the nitrification rate obtained in our study is represented in the **Table** 7.

These resultants allow to plot the **Figure 9** represented the evolution of the nitrification rate compared to the temperature.

In **Figure 9**, we have shown the effect of temperature on the rate of nitrate formation, which becomes the substrate for methanogenic bacteria, allowing the production of biogas.

It shows that the rate of nitrification increases with temperature in mesophyll. According to the results obtained, the values found are higher than those calculated by Bourser. The range of 7.5 to 17.6 mgN/l/h for a temperature that varies between 25°C and 29°C. These values differ from the findings of Yenigün and Demirel's work [21]. Their work shows that the same anaerobic digestion carried out in mesophyll and thermo-phile does not result in the same amount of ammonia in the medium. Specifically, it resulted in 30-mg/liter and 200-20 mg/liter, respectively [21]. Rousseau et al. [22] state that nitrite and nitrate ions cause the pH to change and cause charge balances. Indeed, denitrification allows the production of CO_2 and N_2, which influences the quality of the biogas. At low concentrations, ammoniacal nitrogen in the free, non-ionized NH_3 or ionized NH_4^+ form [23] can neutralize VFAs, which helps maintain a neutral pH. It is generally accepted that a concentration between 50 and 200 mg/l stimulates anaerobic digestion [23]. In this case, the results obtained in our research favor the methanization of sludge. However, ammoniacal nitrogen can be toxic when the concentrations vary between 2 and 3 g/l for the bacterial population [17].

As for the volume production of biogas, the volume production P_v is 0.2 m^3biogaz/m^3 digester.

This result is identical to that found by Markowski et al, which is 0.17 m^3biogaz/m^3 digester [24].

V_{Nti}(T) acid	**37**	**38**	**39**	**42**	**44**	**49**	**43**
V_{Nti}(T) met	18	19	19	21	21	29	21

Table 7.
Variation of the nitrification rate.

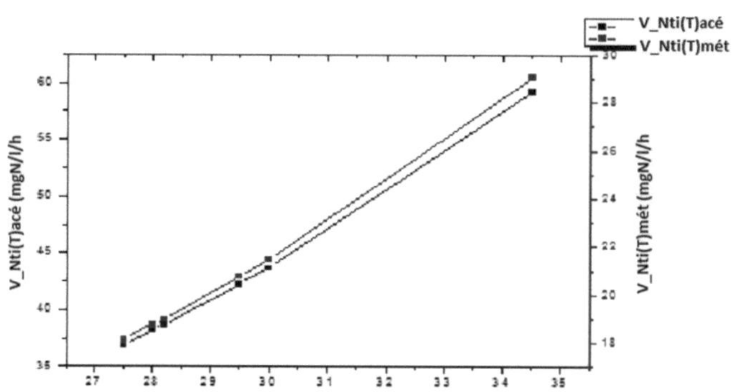

Figure 9.
Nitrification rate.

In the **Figure 8** is shown the evolution of the flow of biogas produced as a function of the acetate respectively without and with inhibition (**Figure 10**).

It is clearly noted that the production of biogas increases with the growth of acetate under normal conditions but decreases with the presence of inhibitor. To confirm these conclusions, we compare the evolution of VFAs and methane flow for these two curves. We noticed an evolution of VFAs which grow up to a certain concentration (3 g/l) and decrease causing a decrease in the flow of biogas thanks to the inhibition of methanogenic bacteria by these VFAs. In addition, the production of methane depends on the growth rate of bacteria.

As part of the viscosity, the **Figure 11** represents the viscosity as a function of pH.

The analysis of the curve allows us to see the change in viscosity as a function of the pH. It is noted that the viscosity increases rapidly with the acidic pH (6) and becomes stable for the pH, which are around neutrality (pH = 7) in accordance with the results found by Culioli et al. [25]. The behavior of effluents within reactors can depend on the solid concentration, the age of the sludge, its type, and the temperature which characterize the pH as it can see in the work done by P. Rousseau et al., [22].

The average value found in our study for viscosity is evaluated at 0.046 Pa.s.. We saw that our result is in the range of those found by those of Laporte [26], Koblan Wilfried et al. [27], and Heléne Caillet [17], which range from 0.0010 Pa.s to 1.29 Pa.s, from 0.018 to 1.3 Pa.s, and from 0.018 Pa.s to 0.445 Pa.s, respectively.

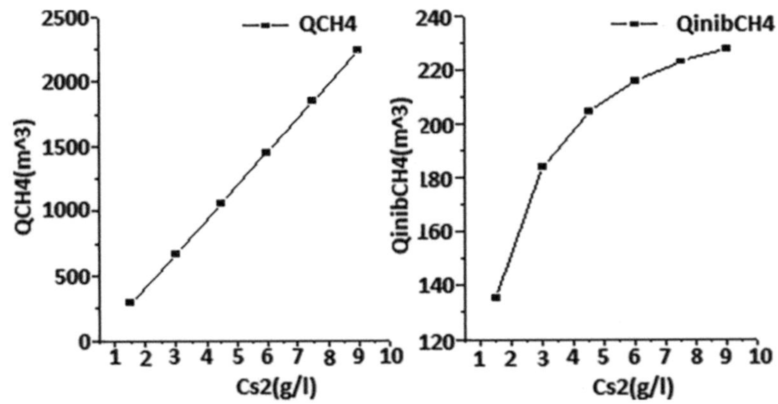

Figure 10.
Evolution of the flow of biogas produced as a function of the acetate, respectively without and with inhibition.

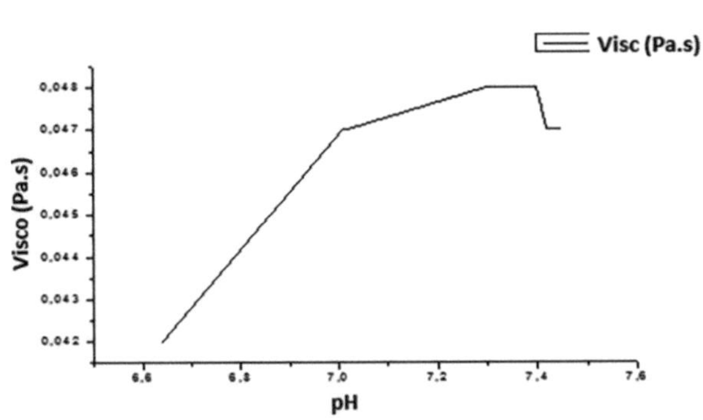

Figure 11.
Representation of viscosity as a function of pH.

About the construction cost of our digester, the total investment cost comes to 1156860FCFA.

4. Conclusion

Throughout this study, we tried to develop the solutions related to the recovery of residues from septic tanks. We also investigated the links between feed, the bacterial activity, and the biogas production using two-step mathematical modeling with and without inhibition to ensure the process stability and regulate the quality of the biogas produced for recovery. Indeed, it has been shown that the value of the parameters influences the rate of biogas produced. Likewise, the increase of the concentrations of the substrates at the inlet of the reactor for a model without biomass mortality (without inhibition) favors the increase of the biogas produced. The study showed that the model without inhibition produces more biogas than the model with inhibition.

The obtained results in our research show clearly that for fixed entry concentrations, there is a maximum level of methane obtained with the weakening of the inhibition of hydrogenotrophic methanogenic bacteria marked by an increase in the concentration of acetoclastic methanogenic bacteria and that of acetate that is quite low.

Upon validation, our models identified a link between feed, digester allkalinity, and biogas quality. In addition, they make it possible to detect inhibition regimes marked by a decrease in the quantities studied.

Author details

Ibrahima Niane[1*], Amadou Diao[2], Fakoro Souleymane Dia[1] and Senghane Mbodji[1]

1 Research Team in Renewable Energies, Materials and Laser of Department of Physics, UFR SATIC – Alioune DIOP University of Bambey, Bambey, Senegal

2 Department of Physics, Faculty of Science and Technology – Cheikh Anta Diop University, Laboratory of Semiconductors and Solar Energy, Dakar, Senegal

*Address all correspondence to: nianeibrahima70@gmail.com

References

[1] Kokoh E. Anaerobic lagooning: modeling combining primary reduction and anaerobic degradation [doctoral thesis]. 2009. p. 128

[2] Lardon L. Modeling of Biofilms of Anaerobic Digestion by Multi-Agent System, Statement of DAA, Agro TIC2001. p. 79

[3] Abunde Neba F, Asiedu NY, Addo A, Morken J, Osterhus SW, Seidu R. A coupled modeling design and investment parameters for optimal operation of methane bioreactors: Attainable region concept approach. Article de revue. Renewable Energy, ISSN: 09601481. 2020;**148**:1054-1064

[4] Amededjihunde Hypolite Hounnou, Chamagne Didier Fréderic Dubas, Antoine Cokou Vianou. Technical-economic dimensioning of penstocks for micro-hydroelectric power stations using generic algorithms. Journal de la Recherche Scientifique de l'Université de Lomé (Togo). 2019;**21**(1):113-122

[5] Lansing S, Botero RB, Martin JF. Waste treatment and biogas quality in small scale agricultural digestors. Bioresource Technology. 2008;**99**: 5881-5890

[6] Nsavyimana G. Modeling of physical and biological processes in septic tanks and routes for recovering sludge drain: Application A Bujumbura-Burundi [thesis]. University of Liège and Faculty of Sciences: "Sanitation and Environment" Unit, 2014-2015

[7] Coulibaly L, Jen-Marie, Oattara P, Tiho S. Potential in agopastoral biogas and human excreta from the watershed of the Sassandra River (Cote d'Ivoire). Journal of Sustanaible development in Africa. 2012;**14**(4)

[8] Kouawa T. Treatment of Faecal Sludge by Drying Beds in a Sudano-Sahelian Climate Other [thesis]. University of Strasbourg; 2016. French. NNT: 2016STRAD018. p.20

[9] Tedjani S, Nouri S. Analytical study and numerical simulation of one-dimensional heat diffusion [thesis for master]. University of Science and Technology Houari Boumediene; 2020. pp. 6-14

[10] Mactar F. Problem of Conduction in Permanent Regime. Thermal Transfer Course. Alioune DIOP University of Bambey UFR Satic. Department of Physics; 2015

[11] Aboubakar, Boli Z, Mbofung CMF. Study of the biogas potential of animal waste: Cattle dung and poultry droppings from a zootechnical center in Maroua - Cameroon. Revue des Energies Renouvelables. 2016;**19**(3):449-450-452-453-454

[12] Tedjani S, Nouri S. Analytical study and numerical simulation of one-dimensional heat diffusion [thesis for master]. University of Science and Technology Houari Boumediene, Faculty of Physics; 2020. pp. 6, 7, 14. Available from: https://www.researchgate.net/publication/341940123

[13] Eynard J. Modeling, dynamic optimization and control of a methanizer by anaerobic digestion [thesis]. 2007. p. 81

[14] Hess J. Modeling the Quality of Biogas Produced by a Methanogen Fermenter and Regulation Strategy in View of its Valuation. University of Nice – Sophia Antipolis – UFR Sciences Doctoral School Information and

Communication Sciences and Technologies. pp. 69-123

[15] Rouez M. Anaerobic Degradation of Solid Waste: characterization, Influencing Factors and Models [thesis order number 2008SAL0061]. National Institute of Applied Sciences of LYON, Doctoral School of Chemistry; 2008. pp. 76-79 95–99

[16] Rousseau P. Study, modeling and optimization of a process for treating pig slurry by coupling anaerobic digestion and biological nitrogen treatment. Environmental Engineering. Doctorate in Chemistry, Rennus University. 2009: 3870

[17] Caillet H. Optimization of the Anaerobic Digestion of Sugar Cane Vinaisse: Modeling and Experimentation at the Laboratory Scale. [doctoral thesis]. University of Reunion, N 542; 2019. p. 243. Available from: https://tel.archives-ouvertes.fr/tel-02894979

[18] Lebranchu A. Study of the Methanation Process in an Anaerobic Digester on a Pilot Scale: Impact Mixing and Gas-Liquid Transfer on the Performance of Biogas Production. Genius of Processes. University of Lorraine; 2017 French. NNT: 2017LORR0336. Tel-01916424

[19] Djaafri M, Kaloum S, Khelafi M, Tahri A, Salem F, Kaidi K, et al. Methanization of organic waste from university residence of the African university of the city of Adrar southwest of Algeria, ISSN: 2028-2508. Journal of Materials and Environmental Science. 2014:2484-2488

[20] Ahmed T, Mohammed D, Mostefa K, Kamel K, Slimane K, Zohra B. Production of biogas from potato waste. In: Conference Paper. 2016

[21] Laura A. Study of scientific and technological obstacles for the understanding and optimization of the discontinuous dry methanization process of by-products of agricultural origin [PhD thesis]. University of Technology of Compiegne Specialty Process Engineering; 2016. pp. 37-38

[22] Rousseau P, Steyer P, Volokey EIP, Bernet N, Béline F. Combined anaerobic digestion and biological nitrogen removal for piggery wastewater treatment: A modelling approach. Water Sciences, Techno. 2008;**5**(2):21-28

[23] Liu T, Sung S. Ammonia inhibition on thermophilic Aceticlastic Methanogenesis. Journal Water Sciences and Technology. 2002;**45**(10):113-120

[24] Markowski M, Treneusz A, Krzemieniewski B. Optimizing low-temperature biogas production from biomass by anaerobic digestion. Renewable Energy. 2014;**69**:219

[25] Culioli J, Bon JP, Maubois JL. Study of the Viscosity of Retentates and Pre-Cheeses Obtained after Treatment of Milk by Ultrafiltration on a Membrane. Vol. 538. Rennes: Laboratory of Dairy Research and Technology, INRA; January 1974. pp. 11-12

[26] M. Laporte. Study of two-phase flow at the millimeter and micrometric scale: applications to foams [doctoral thesis]. 2014

[27] Koblan Wilfried E, Benchabane A, Bekkour karim. Comparative study of the rheological properties of polyethylene oxide (PEO) bentonite mixture of different molecular masses. In: 18th French Congress of Mechanics, Grenoble, 27–31 août 2007, ffhal-03361696. pp. 1-6

Section 3

Sludge Management Systems – The Gutter-to-Good Approaches

Chapter 15

Good Practices in Sludge Management for Non-Annex Mediterranean Countries

Firas Fayssal

Abstract

Sludge management is a critical aspect of wastewater treatment, as it deals with the proper handling, treatment, and disposal of the residual material generated during the process. For non-Annex Mediterranean countries, effective sludge management is vital to protect human health, preserve natural resources, and mitigate environmental impacts. This article examines the challenges faced by these nations in sludge management and highlights the best practices adopted to ensure sustainable and environmentally responsible sludge handling.

Keywords: sludge management, non-Annex Mediterranean countries, energy recovery, solar sludge drying, wastewater treatment

1. Introduction

The Mediterranean region is home to a diverse group of countries, each facing unique challenges in sludge management. Non-Annex Mediterranean countries are those not listed in Annex I of the Barcelona Convention, which includes countries with major economic development. Sludge management in non-Annex Mediterranean countries has often been overlooked, resulting in inadequate practices that pose significant environmental and health risks. This article delves into the good practices that can be implemented to address these challenges effectively.

2. Challenges in sludge management for non-annex Mediterranean countries

1. Limited Financial Resources: Non-Annex Mediterranean countries may face financial constraints, making it challenging to invest in advanced sludge management technologies and infrastructure [1].

2. Institutional Capacity: Inadequate institutional capacity and lack of technical expertise can hamper the development and implementation of robust sludge management strategies [2].

3. Policy and Regulatory Frameworks: The absence of comprehensive policies and regulatory frameworks specific to sludge management may lead to unregulated and improper practices.

4. Awareness and Public Participation: Lack of awareness among the public and stakeholders about the importance of proper sludge management can hinder progress in this field [3].

3. Best practices in sludge management

3.1 Source reduction and pretreatment

One of the fundamental approaches to effective sludge management is source reduction. Implementing pretreatment techniques at the source helps reduce the volume of sludge generated, thereby easing the burden on treatment facilities [4].

3.2 Anaerobic digestion

Anaerobic digestion is a widely adopted practice for stabilizing and reducing the volume of sludge. It involves the breakdown of organic matter in the absence of oxygen, producing biogas that can be utilized as a renewable energy source [5].

3.3 Composting

Composting is another sustainable approach that converts sludge into a nutrient-rich product, suitable for agricultural use. When done correctly, composting can help improve soil quality and reduce the need for chemical fertilizers [5].

3.4 Energy recovery

Efficient sludge management includes harnessing the energy potential within sludge. Various technologies, such as biogas recovery and thermal conversion, can be employed to extract valuable energy resources [6].

3.5 Sludge dewatering

Dewatering is a crucial step in sludge management that reduces its volume and makes it easier to handle and transport [7]. Mechanical, thermal, and chemical methods are commonly used for dewatering purposes [8].

3.6 Nutrient recovery

Sludge often contains valuable nutrients like phosphorus and nitrogen. Implementing nutrient recovery technologies can help extract these nutrients, contributing to a circular economy and reducing dependency on external fertilizer sources [9].

3.7 Safe disposal and reuse

Final disposal of sludge must be carried out responsibly to prevent environmental contamination. Landfilling and ocean dumping should be avoided, and instead, sludge should be appropriately treated or reused in a safe manner [10].

4. Technological advancements

4.1 Mobile sludge treatment units

Mobile sludge treatment units can be particularly beneficial for non-Annex Mediterranean countries with limited infrastructure. These units can be transported to various locations for on-site treatment, reducing logistical challenges [11].

4.2 Decentralized treatment systems

Decentralized treatment systems offer an alternative to large centralized plants, allowing for more efficient sludge management in smaller communities and rural areas [12].

4.3 Sludge-to-energy solutions

Advancements in sludge-to-energy technologies enable the efficient extraction of energy from sludge, contributing to renewable energy generation [13].

5. Policy and governance recommendations

5.1 Develop comprehensive sludge management policies

Non-Annex Mediterranean countries should establish clear and comprehensive policies and regulations specific to sludge management to ensure environmentally responsible practices [14].

5.2 Capacity building and training

Investing in capacity building and training programs for personnel involved in sludge management will improve technical expertise and enhance institutional capacity [15].

5.3 Financial support and incentives

Governments and international organizations should provide financial support and incentives to promote the adoption of sustainable sludge management practices [16].

5.4 Public awareness and participation

Raising awareness about the importance of proper sludge management and involving the public in decision-making processes will foster a sense of ownership and responsibility [17].

6. Case studies

This section presents case studies from various non-Annex Mediterranean countries that have successfully implemented good sludge management practices. It highlights their challenges, strategies, and outcomes, serving as practical examples for others to follow [18].

6.1 Scientific and historical description of non-annex Mediterranean countries

The Mediterranean Sea, enclosed by three continents - Europe, Asia, and Africa, is bordered by numerous countries that share cultural, historical, and ecological ties. Non-Annex Mediterranean countries refer to those nations that are not listed in Annex I of the Barcelona Convention, which includes major economic powers. This scientific and historical description aims to provide an overview of these diverse countries, exploring their geographical characteristics, historical backgrounds, and cultural influences [19].

6.1.1 Geographical features

The Non-Annex Mediterranean countries encompass a wide range of landscapes, from coastal plains to mountainous regions. They share the Mediterranean climate, characterized by hot, dry summers and mild, wet winters. The countries have unique flora and fauna adapted to the region's specific conditions, including numerous endemic species.

- Coastal Areas: The Mediterranean coastline is rich in natural beauty and cultural heritage. Non-Annex Mediterranean countries boast stunning beaches, picturesque harbors, and ancient port cities that have been centers of trade and commerce for millennia.

- Islands: Several Non-Annex Mediterranean countries are home to beautiful islands, each with its distinct history and charm. These islands often showcase unique biodiversity and are popular tourist destinations.

- Mountain Ranges: Many countries in this region feature majestic mountain ranges that have played significant roles in shaping historical events and providing diverse ecosystems and natural resources.

6.1.2 Historical background

The history of the Non-Annex Mediterranean countries is a tapestry woven with threads of ancient civilizations, conquests, and cultural exchanges. The Mediterranean has been a crossroads of civilizations for thousands of years, influencing the development of these countries.

- Ancient Civilizations: The Mediterranean basin is often referred to as the cradle of civilization due to the presence of numerous ancient cultures. From the Phoenicians and Egyptians to the Greeks and Romans, these civilizations left an indelible mark on the region, contributing to art, architecture, philosophy, and governance.

- Medieval and Islamic Period: During the Middle Ages, the Mediterranean witnessed the rise of powerful empires, such as the Byzantine Empire and the Islamic Caliphates. The exchange of knowledge, trade, and cultural practices flourished during this period.

- Maritime and Colonial History: The Mediterranean Sea has long been a hub of maritime activity, with seafaring nations like Venice, Genoa, and Portugal dominating trade routes and establishing colonies in the region and beyond.

- Ottoman Empire and Beyond: The expansion of the Ottoman Empire significantly shaped the history of several Non-Annex Mediterranean countries. Its influence persisted until the 20th century when many countries gained independence and established modern nation-states.

6.1.3 Cultural influences

The cultural diversity of the Non-Annex Mediterranean countries is a result of the intermingling of different civilizations throughout history. These countries have inherited a mosaic of languages, traditions, religions, and culinary delights.

- Language and Linguistic Diversity: Each country in the region boasts its language or regional dialect, reflecting the historical influences of neighboring cultures. Arabic, Turkish, Greek, and other languages enrich the linguistic tapestry of the Mediterranean.

- Religion and Spirituality: Religion has played a vital role in shaping the identity of Non-Annex Mediterranean countries. Islam, Christianity, and Judaism are prominent faiths that coexist and have influenced each other's practices and beliefs.

- Culinary Traditions: Mediterranean cuisine is renowned for its diversity, emphasizing fresh ingredients and unique flavors. Each country offers its culinary specialties, such as Spain's paella, Italy's pasta, and Greece's moussaka.

- Arts and Architecture: The rich cultural heritage of these countries is reflected in their arts and architectural wonders. From ancient temples and amphitheaters to stunning mosques and cathedrals, the Mediterranean is a treasure trove of historical landmarks.

6.1.4 Environmental challenges

Despite the region's natural beauty, Non-Annex Mediterranean countries face numerous environmental challenges, including pollution, overfishing, habitat destruction, and water scarcity. Collaborative efforts are needed to address these issues and promote sustainable development.

The Non-Annex Mediterranean countries are a fascinating amalgamation of diverse cultures, histories, and environments. Their shared geographical features, historical legacies, and cultural influences have shaped the unique identity of each nation. By embracing sustainable practices and preserving their natural and cultural heritage, these countries can build a prosperous and harmonious future for generations to come [20].

6.2 All wastewater legislations regarding Mediterranean Sea with description

The wastewater legislations regarding the Mediterranean Sea primarily revolve around the Barcelona Convention for the Protection of the Marine Environment and the Coastal Region of the Mediterranean. The Convention, adopted in 1976, serves as a key legal framework for cooperation among Mediterranean countries to address environmental issues, including wastewater management. It has been amended and updated over the years to adapt to new challenges and developments in the region [21].

6.2.1 Barcelona convention for the protection of the marine environment and the coastal region of the Mediterranean (1976)

The Barcelona Convention aims to protect and improve the Mediterranean marine environment and coastal areas, prevent and reduce pollution, and promote sustainable development in the region. It sets out general obligations for the contracting parties to take measures and cooperate on environmental protection, including wastewater management. Key aspects of the Convention related to wastewater include:

a. Wastewater Treatment: The Convention encourages the contracting parties to establish and implement wastewater treatment facilities and measures to prevent pollution from domestic, industrial, and other sources.

b. Pollution Reduction: Contracting parties are required to take necessary measures to prevent, abate, and combat pollution from all sources, including domestic wastewater discharges.

c. Monitoring and Reporting: Regular monitoring and reporting of wastewater discharges and their impact on the marine environment are essential components of the Convention to assess the effectiveness of pollution reduction measures.

6.2.2 Mediterranean strategy for sustainable development 2016-2025

The Mediterranean Strategy for Sustainable Development is a regional framework aimed at promoting sustainable development practices in the Mediterranean countries. The strategy covers various aspects, including the management of water resources and wastewater treatment. Its objectives related to wastewater include:

a. Integrated Management of Water Resources: The strategy encourages the integrated management of water resources to ensure the efficient and sustainable use of water, including wastewater reuse and recycling.

b. Capacity Building: The strategy emphasizes the need to strengthen the capacity of Mediterranean countries to implement sustainable wastewater management practices.

c. Promotion of Best Practices: It calls for the exchange of knowledge and promotion of best practices in wastewater management among the countries in the region.

6.2.3 National legislatio

Each country bordering the Mediterranean Sea may have its national legislation and regulations related to wastewater management. These laws often align with the principles and objectives of the Barcelona Convention and regional strategies. National legislation typically covers aspects such as wastewater discharge standards, treatment requirements, monitoring, enforcement, and penalties for non-compliance.

It's important to note that the information provided here is based on the situation as of September 2021. Wastewater legislations and environmental policies are subject to change and continuous improvement, as countries work together to address emerging environmental challenges. For the most up-to-date information on wastewater legislations regarding the Mediterranean Sea, it is recommended to refer directly to official sources from the Barcelona Convention Secretariat and respective national environmental agencies of Mediterranean countries.

6.3 Sludge management in Mediterranean countries

Sludge management in Mediterranean countries is a critical aspect of wastewater treatment, as it deals with the proper handling, treatment, and disposal of the residual material generated during the treatment process. Effective sludge management is essential to protect public health, preserve natural resources, and mitigate environmental impacts. However, sludge management in some Mediterranean countries faces various challenges, while others have made significant progress in adopting best practices [22].

6.3.1 Challenges in sludge management

1. Infrastructure and Technology: Many Mediterranean countries, especially non-Annex countries, may face limitations in infrastructure and technology for proper sludge management. Inadequate treatment facilities and outdated technologies can hinder the efficient handling and treatment of sludge.

2. Regulatory Framework: The absence of comprehensive and well-enforced regulations specific to sludge management can lead to inconsistent practices and improper disposal methods, posing environmental and health risks.

3. Funding and Financial Constraints: Sludge management projects often require significant investments in infrastructure and technology. Some countries may face financial constraints, making it challenging to allocate adequate funds for upgrading and maintaining sludge treatment facilities.

4. Public Awareness and Participation: Lack of awareness among the public and stakeholders about the importance of proper sludge management can result in minimal support for initiatives and investments in improving sludge treatment practices.

6.3.2 Best practices in sludge management

1. Source Reduction and Pretreatment: Implementing source reduction practices can significantly reduce the volume of sludge generated. Pretreatment techniques can also help in the removal of harmful substances, reducing the environmental impact of sludge disposal.

2. Anaerobic Digestion: Anaerobic digestion is a widely adopted practice for stabilizing sludge and producing biogas, a renewable energy source. This process reduces the volume of sludge and provides an additional benefit of energy recovery.

3. Composting: Composting is a sustainable approach that converts organic sludge into nutrient-rich compost suitable for agricultural use. Properly managed composting can help in soil improvement and nutrient recycling.

4. Energy Recovery: Efficient sludge management includes the extraction of valuable energy resources from sludge through processes like biogas recovery and thermal conversion.

5. Sludge Dewatering: Dewatering is essential to reduce the volume of sludge, making it easier and more cost-effective to handle and transport. Mechanical, thermal, and chemical dewatering methods are commonly used.

6. Nutrient Recovery: Sludge often contains valuable nutrients like phosphorus and nitrogen. Implementing nutrient recovery technologies can help extract these nutrients for reuse, reducing dependency on external fertilizer sources.

7. Safe Disposal and Reuse: Final disposal of sludge must be carried out responsibly to prevent environmental contamination. Landfilling and ocean dumping should be avoided, and instead, sludge should be appropriately treated or reused in a safe manner.

6.3.3 Case studies

Several Mediterranean countries have implemented successful sludge management practices:

1. Spain: Spain has made significant progress in sludge management, adopting technologies like anaerobic digestion and promoting the reuse of treated sludge in agriculture.

2. Italy: Italy has invested in advanced sludge treatment facilities and promotes composting and nutrient recovery from sludge to enhance sustainability.

3. Greece: Greece has been working on enhancing its wastewater treatment infrastructure, including sludge management, to comply with EU directives and improve water quality.

Sludge management in Mediterranean countries is a complex and challenging task, but implementing best practices is crucial for sustainable development and environmental protection. By investing in advanced technologies, promoting public awareness, and strengthening regulatory frameworks, Mediterranean countries can ensure responsible sludge management and contribute to a cleaner and healthier marine environment. Collaboration and knowledge exchange among nations will play a key role in overcoming common challenges and achieving effective sludge management practices throughout the region.

6.4 Case study: sludge management in Morocco

Morocco, located on the northwestern coast of Africa, has a diverse geography and is bordered by both the Atlantic Ocean and the Mediterranean Sea. The country faces significant challenges in sludge management due to rapid urbanization, industrial growth, and increased wastewater generation. This case study examines the status of sludge management in the Moroccan Mediterranean coast and explores the initiatives taken to address the associated environmental and health concerns.

6.4.1 Wastewater treatment infrastructure

In recent years, Morocco has made notable progress in improving its wastewater treatment infrastructure, particularly in the coastal regions. Many cities and towns along the Mediterranean coast have established wastewater treatment plants to handle the increasing volume of sewage generated by population growth and economic development.

6.4.2 Challenges in sludge management

Despite efforts to enhance wastewater treatment, sludge management remains a challenging aspect of the process. Some key challenges include:

a. Volume and Composition: The sheer volume of sludge generated can overwhelm existing treatment facilities. Moreover, the composition of sludge can vary depending on the type of wastewater treatment used, making it necessary to tailor sludge management practices accordingly.

b. Disposal Practices: In some areas, improper disposal practices, such as open dumping or land spreading, have been observed, leading to environmental pollution and health risks.

c. Limited Resources: Financial constraints and a lack of advanced technologies can hinder the implementation of effective sludge management practices.

6.4.3 Initiatives and best practices

Despite challenges, Morocco has taken significant steps to address sludge management issues on its Mediterranean coast:

a. Sludge Treatment Facilities: Several wastewater treatment plants along the Mediterranean coast have incorporated sludge treatment facilities, such as anaerobic digesters, to reduce the volume of sludge and recover biogas for energy production.

b. Composting and Land Application: In some regions, treated sludge is composted and used as a soil amendment in agricultural applications, contributing to sustainable agriculture and reducing dependence on chemical fertilizers.

c. Nutrient Recovery: Morocco has shown interest in adopting nutrient recovery technologies to extract valuable nutrients from sludge, which can be recycled for agricultural purposes.

d. Public Awareness and Education: Efforts have been made to raise public awareness about the importance of responsible sludge management and the potential benefits of proper treatment and reuse.

6.4.4 Regional cooperation

As a signatory of the Barcelona Convention, Morocco participates in regional cooperation efforts aimed at protecting the marine environment of the Mediterranean Sea. Collaborative initiatives help share knowledge, best practices, and technological advancements in sludge management among Mediterranean countries.

6.4.5 Results

Sludge management in the Moroccan Mediterranean coast is an ongoing challenge, given the increasing wastewater generation and limited resources. However, Morocco has shown commitment to improving its wastewater treatment infrastructure and adopting best practices in sludge management. Continued efforts in upgrading treatment facilities, promoting sustainable sludge management practices, and enhancing public awareness will be crucial to ensure a cleaner and healthier Mediterranean Sea coastline for both present and future generations.

6.4.5.1 Solar sludge drying project in Nador

Presentation: Grand Nador has a WWTP with a nominal capacity of 20,600 m^3/d and 6800 kg BOD5/d for a total connectable population of 230,000 inhabitants. The capacity is not sufficient to cope with the increase in population and the development of tourist activities in the Atalayoum peninsula, the MR P33 program provides for its extension.

The initial project to extend the STEP plans to increase its capacity to 25,000 m^3/d and 9300 kgBOD5/d. A proposal aimed at replacing this extension with the construction of a second STEP, which would be located near the Atalayoum peninsula and would have a capacity of 7000 m^3/d and 2600 kg BOD5/d, was presented to JICA in April 2016.

Regarding the drying of sludge, the project outlined in the 2012 preparatory study provides for dehydration followed by stabilization with lime and landfill of the sludge, a solution like that implemented in the existing WWTP.

The new project proposed by UNDP consists of replacing the liming of sewage sludge from Grand Nador with solar drying in a greenhouse. If a second WWTP is built in Atalayoum, this sector would be extended to the 2 WWTPs.

Solar sludge drying uses the principle of an agricultural greenhouse, supplemented by a ventilation system and sludge mixing. The greenhouse effect is the walls of the greenhouse and its roof, transparent to solar radiation, allow the heat of the sun to penetrate but do not transmit infrared rays. Solar sludge drying uses this concept to accelerate the process of removing water from dewatered sludge. Heating the surface of the sludge, due to solar radiation and infrared radiation emitted by the transparent material of the greenhouse, increases the water vapor pressure on the surface of the sludge and therefore promotes evaporation. A flow of fresh air generated by a fan and admitted into the greenhouse allows the water vapor resulting from drying to be evacuated.

This technology makes it possible to obtain without the use of fossil energy a tripling of the dryness measured at the centrifugation outlet, with a final sludge dryness rate of 70 to 75%.

The annual volume of sludge produced is 11,000 tons of dehydrated and limed sludge with a dryness rate of 25%, according to the annual operating reports for 2013 and 2014. Future production would be approximately 50% higher than current production.

Justification: The current sludge treatment sector requires significant consumption of gas oil for the evacuation of sludge to the landfill, 22 km, as for the supply of lime. On the other hand, the production of lime is carried out by calcination of limestone at 900°C and therefore also uses fossil fuels. Thus, the sludge treatment process currently used in Nador is not optimized about the endeavor against global warming.

The proposed project will therefore contribute to the reduction of GHG emissions by:

calculation JL 09 09 16		
Indirect CO_2 emissions linked to lime manufacturing		
BOD5 treated	9400	kg/d
Estimation sludge production	10,340	kg MS/d
Estimation sludge production	3774.1	tMS/y
Dryness rate of dewatered sludge	18	%
Lime treatment rate	9.33	%
Consumption of slaked lime	352.2	t/y
CO_2 Eq for slaked lime	0.846	t eq. CO_2/t heat
Total CO_2 Eq	298	t eq. CO_2/y
Total C Eq	81	t eq C/y
Direct CO_2 emission linked to the liming operation (dosing, mixing, etc.)		
Estimation of sludge production to be treated	20,967	t/an
Equi CO_2 for liming (large installation)	0.1219	kg eq CO_2/t sludge
Total CO_2 Eq	3	t eq CO_2/y
Total C Eq	1	t eq C/y
Direct CO_2 emissions linked to the transport of lime		
Tonnage	352.2	t/y
Nb of Trucks	30.31	Per y
Distance 1 way	75	km
Tot CO_2 Eq véhicule full empty	4.9	t equi CO_2/y
Tot C Eq véhicule full empty	1.3	t equi C/y

Table 1.
Impact of liming in terms of CO_2 emissions [19].

- The abandonment of lime in the sludge treatment process;
- Reduction, without recourse to fossil fuels, of the volume of sludge to be evacuated; thus, the final volume of sludge would only represent 30–40% of the current volume and would allow a reduction in the same proportions of the volume of diesel used (**Table 1**).

7. Conclusion

Effective sludge management is vital for non-Annex Mediterranean countries to safeguard public health, protect the environment, and promote sustainable development. By adopting the good practices discussed in this article and addressing the unique challenges they face, these nations can pave the way for a cleaner and healthier future.

Author details

Firas Fayssal
Saint Joseph University – ESIB, Beirut, Lebanon

*Address all correspondence to: firas.fayssal@usj.edu.lb; f_firas@hotmail.com

References

[1] Clauson-Kaas J, Poulsen TS, Jacobsen BN, Guildal T, Wenzel H. Environmental accounting – Decision support tool in WWTP operation. Water Science and Technology. 2001;**44**:25-30

[2] 2019 Energy Balances. US: United Nations Publications; 2021

[3] Clean Development Mechanism. CDM Methodology. US: UN Framework Convention on Climate Change; 2013

[4] Climate Change and Water. UN-Water Policy Brief. UN Water; 2019

[5] Climate Change Information Kit. United Nations Environment Programme. US: UNFCCC & UNEP; 1999

[6] Climate Change MITIGATION. ISO14064 International Organization for Standardization ISO. US: ISO; 2019

[7] Cost Aspects of Wastewater Treatment. United Nations Environment Programme – UNEP. US; 2005

[8] Groundwater Making the Invisible Visible. WWAP – United Nations World Water Assessment Programme; 2021

[9] Report on Susteainable Water and Energy Solutions Addressing Climate Change. United Nations Department of Economic and Social Affairs – UNESCO; 2021

[10] Sustainable Energy Management for WWT Facilities. U.S. Environmental Protection Agency; 2009

[11] The Sustainable Development Goals Report 2021. United Nations SDG; 2021

[12] UN-Water Analytical Brief Water-Use Efficiency. UN Water; 2021

[13] UN-Water Annual Report 2020. UN Water; 2020

[14] UN-Water Country Briefs Project. UN Water; 2013

[15] Wastewater the Untapped Resource. United Nations Educational, Scientific and Cultural Organization – UNESCO; 2017

[16] Water and Energy – UN World Water Development Report 2014. WWAP – United Nations World Water Assessment Programme; 2014

[17] Emissions Gap Report 2021: The Heat Is on –UNEP, UNEP Copenhagen Climate Centre (UNEP-CCC). US: WWAP - United Nations; 2021

[18] Kyoto Protocol to the United Nations Framework Convention on Climate Change – COP3 Kyoto. Japan: UNFCCC; 1997

[19] Forster P, Ramaswamy V. Climate Change 2007: The Physical Science Basis – Fourth Assessment Report – Intergovernmental Panel on Climate Change. UK and USA: Cambridge University; 2007

[20] Convention for the Protection of the Mediterranean Sea against Pollution. Barcelona: Official Journal of the European Communities; 1977 & 2004

[21] Prochaska C, Stavropoulos N. Review of Urban Wastewater Treatment in Greece. Greece: Laboratory of Chemical and Environmental Technology; 2020

[22] Integrated National Energy and Climate Plan for France. EU: The National Low-Carbon Strategy Committee; 2020

Chapter 16

Advancing Sustainable 3D Printing: Harnessing the Potential of Wastewater Sludge Incineration Ash for Composite Material Development and Practical Applications

Dongwon Ki, Shin Young Kang, Se-Won Park and Hong-Kwan Choi

Abstract

This study delves into the unexplored fusion of wastewater sludge incineration ash (WSIA) and 3D printing, uncovering novel possibilities at the crossroads of environmental engineering and advanced manufacturing. The investigation centers on the integration of WSIA within the additive manufacturing framework, specifically material extrusion-fused deposition modeling (FDM). The study focuses on composite materials blending WSIA with commercial thermoplastics like ABS and PC, meticulously analyzing their physical, chemical, and mechanical attributes. Notably, the research highlights the potential for heightened mechanical strength in ABS composite materials, suggesting applications beyond 3D printing. Emphasizing long-term sustainability, the study advocates for the adoption of biodegradable plastics and underscores the importance of continuous research, mass production feasibility, and regulatory adaptations to fully unlock WSIA's potential. This synergy of innovation marries environmental awareness with technological progress, offering a harmonious trajectory to reshape manufacturing paradigms and nurture sustainable development. The study's outcomes present promising avenues for redefining the construction industry and advancing environmental conservation.

Keywords: 3D printing, composite materials, wastewater sludge incineration ash, recycling, sustainable manufacturing, material extrusion-fused deposition modeling

1. Introduction

In the intersection of environmental engineering and cutting-edge technologies, a symphony of possibilities unfolds. This chapter embarks on a journey that delves into the latent potential of wastewater sludge incineration ash (WSIA), unraveling its uncharted connection with the dynamic realm of 3D printing and its implications for the construction industry. As we embark on this exploration, we peel back the layers of complexity surrounding WSIA, unveiling a narrative that embodies both the challenges and prospects embedded within modern environmental concerns.

Metropolitan locales, exemplified by Seoul, house wastewater treatment facilities equipped with sludge incineration units. This incineration process, while crucial for waste management, leads to WSIA as a residual output. Curiously, this seemingly secondary product has evoked interest due to its latent utility. Notably, two wastewater treatment plants (N and S) among four plants annually produce over 12 tons of WSIA each, prompting its incorporation into the production of sturdy cement, thereby reinforcing the infrastructure. Intriguingly, the volume of WSIA and its associated disposal costs have experienced a progressive rise over recent years, attributable in part to a decline in private sector entities involved in its treatment. This confluence of factors underscores the imperative for innovative strategies to navigate the burgeoning landscape of WSIA.

Central to this discourse is the realm of 3D printing, an emerging paradigm within the broader context of environmental engineering and technology. Through an additive manufacturing approach, 3D printing propels a transformative capacity that redefines conventional manufacturing paradigms [1, 2]. This study accentuates the intersection of WSIA and 3D printing, postulating that the integration of WSIA within the additive manufacturing framework may yield pioneering avenues. As we navigate this uncharted terrain, it becomes evident that scant scholarly investigations have delved into the potential of WSIA within the 3D printing domain, except for a preliminary exploratory study conducted by our group [3], which investigated the viability of the binder jet (BJ)-3D printing method.

Intricately interwoven within this investigation is the material extrusion-fused deposition modeling (ME-FDM or FDM) method, a cardinal facet of 3D printing methodology [1, 4, 5]. FDM, revered for its popularity and versatility, predominantly employs thermoplastic materials for its additive manufacturing processes. However, in tandem with evolving needs, novel composite materials have surfaced to enhance the functional attributes of FDM-produced constructs. Remarkably, the use of WSIA within the FDM-3D printing realm remains uncharted. Parameters encompassing filament production, operational requisites spanning printer chambers and nozzles, material interactions, and resultant print quality converge as groundbreaking dimensions shaping the efficacy of FDM-3D printing endeavors [6–8]. This study endeavors to address this void by delving into the synthesis and utilization of WSIA-infused composite materials, with regard to commercial thermoplastics like ABS and PC, within the range of FDM-3D printing.

In sum, the aim of this study was to contribute to new environmental and 3D printing technologies using the potential of WSIA. Ultimately, this research seeks to present new possibilities through the convergence of environmental and technological aspects, suggesting a direction for sustainable development.

2. Review of plastic materials for the development of composite materials using WSIA for FDM-3D printing

The materials used in the FDM-3D printing process are crucial variables at the outset and remain among the most significant factors throughout the entire process [4–8]. In the context of this study, before delving into the development of composite materials using WSIA, it is essential to investigate what materials are available for use. Therefore, a foundational investigation of materials applicable to FDM-3D printing, along with a review of literature related to applied materials, was conducted to ascertain the potential for developing composite materials. The range of materials available for FDM-3D printing is extensive. Primarily, thermoplastic materials (or synthetic resins) in filament form are employed, encompassing not only everyday commodity materials but also engineering materials and high-performance thermoplastic resins. Typically, as the demand for precise and high-performance output increases, the required temperature and associated costs for the printing process also rise [9]. Plastic materials are categorized based on whether they are fossil-based or bio-based and whether they are biodegradable or non-biodegradable. This categorization divides them into four distinct regions, as depicted in **Figure 1** (Sections 1–4). For instance, bio-based and non-biodegradable plastics (Section 2) share nearly identical physicochemical properties with fossil-based and non-biodegradable plastics, differing primarily in production methods and raw materials. Additionally, ongoing research focuses on the development of composite materials beyond the presented individual materials [10–14]. However, a notable absence of research exists regarding the development of composite materials using WSIA.

From the array of plastics corresponding to the four sections (1–4) presented in **Figure 1**, we have selected eight thermoplastic-based materials that can potentially be applied in FDM-3D printing in a composite form when mixed with WSIA. These selections are based on the following criteria: (1) materials that are highly compatible with FDM printing or can be modified to suit printing conditions, (2) materials for which the introduction of WSIA is anticipated to enhance printability or physicochemical properties, and (3) materials that have been researched as composite materials or are deemed to have significant research value. The chosen thermoplastic materials are PP, PE, ABS, PC, PBS, PBAT, PLA, and PHA. Bio-based and non-biodegradable plastics falling within Section 2 (Bio-based & Non-biodegradable) were excluded due to their similarity in physicochemical composition to petroleum-based counterparts.

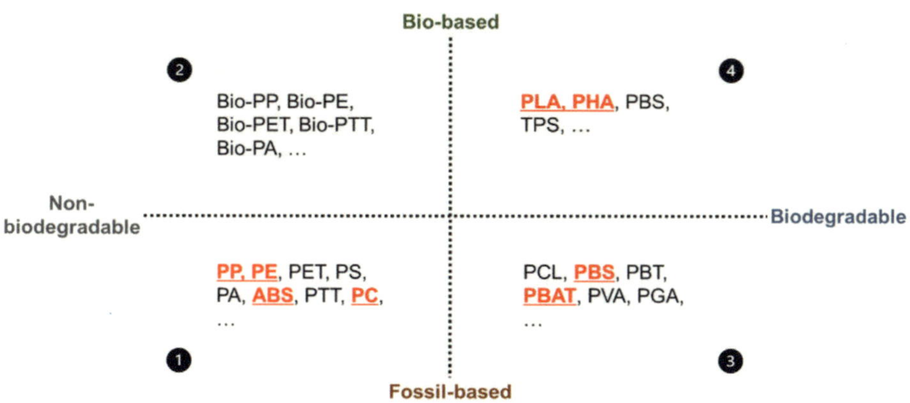

Figure 1.
Classification matrix of plastics based on source (fossil- or bio-based) and biodegradability potential of materials. The red, bold, and underlined ones are selected as potential candidates for composite materials with WSIA. PP, polypropylene; PE, polyethylene; PET, polyethylene terephthalate; PS, polystyrene; PA, polyamide or nylon; ABS, acrylonitrile-butadiene-styrene; PTT, Polytrimethylene terephthalate; PC, polycarbonate; PCL, Polycaprolactone; PBS, polybutylene succinate; PBT, polybutylene terephthalate; PBAT, polybutylene adipate terephthalate; PVA, polyvinyl alcohol; PGA, Polyglycolic acid; PLA, Polylactic acid; PHA, Polyhydroxyalkanoates; TPS, thermoplastic starch.

3. Materials and methods

3.1 WSIA characteristics

Wastewater sludge incineration ash (WSIA) typically appears as a dark brown or gray substance, and its chemical composition may vary depending on additives (e.g., coagulants) used in the wastewater treatment process. In this study, WSIA generated from the N-wastewater treatment plant (N-WWTP) located in Seoul was utilized for the development of composite materials for FDM-3D printing. WSIA samples collected in July 2021 (**Figure 2**) were subjected to analysis of their physical and chemical properties (particle size, oxides, hazardous substances).

The physical properties of the WSIA, specifically the particle size of the ash (powder), were analyzed using a wet analysis method (0.017 ~ 2000 μm) with laser diffraction and scattering analysis using a particle size analyzer (Beckman Coulter Inc., USA). In the process of dispersing the agglomerated ash particles, ultrasonic pre-treatment was carried out using an ethanol-based solvent.

Oxide compounds of the WSIA samples were analyzed using inductively coupled plasma optical emission spectroscopy (ICP-OES, Perkin-Elmer Optima 8300, UK). There were a total of 20 analysis items, but the main nine components were SiO_2, TiO_2, Al_2O_3, K_2O, Na_2O, P_2O_5, MgO, CaO, and Fe_2O_3. Additionally, there were ten trace components including MnO, V_2O_5, SrO, ZrO_2, ZnO, CuO, BaO, Cr_2O_3, La_2O_3, and SO_3, as well as the loss on ignition.

For the analysis of hazardous substances in the WSIA samples, leaching analysis was conducted for six heavy metals (Cr^{6+}, Cu, Cd, Pb, As, and Hg) and cyanide compounds (CN). This analysis was followed by the Korean Official Wastes Test Method [15].

3.2 Composite materials synthesis

In this study, the base materials for composite material fabrication were selected from the aforementioned eight thermoplastic materials based on their suitability for mass production, easy availability, and immediate commercialization. Among the non-biodegradable and fossil-based plastics that fulfill these criteria,

Figure 2.
Wastewater sludge incineration ash (WSIA) sampled from N-WWTP.

acrylonitrile-butadiene-styrene (ABS) and polycarbonate (PC) were employed (ABS – STARLET from Polystar, China; PC – TEIREX from Samyang Corporation, Korea). The WSIA used as a component was also sampled in July 2021 from the N-WWTP.

The synthesis process of composite materials by mixing WSIA with ABS and PC is as follows: (1) mixing of WSIA with ABS and PC plastics, (2) homogenization of the mixed composite material after drying, (3) filament production by feeding the homogenized composite material into an extruder and cooling it, and (4) machining the filament into chip form. Following this sequence, the mixture of WSIA and ABS/PC was created by mixing WSIA into plastics at weight ratios of 0%, 9%, 17%, and 23% for each. The physical mixing was carried out using an overhead stirrer after measuring the weight according to the mixing ratio. The homogenized composite material was dried to a constant weight ratio, then inserted into an extruder, extruded, cooled, and formed into filaments. The FDM printer used in this study requires polymers to be in chip form; therefore, the produced filaments were further processed into chip form for use (**Figure 3**).

3.3 Physical, chemical, and mechanical properties of composite material

For the characterization of the produced composite material, an analysis of its physical, chemical, and mechanical properties was conducted. In terms of physical properties, density analysis was performed, and the synthesized composite material's surface was observed using a Field Emission Scanning Electron Microscope (FE-SEM) for surface characterization. Chemical properties were investigated through Thermogravimetric Analysis (TGA) and Differential Scanning Calorimetry (DSC) for thermal analysis, Fourier-Transform Infrared Spectroscopy (FT-IR) for analysis of chemical bonds within the material, and leaching analysis of heavy metals (according to the waste process testing standards) to assess its harmfulness. Mechanical properties were evaluated by measuring the tensile strength and load deformation temperature, which are fundamental mechanical attributes of plastics.

Figure 3.
Process of fabricating composite materials by mixing WSIA with ABS or PC.

3.3.1 Physical properties of composite material

For density analysis, the method B: Liquid Pyknometer Method from KS M ISO 1183-1 "Plastics - Methods for determining the density of non-cellular plastics" was employed. For surface observation of the composite material, analysis was conducted using an FE-SEM following the guidelines of KS I 0051: 1999 method.

3.3.2 Chemical properties of composite material

For the thermal characterization of the composite material, TGA was conducted to observe the mass loss trend with temperature ramping. The analysis conditions were nitrogen (N_2) up to 600°C, followed by air conditions from 600 to 870°C. To observe the calorimetric changes with temperature in the composite material, DSC analysis was performed. This analysis provides information about parameters like the glass transition temperature (T_g) of polymers and can offer references for 3D printing settings. The analysis conditions involved heating up to 230°C under a nitrogen gas atmosphere. FT-IR analysis was carried out to examine any structural changes in the material resulting from chemical bonding and mixing in the composite material. For component analysis of the composite material, ICP-OES was used to analyze both organic content (ignition loss) and oxides. To confirm the toxicity of the composite material, leaching tests were also conducted to measure the six heavy metals (Cr^{6+}, Cu, Cd, Pb, As, and Hg) and cyanide compounds (CN) according to waste process testing standards aforementioned.

3.3.3 Mechanical properties of composite material

To analyze the mechanical properties of the composite material, specimens were produced using FDM-3D printing, which is owned by 3D Factory (Ulsan, Korea). For the tensile strength specimens, they were fabricated according to the ASTM D638 Type 1 standard, and for the load deformation temperature specimens, they were created with an average width of 2.4 mm in accordance with ASTM D648. For tensile strength analysis, the specimens were tested at a crosshead speed of 50 mm/min following the ASTM D638-14 testing method. The results were obtained as the average of two tested specimens. Load deformation temperature, also known as Heat Deflection Temperature (HDT), was measured using the ASTM D648-18 testing method with Method B, which involves withstanding a light load (0.455 MPa) at elevated temperatures. The average width of the specimens used in the HDT test ranged from 2.1 to 2.7 mm.

4. Results and discussion

4.1 Results of WSIA characteristics

When compared with previous research [3], the nine main components did not show significant differences, with SiO_2 (28.2%), Al_2O_3 (24.4%), P_2O_5 (21.4%), Fe_2O_3 (9.4%), and CaO (7.7%) being the predominant constituents in order. The total content of the 10 trace oxides was 1.18%, and the ignition loss was 2.08% (**Table 1**).

The results of leaching analysis for the six heavy metals (Cr^{6+}, Cu, Cd, Pb, As, Hg) and cyanide compounds (CN), just like the measurements taken in June 2020, showed

Total	SiO_2	TiO_2	Al_2O_3	K_2O	Na_2O	P_2O_5	MgO	CaO	Fe_2O_3
96.64	28.2	0.43	24.4	1.7	0.97	21.4	2.45	7.7	9.39

Table 1.
WSIA's chemical (oxide compounds) composition ratio (%) from N-WWTP.

that all substances except for arsenic (As) were not detected, and even arsenic was found to be below the "designated waste" classification standard (**Table 2**).

The particle size distribution of WSIA is presented in **Figure 4**, with an average particle size (mean size) of 20.0 μm, a median size of 18.4 μm, and a mode of 27.9 μm. The minimum and maximum values were 0.445 μm and 67.5 μm, respectively. When compared to the average particle size of WSIA in the previous study [3], the particle size of this ash is less than half, but there is not a significant difference in the distribution size and proportion.

4.2 Composite materials synthesis

ABS and PC are the most common and widely used materials in the FDM printing method, with numerous related studies being conducted [4, 5]. While ABS is more commonly used, the selection of PC is driven by its qualities as an engineering plastic, offering high mechanical strength, heat resistance, and impact resistance, making it suitable for structures that need to withstand loads.

	Cr^{6+}	Cu	Cd	Pb	As	Hg	CN
"Designated waste" classification standard	1.5	3	0.3	3	1.5	0.005	1
WSIA	N.D.	N.D.	N.D.	N.D.	0.107	N.D.	N.D.

Table 2.
Contents of heavy metals and cyanide compound in WSIA (unit: mg/L).

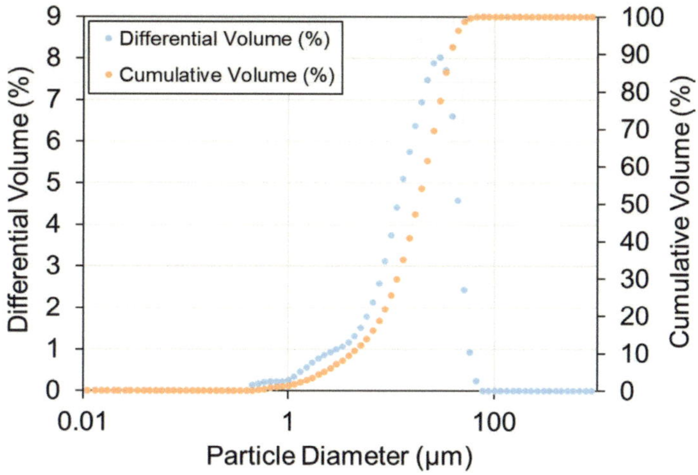

Figure 4.
Particle size distribution of WSIA sampled in July, 2021 from N-WWTP.

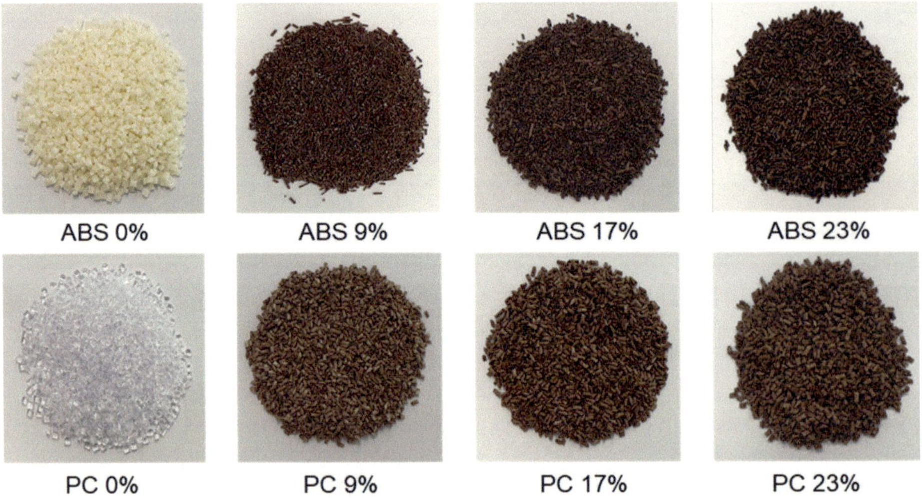

Figure 5.
Composite material chips with varying WSIA mixing ratios (ABS, PC).

The plastic (ABS, PC) and synthesized composite material chips are depicted in **Figure 5**. Pure ABS chips (without WSIA) exhibit an opaque ivory color, gradually transitioning to a darker brown as WSIA is added. Similarly, pure PC chips (without WSIA) appear as transparent white, and as with ABS composite material, the addition of WSIA results in a darker brown hue.

4.3 Physical, chemical, and mechanical properties of composite material

4.3.1 Physical properties of composite material

For ABS without the addition of WSIA, the average density was 1.053 g/cm^3. When WSIA was mixed in at weight ratios of 9%, 17%, and 23%, the measured densities were 1.067, 1.039, and 1.143 g/cm^3, respectively. The density changes after mixing were found to be insignificant. In the case of PC without the addition of WSIA, the average density was 1.193 g/cm^3. With the addition of WSIA at weight ratios of 9%, 17%, and 23%, the densities decreased to 0.736, 0.699, and 0.853 g/cm^3, respectively, representing a maximum reduction of up to 41.4% in density after mixing (**Figure 6**).

For the surface observation of the composite material, an FE-SEM was utilized for analysis. At a low magnification of 500×, the overall sample conditions were observed. The surfaces of pure ABS and PC plastic pellets appeared smooth, while for pellets containing mixed WSIA, a rough surface with WSIA particles of average size 20 μm physically mixed with the polymer was observed (**Figure 7**).

4.3.2 Chemical properties of composite material

Based on the results of thermal analysis (Thermogravimetric Analysis or TGA), for pure ABS (with 0% WSIA content), around 97.9% of the mass decreased around 408°C. As the WSIA content in ABS increased from 9 to 23%, the mass reductions occurred at approximately 90.4%, 78.6%, and 72.5% around 416°C respectively (**Figure 8a**). In the case of ABS composite materials with added WSIA, the

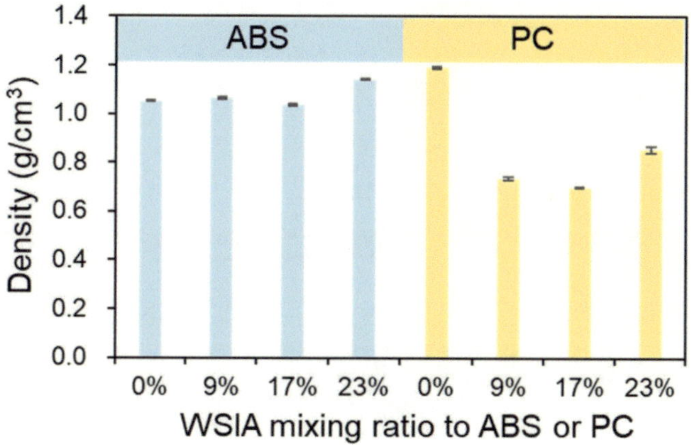

Figure 6.
Density of composite materials with varying WSIA mixing ratio (0–23%).

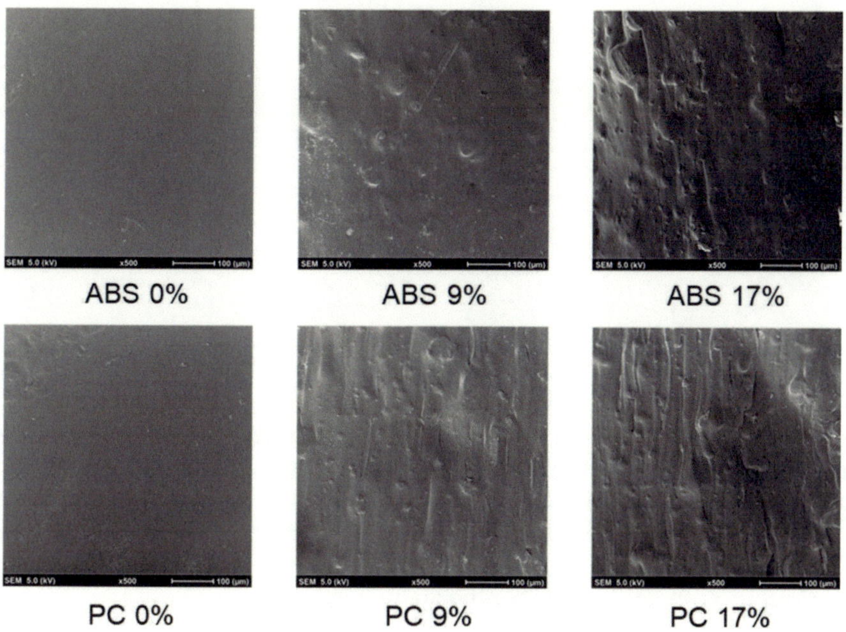

Figure 7.
FE-SEM images of composite materials with WSIA mixing ratio (0–17%).

decomposition temperature increased by about 8°C from 408°C to around 416°C, indicating a slight improvement in thermal stability [16]. For pure PC (with 0% WSIA content), mass reductions of around 98.2% were observed around 514°C and 627°C. When WSIA content in PC was increased to 9%, 17%, and 23%, the mass reductions took place around temperatures ranging from 490 to 500°C and 625°C, resulting in mass reductions of approximately 88.8%, 81.2%, and 77.4% respectively (**Figure 8b**). In the case of PC composite materials with added WSIA, the decomposition temperature decreased from 514°C to the range of 490–500°C, leading to a decrease in thermal stability by a minimum of 13°C to a maximum of 23°C. From the TGA analysis, it was observed that organic polymer materials such as ABS and PC

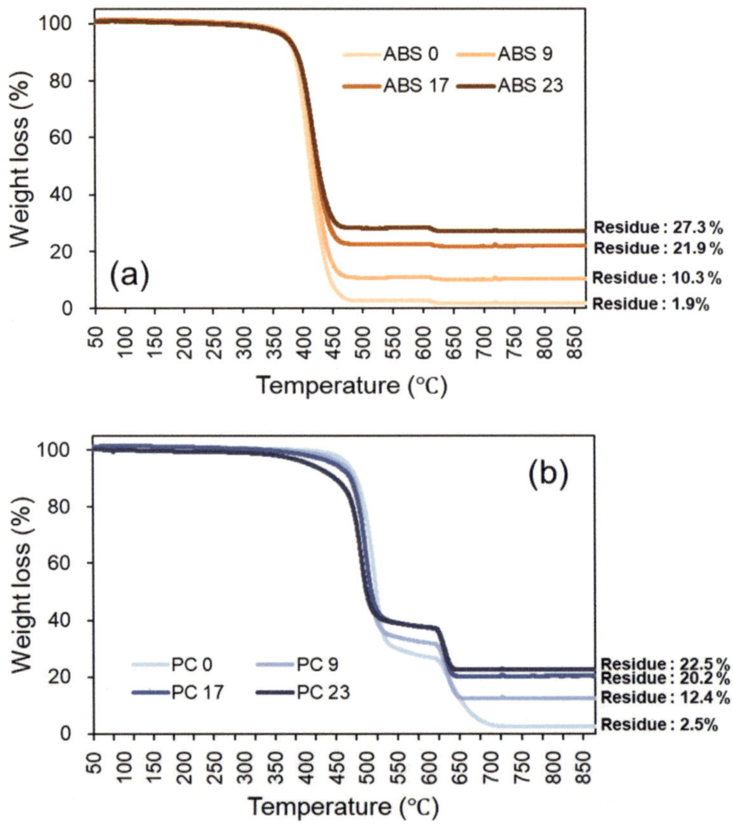

Figure 8.
TGA graphs of composite materials according to WSIA mixing ratios (0–23%) to ABS (a) and PC (b).

undergo thermal degradation at relatively low temperatures below 600°C, while the inorganic WSIA component remains in proportion to the mixing ratio. Additionally, despite the increased decomposition temperature of ABS composite materials after WSIA mixing, PC composite materials exhibited decreased decomposition temperatures. This observation indicates that ABS composite materials have relatively higher thermal stability compared to PC composite materials.

Differential Scanning Calorimeter (DSC) analysis results revealed that in the case of pure ABS, a glass transition temperature (T_g) was observed around 101.1°C (**Figure 9a**) [17]. As the WSIA was mixed in ratios of 9, 17, and 23%, T_g was observed at approximately 102.8°C, 101.2°C, and 100.3°C, respectively. While the main T_g showed no significant change, an additional peak was observed at 75°C, indicating a mixed pattern between the two particles (WSIA and ABS) (**Figure 9a**) [18]. For pure PC, a T_g was observed around 147.2°C (**Figure 9b**) [17], and as the incineration residue was mixed in ratios of 9–23%, T_g values were observed at approximately 141.1°C, 138.9°C, and 136.8°C, respectively. It was confirmed that T_g decreased by 6 ~ 10°C as the WSIA content increased (**Figure 9b**). Through the DSC analysis of T_g influenced by intermolecular interactions, it can be inferred that in the case of ABS composite materials, the main T_g remained relatively unchanged, but an additional peak was observed at the front end, indicating two dispersion patterns between the WSIA and the polymer. Similarly, for PC composite materials, the addition of the WSIA seemed

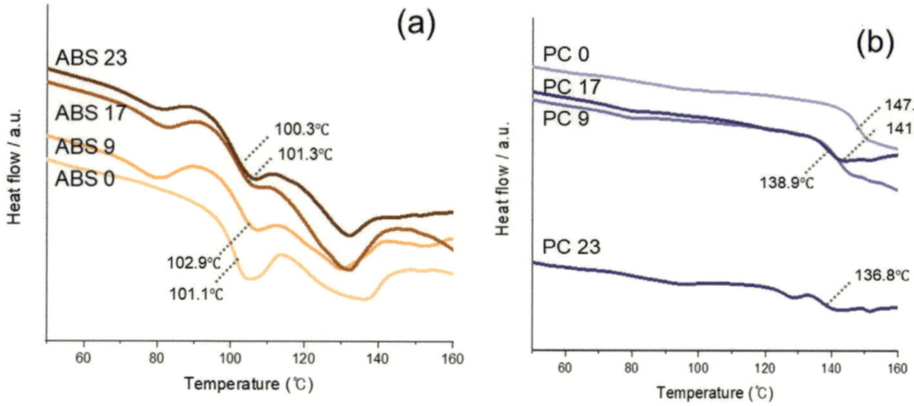

Figure 9.
DSC graphs of composite materials according to WSIA mixing ratios (0–23%) to ABS (a) and PC (b).

to influence the intermolecular interactions between the residue and PC molecules, leading to a decrease in T_g.

The results of FT-IR analysis conducted to assess potential changes in chemical bonding and material structure due to the chemical combination and mixing of WSIA composite materials are shown in **Figure 10**. In the case of pure ABS, distinctive vibrational peaks were observed at 910, 965, 1493, 1639, 2237, 2850, 2920, and 3026 cm^{-1} [19]. When the WSIA was mixed in ratios of 9% and 17%, the same peaks were observed as in pure ABS samples (**Figure 10a**). For pure PC, distinct vibrational peaks were observed at 1220, 1503, 1769, and 2965 cm^{-1} [19]. Similarly, when the WSIA was mixed at 9% and 17%, the same peaks as in pure PC samples were observed (**Figure 10b**). From the FT-IR analysis, it is deduced that the chemical bonding remained unchanged due to the physical mixing of both ABS and PC composite materials with WSIA.

Through compositional analysis of the composite material, it was confirmed that in the case of pure ABS and PC, the result of loss on ignition was determined to be 99.8% and 99.9%, respectively, confirming that they primarily consist of carbon and other organic materials (**Table 3**). As the WSIA was mixed with each plastic (ABS or PC) in the range of 9–17%, an increase in the proportion of inorganic components (such as SiO_2, Al_2O_3, P_2O_5, Fe_2O_3) from the incineration residue was observed. This indicates that the WSIA was appropriately mixed according to the weight ratio. Additionally, the hazardous material analysis results indicated that all parameters of 6 heavy metals (Cr^{6+}, Cu, Cd, Pb, As, and Hg) and CN were below detectable levels, confirming the absence of concerns regarding safety in usage.

4.3.3 Mechanical properties of composite material

The tensile strength analysis results showed that the pure ABS had a strength of 12.8 MPa, while mixing the WSIA from 9 to 23% resulted in strengths of 31.3, 32.1, and 26.5 MPa, respectively, achieving a maximum increase of up to 2.5 times. The highest strength was observed at a mixing ratio of 17%. For pure PC, the initial tensile strength was 32.9 MPa, which was more than twice that of pure ABS. However, as the WSIA was mixed from 9 to 23%, the strengths decreased to 12.7, 8.16, and 9.07 MPa, exhibiting a reduction of over 4 times (**Figure 11a**). In terms of heat deflection temperature (HDT) analysis, pure ABS showed deformation at 90°C. Mixing the ash material from 9 to 23% increased the deformation temperature to 95, 96, and 96°C,

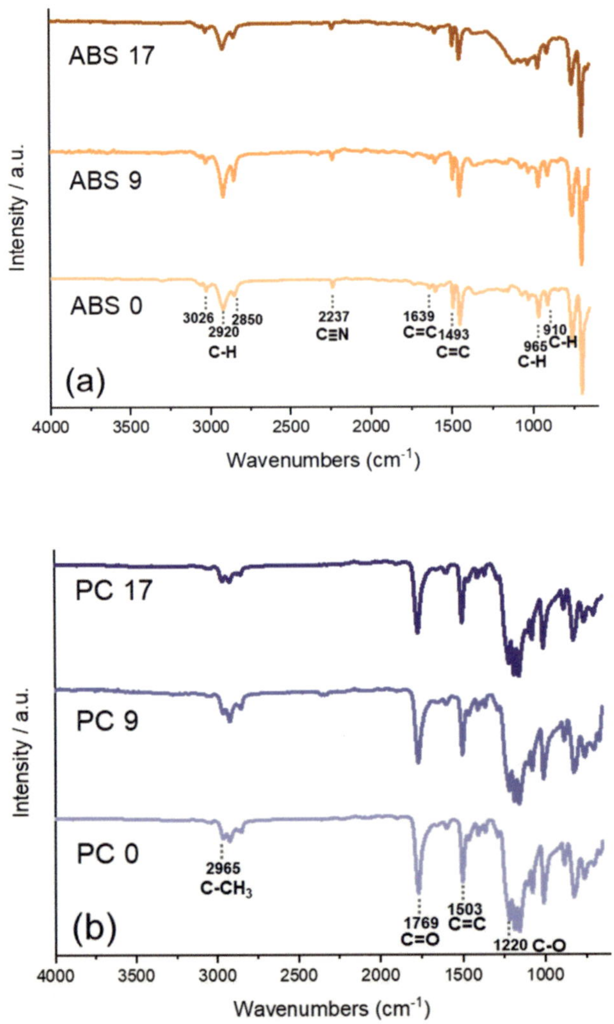

Figure 10.
FT-IR graphs of composite materials according to WSIA mixing ratios (0–17%) to ABS (a) and PC (b).

respectively. On the other hand, pure PC exhibited deformation at 138°C. Mixing the ash material from 9 to 23% led to a gradual decrease in deformation temperature to 124, 116, and 117°C, respectively (**Figure 11b**). Based on the mechanical property analysis results, it was determined that the tensile strength and HDT of the ABS composite material increased as the WSIA was mixed, leading to improved mechanical strength. The optimal mixing ratio was found to be between 9% and 17%. However, in the case of PC, it was observed that the tensile strength and HDT decreased as the WSIA was mixed, indicating that the WSIA composite was not suitable for enhancing the mechanical properties of PC.

4.4 Outlook

In this study, the feasibility of utilizing WSIA as a 3D printing composite material from a material application perspective was examined. However, its applicability extends

		SiO_2	Al_2O_3	P_2O_5	Fe_2O_3	CaO	MgO	K_2O	Na_2O	TiO_2	Others	Ig.loss
ABS	0%	0.06					0.06				0.06	99.8
	9%	1.92	1.6	1.66	0.73	0.52	0.25	0.24	0.08	0.04	0.15	92.7
	17%	4.46	4.47	3.85	1.69	1.25	0.43	0.32	0.15	0.1	0.38	82.9
PC	0%										0.07	99.9
	9%	2.30	2.31	1.94	0.91	0.69	0.21	0.25	0.09	0.06	0.17	91.0
	17%	0.06					0.06				0.06	99.8

Table 3.
Chemical (oxide compounds) composition of composite materials according to WSIA mixing ratios (0–17%) to ABS and PC (unit: %).

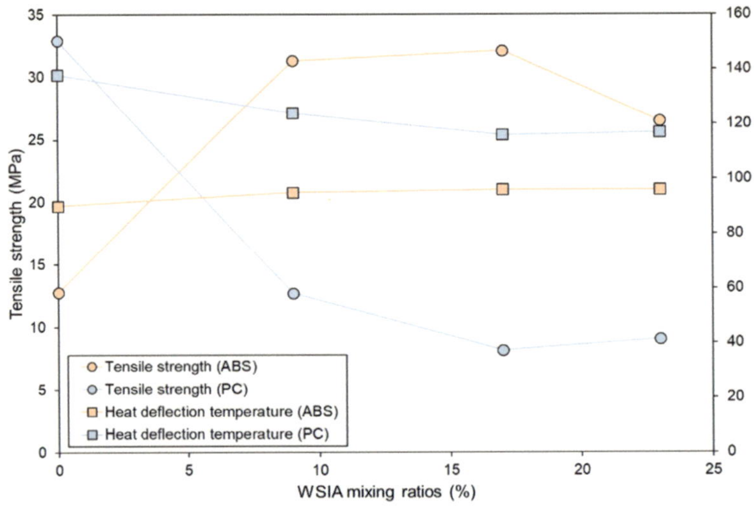

Figure 11.
Results of tensile strength and heat deflection temperature of WSIA composite according to the mixing ratios (0–23%).

beyond 3D printing to general forming processes such as injection molding or extrusion. Moreover, since WSIA can commonly be used as a construction material, its application in the construction field is possible when used as a 3D printing material (e.g., concrete structures, irregular pavilions) [20–22]. Additionally, it can be utilized in outdoor landscaping facilities (benches, pergolas, fences, lighting fixtures, etc.) and other daily life areas (balance bikes, kick scooters, bicycle racks, etc.) (**Figure 12**) [23, 24].

While products using fossil-based and non-biodegradable plastic materials are currently being produced not only through 3D printing but also through general forming processes, the need for research and technological development using more environmentally friendly bio-based/biodegradable plastic materials are increasing. In a long-term perspective, the development of composite materials should also aim at the utilization of biodegradable/bio-based plastics (Section 4 of **Figure 1**). Based on the results of this study, continuous research and development considering mass production and efficient biodegradation are necessary. Furthermore, improvements in the existing legal framework are required to expand the possibilities for WSIA's utilization and commercialization. Initially, this could be achieved by actively leveraging

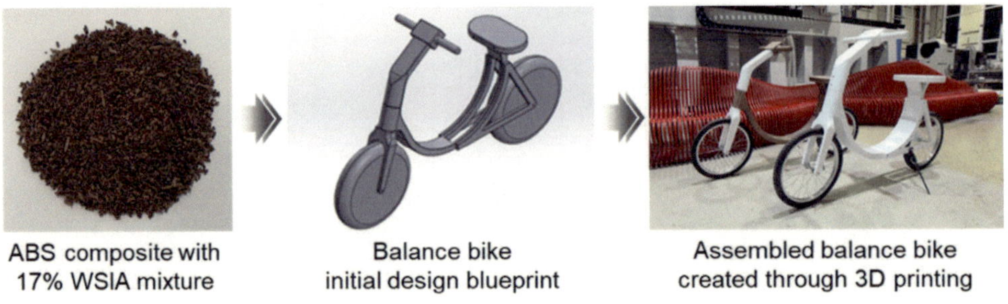

Figure 12.
The initial prototype of balance bike 3D printed using ABS composite material with 17% WSIA mixture (wood color: WSIA composite material, white color: ABS only).

the existing recycling regulations under waste management laws. Subsequently, further improvements could be made to the recognition system for WSIA use within the framework of circular resource management regulations.

5. Conclusions

The exploration of wastewater sludge incineration ash (WSIA) integration with 3D printing uncovers innovative opportunities at the nexus of environmental engineering and advanced manufacturing. The study addresses the uncharted potential of utilizing WSIA in the additive manufacturing framework, particularly within the material extrusion-fused deposition modeling (FDM) method. Focused on composite materials infused with WSIA and commercial thermoplastics like ABS and PC, the investigation meticulously analyzes physical, chemical, and mechanical properties. The findings showcase the potential for enhanced mechanical strength in ABS composite materials, opening avenues for broader applications beyond 3D printing. The study advocates for long-term sustainability by considering the utilization of biodegradable plastics and emphasizes the need for continuous research, mass production viability, and regulatory reforms to fully realize WSIA's potential. In this symphony of innovation, environmental consciousness converges with technological advancement, presenting a harmonious path toward reshaping manufacturing paradigms and fostering sustainable development.

Acknowledgements

This study was supported by the Seoul Institute of Technology (SIT) research projects (2021-AC-002 & 2023-AC-004). The authors extend their gratitude to Tae-Hwan Kim and Chan-Woo Kim from the Water Circulation Safety Bureau, Seoul Metropolitan Government, for their valuable assistance in providing wastewater sludge statistics, incineration ash information, and facilitating connections with wastewater treatment facilities.

Conflict of interest

The authors declare no conflict of interest.

Author details

Dongwon Ki[1*], Shin Young Kang[1], Se-Won Park[1] and Hong-Kwan Choi[2]

1 Seoul Institute of Technology, Seoul, South Korea

2 3D Factory, Ulsan, South Korea

*Address all correspondence to: dongwonk@sit.re.kr

References

[1] Redwood B, Schöffer F, Garret B. The 3D Printing Handbook : Technologies, Design and Applications. Amsterdam: Coers & Roest; 2018

[2] Rosen DW, Seepersad CC, Simpson TW, Williams CB. Design for additive manufacturing: A paradigm shift in design, fabrication, and qualification. Journal of Mechanical Design. 2015;**137**(11):110301. DOI: 10.1115/1.4031470

[3] Ki D, Kang SY, Park K-M. Upcycling of wastewater sludge incineration ash as a 3D printing technology resource. Frontiers in Sustainability. 2021;**2**:697265. DOI: 10.3389/frsus.2021.697265

[4] Dudek P. FDM 3D printing technology in manufacturing composite elements. Archives of Metallurgy and Materials. 2013;**58**(4):1415-1418. DOI: 10.2478/amm-2013-0186

[5] Popescu D, Zapciu A, Amza C, Baciu F, Marinescu R. FDM process parameters influence over the mechanical properties of polymer specimens: A review. Polymer Testing. 2018;**69**:157-166. DOI: 10.1016/j.polymertesting.2018.05.020

[6] Patel R, Desai C, Kushwah S, Mangrola MH. A review article on FDM process parameters in 3D printing for composite materials. Materials Today: Proceedings. 2022;**60**:2162-2166. DOI: 10.1016/j.matpr.2022.02.385

[7] Sharma M, Sharma V, Kala P. Optimization of process variables to improve the mechanical properties of FDM structures. In: Proceedings of the 2nd International Conference on New Frontiers in Engineering, Science & Technology (NFEST); 18-22 February 2019; India. Journal of Physics: Conference Series. 2019;**12400**:012061

[8] Tripathy CR, Sharma RK, Rattan VK. Effect of printing parameters on the mechanical behavior of the thermoplastic polymer processed by FDM technique: A research review. Advances in Production Engineering & Management. 2022;**17**(3):279-294. DOI: 10.14743/apem2022.3.436

[9] 3D HUBS. What is FDM 3D printing? [Internet]. 2023. Available from: https://www.hubs.com/knowledge-base/what-is-fdm-3d-printing/ [Accessed: August 28, 2023]

[10] Roberson D, Shemelya CM, MacDonald E, Wicker R. Expanding the applicability of FDM-type technologies through materials development. Rapid Prototyping Journal. 2015;**21**(2):137-143. DOI: 10.1108/RPJ-12-2014-0165

[11] Geng Y, He H, Liu H, Jing H. Preparation of polycarbonate/poly(lactic acid) with improved printability and processability for fused deposition modeling. Polymers for Advanced Technologies. 2020;**3**(11):2848-2862. DOI: 10.1002/pat.5013

[12] Haryńska A, Janik H, Sienkiewicz M, Mikolaszek B, Kucińska-Lipka J. PLA-potato thermoplastic starch filament as a sustainable alternative to the conventional PLA filament: Processing, characterization ad FFF 3D printing. ACS Sustainable Chemistry & Engineering. 2021;**9**(20):6923-6938. DOI: 10.1021/acssuschemeng.0c09413

[13] Kim H, Ryu KH, Baek D, Khan TA, Kim HJ, Shin S, et al. 3D printing of polyethylene terephthalate glycol–sepiolite composites with nanoscale

orientation. ACS Applied Materials & Interfaces. 2020;**12**(20):23453-23463. DOI: 10.1021/acsami.0c03830

[14] Narancic T, Verstichel S, Chaganti SR, Morales-Gamez L, Kenny ST, De Wilde B, et al. Biodegradable plastic blends create new possibilities for end-of-life management of plastics but they are not a panacea for plastic pollution. Environmental Science & Technology. 2018;**52**(18):10441-10452. DOI: 10.1021/acs.est.8b02963

[15] Ministry of Environment (MOE). Korean Official Test Method on Wastes [Internet]. 2017. Available from: https://www.me.go.kr/gg/web/board/read.do?pagerOffset=0&maxPageItems=10&maxIndexPages=10&searchKey=&searchValue=&menuId=2246&orgCd=&condition.hideCate=1&boardMasterId=228&boardCategoryId=&boardId=788150&decorator= [Accessed: August 28, 2023]

[16] Zheng K, Chen L, Li Y, Cui P. Preparation and thermal properties of silica-graft acrylonitrile-butadiene-styrene nanocomposites. Polymer Engineering and Science. 2004;**44**(6):1077-1082. DOI: 10.1002/pen.20100

[17] Aid S, Eddhahak A, Ortega Z, Froelich D, Tcharkhtchi A. Experimental study of the miscibility of ABS/PC polymer blends and investigation of the processing effect. Journal of Applied Polymer Science. 2017;**134**:44975. DOI: 10.1002/app.44975

[18] Lee MH, Min BH, Kim JH. Morphology and miscibility of PMMA/SMA/clay nanocomposites. Applied Chemistry for Engineering. 2010;**21**(3):252-257. DOI: not available

[19] Li J, Chen F, Yang L, Jiang L, Dan Y. FTIR analysis on aging characteristics of ABS/PC blend under UV-irradiation in air. Spectrochimica Acta Part A: Molecular and Biomolecular Spectroscopy. 2017;**184**:361-367. DOI: 10.1016/j.saa.2017.04.075

[20] Na O, Kim K, Lee H, Lee H. Printability and setting time of CSA cement with Na_2SiO_3 and gypsum for binder jetting 3D printing. Materials. 2021;**14**(11):2811. DOI: 10.3390/ma14112811

[21] Jipa A, Dillenburger B. 3D printed formwork for concrete: State-of-the-art, opportunities, challenges, and applications. 3D Printing and Additive Manufacturing. 2022;**9**(2):84-107. DOI: 10.1089/3dp.2021.0024

[22] Han D, Yin H, Qu M, Zhu J, Wickes A. Technical analysis and comparison of formwork-making methods for customized prefabricated buildings: 3D printing and conventional methods. Journal of Architectural Engineering. 2020;**26**(2). DOI: 10.1061/(ASCE)AE.1943-5568.0000397

[23] Hyundai Engineering & Construction creates irregular architectural structures using 3D printing [Internet]. 2021. Available from: https://www.hdec.kr/kr/company/press_view.aspx?CompanyPressSeq=39#.YZXjeGBByUk [Accessed: August 28, 2023]

[24] Carlota V. Recycling plastic to 3D print furniture in cities [Internet]. 2019. Available from: https://www.3dnatives.com/en/recycling-plastic-to-3d-print-furniture-in-cities/#! [Accessed: August 28, 2023]

Chapter 17

Future of Sludge Management

Massahi Tooraj, Hossini Hooshyar, Parnoon Kimya, Atashkar Sheida, Sharafi Maryam, Massahi Dariush and Ahmadi Borhan

Abstract

Managing sludge is a crucial component of wastewater treatment plants. As the amount of sludge produced continues to grow, it is necessary to explore future strategies and technologies for sustainable sludge management. This chapter aims to provide an overview of the future of sludge management, with an emphasis on emerging technologies and sustainable practices. The chapter stresses the significance of environmentally sustainable and cost-effective sludge management and extensively evaluates the technological, economic, and environmental criteria for assessing sludge management strategies. The chapter explores innovative methods for handling sewage sludge and its reuse, as well as emerging technologies for sustainable sludge management. It assesses the economic feasibility of advanced treatment technologies and sludge-based products, considering beneficial sludge utilization such as land application, energy recovery, alternative fuel sources, construction materials, and resource recovery. Lastly, the chapter discusses future developments in sludge management, including issues and trends. Recent research findings are utilized to provide insights into the future of sludge management that can guide decision-making and policy development in the field.

Keywords: emerging technologies, future developments, sludge management, sustainable sewage sludge management, beneficial sludge utilization

1. Introduction

Effective sludge management is a crucial component of wastewater treatment and requires sustainable and cost-efficient solutions. Innovative technologies have transformed sludge management into a sophisticated research and development sector, maximizing the efficiency of sewage treatment plants and minimizing their carbon footprint. Sewage sludge is no longer considered waste but rather a renewable resource for energy and material recovery. Consequently, legislation regarding sewage sludge management now encompasses environmental protection, public health, climate change impacts, and socio-economic benefits.

The annual global production of municipal sewage sludge is rapidly increasing and is influenced by factors such as urbanization, development level, and lifestyle [1]. As conventional techniques like landfilling and incineration are now considered unacceptable, it is crucial to explore sustainable and economically viable alternatives for sludge management. Recent studies have evaluated the economic feasibility of various sewage sludge management technologies, including solar drying systems and the material recovery of sewage sludge through the struvite production pathway [2]. These studies emphasize the importance of considering the local context, system size, and environmental impact when assessing the economic feasibility of these technologies. In light of these findings, this chapter aims to provide an overview of the future of sludge management, covering the latest technological and policy advancements, as well as exploring the challenges and opportunities in this field. Additionally, the chapter will investigate the potential for resource recovery from sewage sludge and the role of circular and bioeconomy in sludge management.

2. Major trends in sludge management: Past, present, and future perspectives

Sludge management is a complex issue that has significant implications for various fields, including water and sewage management, waste management, agriculture, and energy [1]. A review of literature published in the last 30 years shows that while the range of unit processes for sludge treatment has greatly expanded, there are still only three primary disposal options available for treated sludge: land application, landfill disposal, and incineration. However, it's important to note that none of these disposal methods are inherently good or bad; their suitability depends on specific circumstances. The choice of a sludge management strategy is susceptible to prevailing trends, which may introduce subjective criteria instead of being based on an objective assessment of each project's needs [3].

The status of sludge management and disposal varies across countries. For example, in China, as of the end of 2019, there were 5476 operating municipal wastewater treatment facilities, which produced an annual 39.04 million tons of sludge (with 80% water content). The most commonly used techniques for disposal were composting (45.11%), incineration (29.24%), and sanitary landfills (23.04%) [1]. However, to ensure sustainable sludge management in the future, technological advancements and research and development are necessary. Poor management of sewage sludge can lead to higher costs of wastewater treatment, while advanced technologies offer the potential to recover energy and raw materials. Potential solutions could involve burning with the recovery of energy and raw materials from the ashes or organic recycling, such as composting with the recovery of energy and heat, and utilizing the resulting fertilizer [1].

In summary, notable trends in sludge management, both past and present, have involved the use of diverse unit processes for sludge treatment, the three primary disposal options for treated sludge, and the impact of current trends on decision-making in sludge management practices. Future trends in sludge management will require technological advancements, research and development, and the implementation of modern technologies to recover energy and raw materials from sludge.

3. Importance of environmentally sustainable and cost-effective sludge management

Effective and environmentally friendly management of sludge produced by wastewater treatment plants is crucial to minimize the environmental impact of sludge disposal. Sustainable and cost-effective sludge management is of significant importance for several reasons:

3.1 Environmental impact

Proper management of sludge in wastewater treatment plants is crucial to reduce its environmental impact. If not managed effectively, sludge can contain harmful substances and microorganisms that may pollute soil and water bodies, leading to damage to ecosystems and potential health risks. Recent research has highlighted the need to consider the environmental effects of sludge management practices, assess the sustainability of different treatment and disposal methods, and explore innovative approaches such as hydrodynamic cavitation to effectively stabilize sludge. By adopting efficient strategies for managing sludge, the environmental impact of wastewater treatment plants can be minimized, which will help safeguard ecosystems and protect public health [4, 5].

3.2 Resource recovery

Sludge contains valuable resources such as organic matter, nutrients (including nitrogen and phosphorus), and energy. By adopting sustainable sludge management practices, these resources can be extracted and utilized, reducing the need for external inputs and promoting circular economy principles. Waste sludge can be used as a renewable resource for energy recovery, which is a practical solution for managing the ever-increasing waste sludge [6].

3.3 Emissions reduction

Innovative sludge management technologies can assist in reducing greenhouse gas (GHG) emissions. The treatment and disposal of sewage sludge is a significant source of GHG emissions due to environmentally damaging traditional methods that require a lot of energy. However, recent research has explored novel methods for recovering resources and energy from sewage sludge. One such approach involves utilizing sludge as a feedstock for energy production, which can replace fossil fuels and result in significant reductions in emissions [7].

3.4 Cost savings

The adoption of environmentally sustainable sludge management practices can lead to long-term cost savings. Wastewater treatment plants can reduce operational costs and potentially generate income by extracting resources such as energy and nutrients from sludge. Sustainable sludge management involves utilizing efficient and sustainable methods to recover the maximum amount of valuable resources or energy from sludge. Achieving sustainable management of sludge requires a thorough analysis of the environmental impacts resulting from various sludge treatment and disposal methods. By considering sustainable and cost-effective options for managing sewage sludge, wastewater treatment plants can recover and utilize valuable resources, contributing to the development of efficient and sustainable practices for sewage sludge management [8].

3.5 Land and soil remediation

Using sludge as a fertilizer for agricultural and degraded lands is a sustainable approach that can help remediate such lands and improve soil fertility. This method has been proven to reduce the need for chemical fertilizers, which can cause adverse environmental impacts, and promote the rehabilitation of degraded ecosystems. Applying sludge to these lands can enhance soil quality and support plant growth by utilizing the nutrients and organic matter present in the sludge. Furthermore, this approach can decrease the amount of waste that would otherwise end up in landfills. In summary, using sludge as a fertilizer is a promising approach for achieving both environmental and agricultural sustainability objectives [9].

Achieving environmentally sustainable sludge management is a critical objective for wastewater treatment systems. To accomplish this goal, it is essential to consider a range of indicators and criteria that encompass technical, social, environmental, and economic aspects. These indicators can be utilized to evaluate the sustainability of various sludge management technologies and support decision-making processes. By considering these indicators, stakeholders can assess the environmental impact, social acceptability, technical feasibility, and economic viability of different sludge management options. Ensuring environmentally sustainable and cost-effective sludge management is crucial to wastewater treatment. It is a vital element that can contribute to a more sustainable and eco-friendly future. With effective sludge management, it is possible to safeguard the environment, recover valuable resources, reduce emissions, and promote circular economy principles. Therefore, it is essential to prioritize sustainable and cost-effective sludge management practices in wastewater treatment systems.

4. Techno-economic-environmental criteria for evaluating sludge management strategies

Techno-economic-environmental criteria are essential for evaluating sludge management strategies. Recent research studies have proposed various indicators and criteria for sustainable and innovative sewage sludge treatment technologies [7, 8]. These indicators are categorized into four main categories: technical, social, environmental, and economic. Specific indicators are explained, followed by a description of their impact on sustainability. For instance, technical indicators may include the efficiency of sludge dewatering, while social indicators may involve public acceptance of sludge management technologies [7].

A case study utilizing the LEAP tool was conducted to evaluate the potential for emissions reduction by utilizing sewage sludge as an energy feedstock to replace existing fossil fuels in the energy mix in several scenarios. The results showed a significant decrease in emissions when sludge was utilized [7].

The PROMETHEE method was utilized to analyze technologies for the sustainable management of sewage sludge in Croatia. The model indicated that the most favorable solution is the material recovery of sewage sludge, utilizing the struvite production pathway through the analysis of selected criteria [8].

A life cycle assessment (LCA) approach was proposed to address the issue of the significant amount of sludge generated by wastewater treatment plants by converting it into granular activated carbon (GAC) for use in water pollution control. The paper reviews various approaches and techniques for drying, pyrolysis, and activation, as well as the optimal operating conditions for sludge-based activated carbon (SBAC) to achieve optimal pollution control [10].

A recent study proposes a solar drying system as a suitable and cost-effective alternative to mechanical drying for refining sewage sludge with the aim of minimizing environmental impacts and promoting the reuse of sewage sludge. The solar drying system demonstrated high efficiency, with all three samples attaining a dry solid content of 96.96% for R.S. (raw sludge collected during winter), 96.75% for D.S.1 (digested sludge collected during winter), and 93.60% for D.S.2 (digested sludge collected during summer) after undergoing solar drying. The pollution index calculations revealed that all three types of sludge presented a low potential ecological risk [11].

In summary, evaluating sludge management strategies requires considering multiple criteria that encompass technological, economic, and environmental factors. These criteria can aid in assessing the sustainability of various sludge management technologies and guide decision-making processes.

5. Sustainability in sludge management: Balancing present needs and future generations

Sustainability in sludge management involves meeting current societal requirements while also ensuring that future generations can meet their needs. Sludge management encompasses the handling and disposal of residues produced during wastewater treatment, which may pose a significant threat to the environment if not adequately managed. Achieving a balance between present and future needs in sludge management requires adopting practices that are environmentally sustainable, socially acceptable, and economically viable.

An approach to ensure sustainable sludge management is to view sludge as a valuable resource rather than a waste product. This approach involves processing sludge to generate biogas that can serve as a renewable energy source or utilizing sludge as a fertilizer in agriculture. These methods can mitigate the environmental impact of sludge management while also providing economic benefits.

Another critical aspect of sustainable sludge management is to ensure that the management practices are socially acceptable. This involves engaging stakeholders, including local communities, in decision-making processes and addressing their concerns to establish trust and support for sludge management practices. This approach is essential for the long-term sustainability of such practices.

In summary, achieving sustainable sludge management requires a comprehensive approach that considers environmental, social, and economic factors. By balancing the needs of present and future generations, sustainable sludge management can ensure that the benefits of wastewater treatment are achieved without compromising the ability of future generations to meet their own needs [12, 13].

6. Innovative approaches for sewage sludge treatment and reuse

Conventional methods of treating and disposing of sludge can be costly, energy-intensive, and have adverse environmental effects. However, innovative approaches to treating and reusing sewage sludge have garnered increasing attention in recent times. These approaches prioritize sustainability, resource recovery, and environmental impact, facilitating the development of effective and eco-friendly solutions for managing sewage sludge.

The reuse of water and sludge in agriculture involves treating wastewater and sewage sludge to extract the maximum amount of water and avoid waste. The treated water and sludge can then be utilized as fertilizer or for irrigation in agriculture. This technique has the potential to improve water efficiency, minimize waste, and enhance crop production. Recent research highlights the benefits of using reclaimed water and sludge in agriculture, which includes reducing costs, creating a dependable local water supply, and promoting the circular economy. By adopting new treatment technologies, the agricultural sector can play a significant role in sustainable resource management and contribute to a more efficient and environmentally friendly approach to water and sludge management [14].

The article titled "Wastewater and Sludge Reuse: Selected Case Studies Across the Globe" presents a descriptive analysis of the current state of wastewater and sewage sludge reuse in different countries. The research focuses on case studies from nations such as Israel, the Philippines, and others, highlighting effective practices such as using nanotechnology to improve water quality, reusing domestic and food-producing wastewater, and repurposing distillery-based wastewater. The review identifies the potential for further expansion of sewage and sludge reuse in energy production and agricultural applications. In summary, the study provides insights into the global status of wastewater and sludge reuse, showcasing successful case studies and emphasizing the potential for increased utilization in various sectors [14].

The process of sludge stabilization with clay minerals and biochar involves adding these materials during sewage sludge stabilization. This approach has the potential to reduce microbial load, control heavy metals, and promote efficient nutrient reuse. Research shows that stabilized sewage sludge can be a safe soil conditioner and fertilizer. Using clay minerals and biochar for sewage sludge stabilization is an eco-friendlier method for agricultural purposes. Further research is currently exploring the effectiveness of this approach in enhancing soil properties and its potential as a safe soil conditioner and fertilizer. By utilizing these integrated methodologies, the treatment and utilization of sewage sludge can be made more sustainable and environmentally friendly [2].

Hydrodynamic cavitation (HC) is a technique that can improve sewage sludge stabilization and promote safe and efficient reuse in agriculture. HC involves the gradual breakdown of activated sludge and the dissolution of dissolved organic matter using cavitation bubbles, which can also reduce the size of suspended sludge particles. Recent research has demonstrated the potential of HC in enhancing sewage sludge stabilization in both anaerobic and aerobic digestion units, making it easier to manage and dispose of [1, 14, 15]. This eco-friendly and cost-efficient pretreatment method shows promise in enhancing the reuse of sewage sludge for agricultural applications [5].

The technique of utilizing sewage sludge in the production of fired bricks involves using urban sludge as a raw material in the manufacturing of clay bricks. This process reduces the amount of sludge and converts it into useful materials. Recent research has assessed the characteristics of bricks made from sewage sludge, resulting in an efficient elimination of urban sludge. According to the studies, incorporating 10% of sludge allows for the production of effective bricks, highlighting the potential of this valorization method. This approach has the potential to reduce the amount of sludge and convert it into valuable materials, thereby promoting sustainable waste management practices [16].

Overall, the available resources emphasize the importance of sustainable sewage sludge management and showcase innovative techniques for its treatment and reuse. These techniques consider factors such as sustainability, resource recovery, and environmental impact, resulting in the development of sustainable and eco-friendly solutions for sewage sludge management.

7. Emerging technologies for sustainable sludge management

Emerging technologies for sustainable sludge management are being developed to address environmental and health concerns associated with traditional methods. The following are some of the identified emerging technologies:

7.1 Anaerobic digestion

Anaerobic digestion is a process that decomposes organic matter in the absence of oxygen, resulting in biogas and a nutrient-rich digestate. This technology can be used to manage sewage sludge and generate renewable energy in the form of biogas. Anaerobic digestion provides a sustainable solution for sewage sludge management by recovering valuable resources and reducing the environmental impact of sludge disposal. The process involves the breakdown of organic matter by microorganisms in an oxygen-free environment, which produces biogas and a nutrient-rich digestate that can be used as fertilizer. The biogas produced can serve as a renewable energy source, reducing the need for fossil fuels and contributing to the development of a circular economy. Overall, anaerobic digestion is a promising technology for achieving sustainable sludge management and promoting a more eco-friendly future [17].

7.2 Dewatering and drying

Dewatering and drying technologies are crucial for managing sewage sludge and reducing its moisture content to enhance handling and transportability. These technologies also contribute to sludge stabilization and volume reduction, making it more suitable for disposal or reuse. Recent advancements in sludge dewatering and drying have focused on minimizing sludge volume, improving transportability, enhancing calorific value, and reducing leachate production in landfills. The dewatering process involves addressing influencing factors and employing conditioning methods to improve performance. Dewatering technologies are classified based on the degree of moisture content, including common, deep, and high-dry dewatering. On the other hand, the drying method is divided into convective drying, conduction drying, radiation drying, and other new drying methods [18]. The selection of a dewatering technique depends on the characteristics of the sludge, the need for chemical conditioning, and the final method of sludge disposal [19]. Technological comparisons and system modeling of sewage sludge treatment are critical for sustainable development and climate change mitigation. Dewatering and drying processes are important for reducing volume and are often a requirement for further sludge treatment. Inventory data on mass transfers, material and energy consumption, and other factors are crucial for improving and understanding sludge management systems. However, limited data availability hinders the ability to identify detailed relationships between inputs and outputs for different technologies [20]. In addition to conventional methods, there are innovative approaches to sludge drying. For example, the use of solar greenhouse dryers has been studied for reducing the moisture content of dewatered sludge. Solar drying technologies, both open and closed systems, have been employed, and their performance recorded for different seasons [21]. These technologies offer potential for volume reduction of sewage sludge. In conclusion, dewatering and drying technologies are essential for sustainable sludge management. They contribute to reducing sludge moisture content, enhancing handling and transportability, and stabilizing sludge for disposal or reuse. The appropriate technology selection depends on the sludge's characteristics and the desired method of sludge disposal. Continued research and development in this field is crucial for advancing sustainable sludge management practices.

7.3 Pyrolysis and gasification

Pyrolysis and gasification are thermal techniques that can be utilized for sustainable sludge management. These technologies transform organic materials, such as sewage sludge, into biochar, syngas, and other beneficial products. Pyrolysis and gasification can help to reduce the volume of sludge, recover energy, and produce biochar that can be used as a soil amendment. Pyrolysis is a thermal process that involves heating organic materials without oxygen to create biochar and other valuable products. In contrast, gasification is a process that involves heating organic materials with a limited amount of oxygen to produce syngas, which can be used as fuel. Both processes can be employed for sustainable sludge management, as they can help to decrease the volume of sludge and recover valuable resources. One of the primary benefits of pyrolysis and gasification is the production of biochar, which can be used as a soil amendment. Biochar is a stable form of carbon that can enhance soil fertility, water retention, and nutrient availability. Additionally, biochar can aid in carbon sequestration in the soil, which can reduce greenhouse gas emissions. Overall, pyrolysis and gasification are promising technologies for sustainable sludge management. These processes can help to reduce sludge volume, recover energy, and produce beneficial products such as biochar and syngas [19, 20, 22–24].

7.4 Nutrient recovery

The recovery of valuable nutrients, such as phosphorus and nitrogen, from sewage sludge for reuse as fertilizers or other applications, is becoming increasingly important in sustainable sludge management. Nutrient recovery technologies aim to extract these nutrients, reduce the environmental impact of sludge disposal, and promote circular economy principles. Several emerging nutrient recovery technologies have been developed to recycle finite natural resources. These technologies include acid-mediated hydrothermal treatment, which allows for the concentration of phosphorus and nitrogen in the process water [25]. Other technologies, such as anaerobic digestion and nutrient recovery and reuse, have been successfully implemented in full-scale installations throughout Europe [26].

7.5 Struvite production

The process of struvite production involves extracting phosphorus from sewage sludge and converting it into a slow-release fertilizer known as struvite. This technology provides several benefits in terms of sustainable nutrient management and reducing dependence on mined phosphorus. Firstly, struvite production reduces reliance on mined phosphorus, offering an alternative source of this essential nutrient and promoting sustainable nutrient management practices. Secondly, the slow-release fertilizer produced through struvite production can be used to replenish phosphorus in agricultural soils, recycling and reusing nutrients in a sustainable manner. This reduces the need for chemical fertilizers and minimizes the environmental impact associated with their production and use. The significance of struvite production in sustainable sludge management cannot be overstated. By recovering phosphorus from sewage sludge and converting it into a valuable fertilizer, this technology provides a promising solution for reducing dependence on mined phosphorus and promoting sustainable nutrient management practices. It is crucial to continue researching and implementing struvite production technologies to ensure efficient and environmentally sound sewage sludge management [8, 24, 27].

7.6 Material recovery

Material recovery technologies aim to extract valuable materials, including metals or organic compounds, from sewage sludge for reuse or recycling. These technologies help recover resources from sludge and minimize the need for raw materials. The primary objective of material recovery technologies is to retrieve valuable resources from sewage sludge, such as metals and organic compounds, that can be reused or recycled, and to enhance resource efficiency by maximizing the extraction of valuable materials. Material recovery from sewage sludge offers numerous advantages, such as resource conservation, reduction in the demand for raw materials, and environmental sustainability [8].

It is worth noting that these emerging technologies are still under development and may have limitations or challenges that need to be addressed. However, they offer promising potential for more sustainable and efficient management of sludge in the future.

8. Economic viability of advanced treatment technologies and sludge-based products

The economic feasibility of advanced treatment technologies and sludge-based products is a crucial factor to consider for their widespread adoption. Recent studies have shed light on the economic aspects of these technologies. In a 2018 study, the economic feasibility of decentralized wastewater treatment systems was evaluated, and it was found that they can be more cost-effective than centralized systems in certain contexts. However, this study did not specifically focus on sludge-based products. Another study in 2018 highlighted the challenging economic issues related to sewage sludge processing, reuse, and disposal, emphasizing the need for sustainable and cost-effective solutions for sludge management. In the same year, a study evaluated the life cycle costs of advanced treatment techniques for wastewater reuse and resource recovery from sewage sludge. The findings revealed that these techniques can be economically viable, but their viability depends on factors such as the scale of the system and the local context. Similarly, another study utilized a life cycle sustainability assessment to evaluate the economic, technical, social, and environmental criteria of advanced treatment techniques for urban wastewater reuse and sewage sludge resource recovery. The study concluded that the most economically viable option varied depending on the specific context. Additionally, a study conducted on the economics of advanced technologies for wastewater treatment in the pulp and paper industry found that biological treatment methods were more economically viable than chemical treatment methods. In Croatia, two recent studies assessed the economic feasibility of different sewage sludge management technologies. Both studies concluded that material recovery of sewage sludge through the struvite production pathway was the most cost-effective option when considering economic and environmental criteria. These studies provide valuable insights into the economic feasibility of advanced treatment technologies and sludge-based products, which can inform decision-making and policy development in the field of sludge management. While sludge-based products offer potential for a circular economy, current methods for treating sewage and wastewater face challenges such as high energy requirements, unstable treatment effects, lengthy processes, carbon emissions, excess sludge discharge, and wastage of recyclable resources. In summary, recent studies suggest that advanced treatment technologies and sludge-based products can be economically feasible, but their feasibility depends on factors such as the scale of the system, local context, and specific technology used. Material recovery of sewage sludge through the struvite production pathway has been identified as a particularly promising and cost-effective option [8, 28–32].

9. Beneficial sludge utilization: Land application, energy recovery, alternative fuel source, construction material, and resource recovery

Sludge, a byproduct of wastewater treatment, holds the potential for both environmental and economic benefits. By utilizing sludge, we can address the environmental concerns linked to its disposal in landfills or through incineration, while also reaping economic advantages through resource recovery and energy generation. In this section, we will delve into various methods of utilizing sludge for its beneficial properties.

Land application is a well-established approach for the beneficial utilization of sludge. Sewage sludge can be used as an organic fertilizer to improve soil physical, chemical, and biological properties, thereby promoting plant growth and development. This is achievable because sludge contains vital plant nutrients like nitrogen, phosphorus, and potassium, as well as organic matter that enhance soil structure and water-holding capacity. However, it is crucial to assess the characteristics of sewage sludge before utilizing it to prevent any adverse effects on the soil environment and human health. The application of sludge to land is governed by stringent guidelines to ensure its safe and effective utilization [33, 34].

With the increasing demand for phosphorus and nitrogen, the importance of resource recovery from sludge is growing. Sludge can undergo processing to extract valuable resources, including phosphorus, nitrogen, and organic matter. One promising technology for reducing sludge moisture content and maintaining high organic matter content suitable for composting treatment is deep dewatering of sludge using chemical conditioning reagents. The recovered resources can then be used as fertilizers or soil amendments, reducing the reliance on chemical fertilizers and enhancing soil health [34–36].

Another noteworthy method for the beneficial utilization of sludge is energy recovery. Through anaerobic digestion, the organic matter in sludge undergoes a process that converts it into biogas. This biogas can then be utilized for generating electricity or heat. The biogas produced from anaerobic digestion is a renewable energy source, contributing to the reduction of greenhouse gas emissions and reducing dependence on fossil fuels. The utilization of biogas as a fuel source is gaining traction, particularly as the production cost of renewable energy decreases [6, 36, 37].

Pyrolysis is an additional method for converting sludge into a valuable fuel source. This process involves heating the sludge in the absence of oxygen to produce biochar, which is a solid fuel capable of replacing coal. The utilization of biochar as a fuel source offers the potential to reduce greenhouse gas emissions and decrease reliance on fossil fuels. However, it's important to note that the production of biochar from sludge is currently in the experimental phase, and further research is required to assess its feasibility and economic viability [6, 35, 37].

Lastly, sludge can be utilized as a construction material in the form of bricks or blocks. This involves mixing sludge with other materials like clay and cement to create a durable and lightweight material. This approach offers the potential to mitigate the environmental impact associated with conventional building materials such as concrete and bricks, which require substantial energy for production. However, it's important to note that the use of sludge as a construction material is currently in the experimental phase, and further research is required to assess its safety and feasibility [36, 37].

In conclusion, the utilization of sludge offers numerous environmental and economic advantages. The various methods of beneficial sludge utilization, including land application, resource recovery, energy recovery, alternative fuel sources, and construction materials, help address the environmental concerns associated with sludge disposal while also yielding economic benefits through resource recovery and energy generation. However, it is crucial to conduct thorough evaluations of sewage sludge characteristics prior to utilization in order to prevent any adverse effects on the environment and human health. Further research is necessary to determine the safety and feasibility of newer sludge utilization methods, such as utilizing sludge as a fuel source and construction material.

10. Future developments in sludge management: issues and trends

Sludge management is a critical issue that requires careful consideration of available options. Over the last 30 years, the diversity of unit processes for sludge treatment has increased dramatically. This trend is likely to continue as new technologies are developed to improve the efficiency and effectiveness of sludge treatment. Despite the increase in unit processes, there are still only three basic ultimate disposal routes for processed sludge: land application, landfill, and incineration [38]. However, the selection of a sludge management option should be based on a critical evaluation of the specific project's needs rather than popular trends.

The land application of sludge is experiencing a well-deserved period of growth based on good science, cleaner sludges, and proactive public education [38]. This trend is likely to continue as more research is conducted on the benefits of using sludge as an organic fertilizer. Resource recovery from sludge is becoming increasingly important as the demand for phosphorus and nitrogen increases [39]. Sludge can be treated to recover valuable resources such as phosphorus, nitrogen, and organic matter. Energy recovery from sludge is also an important trend in sludge management. Anaerobic digestion is a process that converts organic matter in sludge into biogas, which can be used to generate electricity or heat [38].

The use of sewage sludge as fertilizer for agricultural production is becoming an urgent issue due to the increasing demand for sustainable development and the rising price of chemical fertilizers [39]. However, it is crucial to carefully assess the characteristics of sewage sludge before its application to avoid any adverse effects on the soil environment and human health. Deep dewatering of sludge with chemical conditioning reagents is a promising technology for reducing the moisture content of sludge and maintaining its organic matter content at a high level, which is beneficial for composting treatment. LCA is a method that can be used to evaluate the environmental impact of sludge management technologies [40]. This approach can help identify the most sustainable and environmentally friendly options for sludge management.

In conclusion, future developments in sludge management are likely to focus on resource recovery, energy recovery, and safe utilization in agriculture. These trends will require the development of new technologies and the careful evaluation of the environmental impact of sludge management options. The selection of a sludge management option should be based on a critical evaluation of the specific project's needs rather than popular trends.

Acknowledgements

We would like to express our gratitude to the Research Council of Kermanshah University of Medical Sciences for their support.

Conflict of interest

The authors declare no conflict of interest.

Note

It should be noted that AI tools were used here to enhance the grammar of the text.

Author details

Massahi Tooraj[1*], Hossini Hooshyar[1], Parnoon Kimya[1], Atashkar Sheida[2], Sharafi Maryam[1], Massahi Dariush[3] and Ahmadi Borhan[1]

1 Department of Environmental Health Engineering, School of Public Health, Kermanshah University of Medical Sciences, Kermanshah, Iran

2 Department of Environmental Health Engineering, School of Public Health and Safety, Shahid Beheshti University of Medical Sciences, Tehran, Iran

3 Department of Pure Chemistry, Razi University, Kermanshah, Iran

*Address all correspondence to: toorajmassahi@gmail.com

References

[1] Baum MA et al. The State of the Nation: A 50-State Covid-19 Survey Report #2. 2020

[2] Gruber A et al. Evolution of the ESA CCI soil moisture climate data records and their underlying merging methodology. Earth System Science Data. 2019;**11**:717-739

[3] Roth AJ. The Research Paper. Writing the Research Paper. 2022

[4] Obaideen K et al. The role of wastewater treatment in achieving sustainable development goals (SDGs) and sustainability guideline. Energy Nexus. 2022;**7**:100112

[5] Mancuso G et al. Effect of hydrodynamic cavitation on flocs structure in sewage sludge to increase stabilization for efficient and safe reuse in agriculture. Bioremediation Journal. 2022;**26**(1):41-52

[6] Ahuja N et al. Lessons learned from start-up of a codigestion process. Florida Water Resources Journal. 2017;**4**:34-38

[7] Đurđević D, Žiković S, Čop T. Socio-economic, technical and environmental indicators for sustainable sewage sludge management and LEAP analysis of emissions reduction. Energies. 2022;**15**:6050

[8] Đurđević D, Žiković S, Blecich P. Sustainable sewage sludge management technologies selection based on techno-economic-environmental criteria: Case study of Croatia. Energies. 2022;**15**(11):3941

[9] Nicorescu V et al. Sustainable sludge management-its use for remediation of degraded lands. Romanian Journal of Ecology & Environmental Chemistry. 2020;**2**:193-201

[10] Mu'azu ND et al. Life cycle assessment approach to sustainable sewage sludge management for water pollution control. Journal of Physics: Conference Series. 2019;**1349**:012145

[11] Dramé SI, Oba OA, Pasaoglulari Aydinlik N. Environmental and ecological risk assessment of municipal sewage sludge management using a sustainable solar drying system. Water Science and Technology. 2023;**88**(1):151-168

[12] Colella M, Ripa M, Cocozza A, Panfilo C, Ulgiati S. Challenges and opportunities for more efficient water use and circular wastewater management. The case of Campania Region, Italy. Journal of Environmental Management. 1 Nov 2021;**297**:113171

[13] Tyagi VK, Lo SL. Sludge: A waste or renewable source for energy and resources recovery?. Renewable and Sustainable Energy Reviews. 1 Sep 2013;**25**:708-728

[14] Fernandes SC et al. Reuse of water and sludge in agriculture: New technologies for the treatment. Research, Society and Development. 2023;**12**:e30012128262

[15] Zhang J, Ma L. Comparative environmental sustainability study of an improved sewage sludge treatment and sludge reuse system based on emergy analysis in China. Applied Water Science. 2021;**11**:146

[16] Minh Trang NT, Dao Ho NA, Babel S. Reuse of waste sludge from water treatment plants and fly ash

for manufacturing of adobe bricks. Chemosphere. 2021;**284**:131367

[17] Spinosa L et al. Sustainable and innovative solutions for sewage sludge management. Water. 2011;**3**(2):702-717

[18] Rao B, Wang G, Xu P. Recent advances in sludge dewatering and drying technology. Drying Technology. 2022;**40**:3049-3063

[19] Varjavandi J. Dewatering and Disposal of Sewage Sludge. New Zealand: Chemical Engineering Group; 1990

[20] Chang H et al. Mini-review of inventory data for the dewatering and drying of sewage sludge. Waste Management & Research. 2023;**41**(6):1081-1088

[21] Raval N. A review on solar greenhouse aided drying of the sewage sludge. International Journal of Latest Technology in Engineering, Management & Applied Science (IJLTEMAS). 2018;7:182-185

[22] Jagaba AH, Kutty SR, Lawal IM, Birniwa AH, Affam AC, Yaro NS, et al. Circular economy potential and contributions of petroleum industry sludge utilization to environmental sustainability through engineered processes - A review. Cleaner and Circular Bioeconomy. 14 Oct 2022:100029

[23] Bhattacharjee P, Sengupta T, Ghosh K. 5-sewage sludge to biofuel: Emerging technologies for a sustainable environment. In: Shah MP et al., editors. Development in Waste Water Treatment Research and Processes. USA: Elsevier; 2022. pp. 63-89

[24] Dichtl N, Rogge SMJ, Bauerfeld K. Novel strategies in sewage sludge treatment. Clean-Soil Air Water. USA. 2007;**35**:473-479

[25] Sarrion A et al. Acid-mediated hydrothermal treatment of sewage sludge for nutrient recovery. Science of the Total Environment. 2022;**838**(Pt 4):156494

[26] Orner KD, Smith SJ, Breunig HM, Scown CD, Nelson KL. Fertilizer demand and potential supply through nutrient recovery from organic waste digestate in California. Water Research. 1 Nov 2021;**206**:117717

[27] Shaddel S et al. Sustainable sewage sludge management: From current practices to emerging nutrient recovery technologies. Sustainability. 2019;**11**:3435

[28] Xu G. Analysis of sewage sludge recovery system in EU-in perspectives of nutrients and energy recovery efficiency, and environmental impacts. Norway: Institutt for Energi-og Prosessteknikk; 2014

[29] Di Giacomo G, Romano P. Evolution and prospects in managing sewage sludge resulting from municipal wastewater purification. Energies. 2022;**15**(15):5633

[30] Christodoulou A, Stamatelatou K. Overview of legislation on sewage sludge management in developed countries worldwide. Water Science and Technology. 2016;**73**(3):453-462

[31] Rakotonimaro TV, Neculita CM, Bussière B, Benzaazoua M, Zagury GJ. Recovery and reuse of sludge from active and passive treatment of mine drainage-impacted waters: A review. Environmental Science and Pollution Research. Jan 2017;**24**:73-91

[32] Salkunić A, Vuković J, Smiljanić S. Review of technologies for the recovery of phosphorus from waste streams. Chemical and Biochemical Engineering Quarterly. 2022;**36**(2):91-116

[33] Campbell H. Sludge management–future issues and trends. Water Science and Technology. 2000;**41**(8):1-8

[34] Muter O. Economic and technological principles of safe use of sewage sludge in agriculture. Economic Innovations. 2022;**24**(2 (83)):139-157

[35] Jatav SS et al. Characterization of sewage sludge for quality assessment and its safe utilization in agriculture. Current Journal of Applied Science and Technology. 2021;**40**(25):28-35

[36] Rhyner CR et al. Waste Management and Resource Recovery. England: CRC Press; 2017

[37] Ward JE. Wood Wastes and Residues Generated Along the Colorado Front Range as a Potential Fuel Source. USA: US Department of Agriculture, Forest Service, Rocky Mountain Research Station; 2004

[38] Andreoli CV, Von Sperling M, Fernandes F. Sludge Treatment and Disposal. England: IWA Publishing; 2007

[39] Li L et al. Worldwide research progress and trend in sludge treatment and disposal: A bibliometric analysis. ACS ES&T Engineering. 2023;**3**:1083-1097

[40] Wei L et al. Development, current state and future trends of sludge management in China: Based on exploratory data and CO_2-equivaient emissions analysis. Environment International. 2020;**144**:106093